EXONS, INTRONS, AND TALKING GENES

EXONS, INTRONS, AND TALKING GENES

The Science Behind the Human Genome Project

CHRISTOPHER WILLS

BasicBooks
A Division of HarperCollins*Publishers*

Credits can be found on page xv.

Library of Congress Cataloging-in-Publication Data
Wills, Christopher.
Exons, introns, and talking genes / Christopher Wills.
p. cm.
Includes bibliographical references and index.
ISBN 0-465-05020-4
1. Medical genetics—Popular works. 2. Molecular genetics—Popular works. 3. Exons (Genetics)—Popular works. 4. Introns—Popular works. I. Title.
RB155.W54 1991 91-70062
616'.042—dc20 CIP

PRINTED IN THE UNITED STATES OF AMERICA
Designed by Ellen Levine
91 92 93 94 CC/HC 9 8 7 6 5 4 3 2 1

To the memory of
KEVIN,
and all the others

The web of our life is of a mingled yarn, good and ill together.

—William Shakespeare
All's Well That Ends Well

Contents

PREFACE

Not long ago, a mother of a Down syndrome child told us that when her daughter, Mindy, was born in 1973, her husband, desperate for information about the condition, pulled the 1970 Encyclopedia Britannica *from their home library shelf and began to search it for information. . . . Looking back on that incident, they can both laugh, although sardonically, about what he found: Down syndrome was described under the unfortunate heading of "Monster."*

—John E. Rynders and Siegfried M. Pueschel

The fusion of sperm with egg brings together half the mother's genes and half the father's. The cell that results, called a zygote, briefly embodies all the potential of a human being and none of the realization. For each of us, this moment established the pattern of DNA that makes up our unique *genome,* or set of genes. The ten trillion cells of our adult bodies have all been shaped by this pattern.

There is far more to the story than this, however. What happens to the zygote on its way to producing an adult human is the result of complex interactions between its genome and the environment. When the environment changes, so does the result.

For example, the genetic disorder Down syndrome is caused by a faulty distribution of genetic material, resulting in an extra copy of chromosome 21. A third of the cases of mental retardation in Western countries can be attributed to Down syndrome, and only a few decades ago people with this disorder were often locked away in institutions to spare society embarrassment. Many succumbed before the age of ten to respiratory infections or to the cardiac problems that accompany the syndrome. Not until the 1960s was there an organized effort in the United States to remove retarded children from institutions and try to integrate them into society. In the 1970s state and federal laws were passed mandating appropriate education for these children.

The process of changing the attitudes of parents, schools, and U.S. society toward Down syndrome has been a lengthy one, and it continues to this day. In most of the world it has barely begun. Currently, diagnostic tests can inform pregnant women if their fetuses carry the extra chromosome, allowing them the option of abortion. The fact that many parents choose this option has helped fuel the controversy surrounding both abortion and prenatal diagnosis.

Changing societal attitudes and the power of communications to shape them are illustrated by the American television series "Life Goes On," which stars a young actor with Down syndrome, Chris Burke. His interactions with the people around him are explored in a glamorized and yet ultimately illuminating way, one that can help educate the public about its own biases. At the same time, reality is in danger of being supplanted by fiction. The portrayal of this young man's life as happy and fulfilling has been cited by some prospective parents as their reason for continuing a pregnancy in which the syndrome has been diagnosed.

Despite continuing social changes, the effects of an extra chromosome 21 remain irreversible. This book will explain how our increasing knowledge of the human genome may eventually allow us to reverse the effects of this disorder and of many others, opening up an ever more extensive variety of choices to confused parents, doctors, and politicians. The laws that arise from genetic issues continue to change rapidly (though not as rapidly as the discoveries that make the legal changes necessary). But the changing patchwork of laws and attitudes worldwide can be better followed in the newspapers than in a book.

Indeed, the pattern is not a particularly logical one. At the moment, for example, Norway, unlike the United States, absolutely forbids any experimentation with human embryonic tissue. (The United States only forbids government funding of such research.) At the same time, and again in vivid contrast to the United States, the Norwegian government can and does step in to terminate a pregnancy in which the fetus is known to be defective. All this may have changed by the time you read this, as people and institutions scramble to keep up with new developments in the field.

The story I have to tell thus concentrates on the science, not the laws, and it may at times seem bewildering. As we venture into the genomic thickets, it will be necessary to look for signposts and patterns as guides. Some of the patterns are already well-known to human geneticists. Others are not so obvious. When they are revealed, you will see that they have the capacity to destroy old prejudices that we hold about our fellow humans. Please, if you have only a glancing acquaintance with genetics or none at all, start at the beginning and follow patiently along. I have designed this book to introduce you step by step to the amazing world of molecules and genes, while doing my best not to bore readers who know a fair amount already. By the end you will be quite comfortable with such esoterica as gene mapping, RFLPs, and introns.

Assiduous readers will find at the end of the book a set of notes giving further references for each chapter. And beginning on page 341, there is a glossary of terms for those who have forgotten after many pages what a Southern blot or an allele might be. I hope you find it useful; I have often wished that Tolstoy had done the same to enable us to keep track of the characters in *War and Peace!* Finally, a chapter called "Stop Press" covers recent developments. Written just before the presses began to roll, this was an effort to respond to the fast-moving world of human genetics. It updates some of the topics covered in the book and gives a glimpse of the latest directions of research.

Many of my fellow scientists have contributed to this book, in some cases unwittingly. So have the students I have taught over the years—the thousands of undergraduates in many different courses and the hundreds of budding doctors who have taken my human genetics course at the Medical School of the University of California, San Diego. They all have a boundless capacity for questioning the established order of things. I am also most grateful for the time taken by a number of children and adults with genetic diseases to explain to me the impact these conditions have had on their lives.

I particularly want to acknowledge help from and discussions with Rod Balhorn, Alan Beggs, Tony Beugelsdigt, Walter Bodmer, Morton Bradbury, Elbert Branscomb, Howard Cann, Charles Cantor, Jean Dausset, Larry Deaven, Peter de Jong, Joseph de Verna, Renato Dulbecco, Dick Dutton, Lisa Foreman, Walter Gilbert, Joe Gray, Hugh Gurling, Michael Harrington, Ivan Harwood, Ed Hildebrand, Eric Hoffmann, Lee Hood, Tim Hunkapiller, Jim Jett, Michael Kaback, Rob Kaiser, Jo Kiernan, Mary Claire King, John Kjekshus, Ben Kope, Louis Kunkel, Hans Lehrach, Dick Lewontin, Mark Martin, Hans Mohrenweiser, David Nelson, Debbie

Nickerson, Anne Olson, the late Kevin Riley, Neil Risch, Johanna Rommens, Leslie Russell, Jay Seegmiller, Steve Shepherd, Robert Sinsheimer, Cassandra Smith, Dan Sulzbach, David Torney, Marvin Van Dilla, Ron Walters, Rick Wilson, John Yates, Alice Yu, and Bruno Zimm.

I am particularly grateful to John Cairns, Tony Carrano, Ted Case, Charles DeLisi, Russ Doolittle, Ted Friedmann, Trevor Price, Fred Sanger, and Lap-Chee Tsui, who read parts of the manuscript at various stages and made many valuable comments. Susan Rabiner, my editor at BasicBooks, made many sensible suggestions and shot down some of my wilder flights of fancy. My mistakes are my own, of course, and occurred in spite of all their help. So are my conclusions, along with the assorted prejudices that will be on display in the course of the book.

ACKNOWLEDGMENTS

Grateful acknowledgment is made to the following for their permission to reproduce the figures in this book:

Figure 2–3 from *Molecular Biology of the Gene,* vol. 1, 4th ed., by J. D. Watson et al. (Menlo Park, Cal: Benjamin/Cummings, 1987), p. 283. Reprinted by permission.

Figure 5–1 from *Single Lens* by Brian J. Ford. Copyright © 1985 by Brian J. Ford. Reprinted by permission of HarperCollins Publishers.

Figure 5–2 from J. J. Yunis, "The Origin of Man: A Chromosomal Pictorial Legacy," *Science,* 19 March 1989, pp. 1525–29. Copyright © 1982 by the AAAS.

Figure 5–3 reproduced from R. Nave, D. O. Furst, and K. Weber, "Visualization of the Plarity of Isolated Titin Molecules," *Journal of Cell Biology* 109 (1989): 2177–87. Reprinted by permission of Rockefeller University Press.

Figure 5–4 from Richard E. Dickerson and Irving Geis, *The Structure and Action of Protein* (New York: Harper & Row, 1969).

Figure 6–1 from Y. Nakamura, et al., "A Mapped Set of DNA Markers for Human Chromosomes, *Genomics* 2 (1988): 302–9.

Figure 6–2 from Thomas Keneally, "Picture of the Simpson Desert," *Outback* (London: Rainbird, 1983). Photograph copyright © Gary Hansen, 1982.

Figure 6–3 from M. A. Van Dilla et al., "Human Chromosome-Specific DNA Libraries," *Bio/Technology* 4(1986): 537–52.

Figure 6–6 courtesy of David Ward, Yale University, and Peter Lichter, Max Planck Institute.

Figure 7–1 courtesy of Rod Balhorn, Lawrence Livermore Laboratory.

Figure 8–1 from "A Sensitive New Prenatal Test for Sickle-Cell Anemia," by J. C. Chang and Y. W. Kan. Reprinted with permission from *The New England Journal of Medicine* 307 (1982): 30–32.

Figure 8–2 courtesy of Cellmark Diagnostics.

Figure 12–1 from A. Bassett et al., "Partial Trisomy Chromosome 5 Cosegregating with Schizophrenia," *Lancet* (April 9, 1988): 799–801. Copyright © The Lancet Ltd.

Figure 12–2 from R. Sherrington et al., "Localization of a Susceptibility Locus for Schizophrenia on Chromosome 5." Reprinted by permission from *Nature* 336, pp. 164–67. Copyright © 1988 Macmillan Ltd.

Figure 12–3 courtesy of Charles Rick, University of California, Davis.

Figure 13–2 from B. Schwendowius and W. Dömling, ed., *Johann Sebastian Bach* (Barenreiter: Verlag, 1984), p. 53.

Figure 13–3 adapted from *The Trend of Scottish Intelligence* (London: University of London Press). Copyright © 1949, The Scottish Council for Research in Education, Edinburgh.

Figure 13–5 from *Architettura del Renascimento* by Peter Murray (Milano: Electra, 1971). Photo copyright © 1971 by Pepi Merisio.

INTRODUCTION

How Brave a New World?

"And now," Mr. Foster went on, "I'd like to show you some very interesting conditioning for Alpha-Plus Intellectuals. We have a big batch of them on Rack 5. First Gallery level," he called to two boys who had started to go down to the ground floor. . . .

But the Director had looked at his watch. "Ten to three," he said. "No time for intellectual embryos, I'm afraid. We must go up to the Nurseries before the children have finished their afternoon sleep."

Mr. Foster was disappointed. "At least one glance at the Decanting Room," he pleaded.

"Very well, then." The Director smiled indulgently. "Just one glance." —Aldous Huxley, *Brave New World* (1932)

It is the year 2025, and you are a genetic neurosurgeon working in a large big-city hospital. This morning you are confronted with a particularly urgent case: a newborn baby who has just been brought in by helicopter from a hospital serving a poor part of town. The baby's mother had been using super-crack, the latest addictive drug. Like cocaine, super-crack in a pregnant woman's system can cause the placenta to separate early, starving the fetus of needed oxygen during critical stages of development. The baby is small, under four pounds, and moves feebly. Its eyes have a fixed glassy stare; its skin is yellow with infantile jaundice. A CAT scan shows that the brain is also undersized and oddly smooth, the convolutions of the cerebral cortex less obvious than in a normal brain.

The baby is quickly anesthetized. Using the portable CAT scanner as a guide, you punch a large syringe through the soft spot at the top of the baby's head, between the hemispheres of the brain and far down into the fourth ventricle, where the fluid that bathes the brain joins that of the spine. You convert the needle swiftly into a catheter, and attach the first of a series of small plastic pouches. Inside the pouches, ready to diffuse into both the brain and the central canal of the spine, are a remarkable series of products of the geneticist's art. A set of twenty harmless viruses carry the genes that must be present in order for the mature neurons, or nerve cells, of the brain to dedifferentiate—to turn back into cells capable of dividing again. These viruses are highly specific; the cells of the brain stem that control breathing and other vital functions are unaffected by them. Other viruses carry genes that stimulate the growth of blood vessels, supplying needed oxygen to the baby's tiny brain as it begins to grow. All these viruses have been engineered so that they will destroy themselves if they escape from their container or the baby's body.

As a result of the introduction of these genes, over a period of weeks the baby's cerebral cortex will soften, diffuse, and grow as the cells round up and begin to multiply. Later, genes carried by other viruses will be introduced that will reverse this process, allowing the neurons to differentiate again. Many of the new neurons will die, a quite normal process. You will then introduce a carefully adjusted cocktail of nerve growth factors that will stimulate the formation of axons and dendrites on the neurons. These in turn will link up through synapses. The eventual result will be a complex neural network that will function quite well, though not quite as well as if the brain had developed normally throughout gestation.

The procedure is still crude, though it is constantly being refined. It works because, given the right push, the baby's own brain will carry out most of the healing process itself. With luck, after a month or two the baby's brain will be restored to near-normal function. You have had to carry out the procedure swiftly, for too much delay will cause many of the genes in the neurons to become irreversibly altered, making the cells unable ever to multiply again.

Astonishing as this procedure is, you know that it is only the start of what will one day be feasible. About a third of the hundred thousand or so genes of a human being are responsible for the development of the nervous system, and while all of them have been sequenced by this time as a result of the Human Genome Project, only a tiny fraction of them are even dimly understood. As you work, you think about the many other procedures that are under development, some designed to reverse the effects of brain injury, others to halt the degeneration that results in Alzheimer's disease. Such procedures are far more difficult to devise, since the adult brain cannot be manipulated as easily as that of a newborn.

The baby is attached to an IV for nourishment and put in an incubator to await what is nothing less than the renewal of its brain. You turn to your next case, a girl baby born with Tay-Sachs disease. Tay-Sachs results from a mutation in a stretch of DNA on chromosome 15. The normal, unmutated gene specifies an enzyme that breaks down a fatty substance inside nerve cells. In people with Tay-Sachs, this enzyme never gets made, and the fatty substance accumulates, eventually killing the cells. A baby with the disorder appears normal at birth, but without treatment will inevitably die within a year or two as a result of progressive neurological damage.

Most people carry two copies of the normal *allele* (the technical word for a particular form of this gene). One allele is received from each parent. A few people are *heterozygous* for the mutant, meaning that they carry one normal and one mutant allele. In these cases, the mutant allele is quite harmless. It is *recessive* while the normal allele is *dominant,* meaning that the normal allele is able to carry out the gene's function even if a mutant allele is also present. For many years, most of the people who are heterozygous for Tay-Sachs have been alerted to this fact by their physicians. Tracking down the hidden mutant alleles has been made easier by the fact that many of them are found among Ashkenazi Jews.

Much rarer than these heterozygotes are those who, like your tiny patient, are *homozygous* for the mutant allele. For this to happen, both parents must be heterozygous, and even so the child of such a marriage has only a 25 percent chance of being homozygous. If two heterozygotes do marry and a homozygous baby is conceived, the problem can be detected early in development, provided that the appropriate test is carried out. Until recently, most such fetuses were aborted, but since the beginning of the twenty-first century it has been possible to do gene surgery early in development and correct the defect.

This child has slipped through the net. Her mother was known to be heterozygous for the mutant gene, but her father appeared to be normal. Unfortunately, a mutation had occurred in one of the father's cells during his early development, making some of his sperm-producing cells heterozygous. Most of the cells in his body were normal, so this event was not detected through the standard screening. By chance he has passed this new mutant gene on to his child. The pregnancy appeared to be normal, but the baby's condition was picked up shortly after birth, when a routine retina print showed the damaged retinal nerve cells typical of the earliest stages of Tay-Sachs.

Curing the baby's mutant Tay-Sachs gene was straightforward. A specifically tailored virus, carrying a healthy piece of gene to replace the mutant one, was introduced into the affected brain cells. But undoing the damage done during the nine months of gestation has been far more difficult, because the excess material that collects in the cells of Tay-Sachs victims is made up of molecules that normally form an integral part of the cells themselves. In your repair efforts, you have been forced to draw a fine line between breaking down the excess material and breaking down the cell. Luckily, the extra fatty material in each cell has become chemically modified. This fact enables you to introduce into the baby's nerve cells other genes, derived from bacteria, that can recognize and break down only the modified material.

But homozygosity for the Tay-Sachs allele has caused more problems. Untreated, the baby's retinal damage will lead to blindness, so you have had to use techniques similar to those used on the brain-damaged baby. To rebuild the Tay-Sachs baby's retinas, you have diffused a series of viruses into the jellylike vitreous humor of the baby's eyes. The viruses carry genes that dedifferentiate retinal nerve cells, allow them to multiply, then differentiate them again. This morning you give the baby, who is healthy and active, eyedrops containing the last of the series. The infant's retinas will soon be healed.

Your next case is a far more difficult one. A worried woman, pregnant with her first child, arrives for a conference. Her husband has recently been diagnosed as a schizophrenic, his condition made even worse by severe periodic depressions. By this time, geneticists have managed to track down three alleles—at three different genetic *loci,* or locations on the chromosomes—that are known to contribute to schizophrenia. Her husband carries two of them. A test of the unborn baby's genes shows that it carries one. Your patient is distraught over the news, and wants to know if she should terminate the pregnancy. You tell her that there are no easy answers. You can give her more definite numbers than would have been possible a generation ago, but even with all the knowledge that has been gained about the human genome in the last few decades, mental illness is still a difficult problem.

There happens to be a good deal of hope for this particular case. First of all, there is no schizophrenia elsewhere in the father's family. This means that there may be other genes in his side of the family helping to counteract the effects of the schizophrenia alleles. You tell your patient that a child with such a family history has only a 5 percent chance of developing schizophrenia, compared with a 1 percent chance for the general population. Second, excellent drug therapies are now available for schizophrenia. Her husband is unlucky, since his case is so severe and has been complicated by depression. Only about 5 percent of schizophrenia cases do not, like his, respond to drug therapy. So there is only a 5 percent of a 5 percent chance that the child will develop an untreatable case of schizophrenia—one chance in 2,500. The gene the child carries is a worry, but simply carrying a schizophrenia gene does not automatically mean that schizophrenia will result.

The mother-to-be is relieved, but still uncertain. She asks your opinion on whether she should abort. You tell her that the pregnancy appears normal and that there is an excellent chance her child will be fine. Most importantly, given her family situation, the chances of schizophrenia would be about the same even if the child did not carry any of the known schizophrenia alleles. Aborting this child and having another will not change the odds appreciably. This information leaves the mother somewhat reassured.

After a hurried lunch, you attend an exciting seminar on advances in the study of aging. This is now a very fast-moving field, and the seminar speaker is in the thick of the excitement. He holds the large audience in thrall. They know that in the last few years a number of genes that can lengthen the life span have been isolated in mice. He announces that the same genes have now been found in humans, and the relatively rare alleles that lengthen the life span in mice have turned out to be very similar to the predominant alleles in the human population. In other words, the speaker points out, the lengthening of the life span that has occurred over the course of human evolution has drawn in part upon genetic variation that must have been present in the common ancestors of humans and mice.

Now, one of these genes has been modified in the laboratory and inserted into mice, with startling effects—the new gene doubles its life span. The next step will be to make a similar modification to the corresponding gene in chimpanzees, to see the effect. Years of study will be needed to determine whether the modification is safe, or whether it interacts with other genes to produce unexpected results. But it could in theory allow humans to live to an age of 150 or more.

You leave the seminar with your brain spinning. In mice, the new gene has been found to slow down all the stages of development and senescence about equally. If such a modification could be made to work in humans, would people be condemned to a difficult, drawn-out old age? What would this do to the Social Security system and to the system of geriatric care, both of which are already on the ropes? Sixty million people in the country—a fifth of the population—are now over sixty-five, and the burden on the working young is becoming insupportable. On the other hand, the ever-earlier onset of puberty due to improved nutrition over the last two centuries has contributed to tremendous societal problems. Would lengthening the time of childhood give children a chance to attain greater intellectual maturity before puberty begins, so as to withstand better the emotional onslaught of adolescence?

Still wrestling with these problems, you go to your next appointment. It is a meeting with other specialists about the child of the addicted mother whom you treated in the morning. There are many complications. One debate concerns just when the regrowth of the baby's cerebral cortex should be halted. Studies have shown that if it is stopped early, the baby will have reduced intelligence. If it is stopped late, the baby's intelligence will be closer to normal, but there will be a much greater chance of strokes later on in life because the newly grown brain tissue will tend to block off blood vessels. It is an agonizing choice to have to make, and all you can do is try to pick what you hope will be the optimal moment to stop the regrowth.

The second problem has to do with the mother. She has been provided with a court-appointed lawyer who scents an opportunity to create a legal precedent. He is claiming that the baby was taken from her and genetically altered without her permission. The hospital's legal branch has countered by seeking a ruling declaring the mother to be temporarily unfit. The law, which always follows far in the wake of scientific developments, is very unclear on the mother's rights.

As a doctor, you are impatient with all this legal wrangling. The important thing, you point out, was to save the baby from becoming a vegetable. You suggest facetiously that because the baby will have a new personality along with its new brain, it might no longer be considered the child of its biological mother. The lawyers at the meeting are not amused by this argument.

There may be a way around the difficulty. Children whose brains have been regrown in this fashion often have behavioral problems. Their attention spans are shortened, and they may not respond appropriately to external stimuli. Untreated, such a child may show aberrant behavior all its life. The best way to deal with these problems is to put the child in a course of intensive therapy, as early as possible, to compensate for these deficits. Started early enough, a well-designed course of therapy can actually improve the brain's function. Special classes have been established for children who have problems of this type that have been traced to genetic defects. Some of these children, already in their teens, are far better adjusted than an untreated control group. You suggest that the mother be persuaded to enroll the child in such a program. She cannot be forced to do so, but you promise to meet with her and bring all your persuasive powers to bear.

The Tay-Sachs baby also presents problems. While the parents are happy that their child has been effectively cured of the physical manifestations of this dreadful disease, they want her to be genetically cured as well. This would require replacing the defective alleles with normal alleles, not just in her nerve cells but in the cells of her ovaries as well. Some of the baby's ovary cells have already partially matured into the egg cells that will give rise to her own babies, making up the so-called germ line. If these germ-line cells could be repaired, she would then be able to grow up and have children, secure in the knowledge that she would not be passing the Tay-Sachs allele on to her own offspring. With only her nerve cells repaired, she can still pass the mutant gene on, because the repaired cells will never turn into germ-line cells.

The problem is difficult for a number of reasons. It is not yet certain that the genetic surgery needed to cure the germ-line cells is sufficiently precise. Unwanted genes may be put into these cells by accident, and passed on to future generations. You are confident that current techniques are good enough to make this very unlikely, but you admit that the risk exists. In addition, a dangerous precedent might be set. Germ-line gene surgery has been carried out in other countries, but never so far in the United States. (If it has, it has not been revealed). Strong political sentiment has been rallied against it, led by descendant organizations of the Right-to-Life and Green movements of the twentieth century. With the development of new methods of birth control in the late 1990s, the number of elective abortions in the country dropped precipitously. The Right-to-Life movement largely disappeared, to be replaced by a strong and vocal lobby against the alteration of any germ-line genes by genetic surgery. The lawyers at the conference are adamant that the hospital's resources should not be squandered on what promises to be a long and expensive courtroom battle.

You agree regretfully that the parents should be told that germ-line surgery is simply not feasible at the present time. Privately, you feel strongly that ridding the population of an allele like Tay-Sachs would be a very good thing to do.

That evening you talk to your husband, who is a teacher in a school with a large number of disadvantaged children. Many of the children have just participated in an exciting genetic test carried out by one of your colleagues. One consequence of the complete mapping of the human genome has been a great upsurge of interest in determining whether particular alleles at various gene loci are common among people who exhibit strong talents. An early success in this search has been the discovery of an allele that is found more commonly among mathematicians than in the general population. The gene at this locus has several alleles, only one of which shows an association with mathematicians, and its function is a complete mystery. The allele, whatever it is, has been found in about 20 percent of mathematicians, but in only about 10 percent of the population at large. Your colleague has surveyed the children in your husband's school and has found the allele in about 12 percent of them. This news has been greeted with excitement by both children and teachers, who take it for what it is—a direct genetic demonstration of racial equality, at least at this gene locus. They are anxious to start an advanced math program in the school; the question is who should be enrolled. Your colleague is out of town, so your husband asks you to address a meeting at the school to try to explain the genetic complexities of the situation.

After dinner you go to the school and talk to a group of parents, students, and teachers. You explain that not all mathematically talented children will have the allele, and not all those with the allele will be mathematically talented. After all, 80 percent of mathematicians do *not* have the allele, compared with 90 percent of people in the general population. And lots of people in the general population have the allele but show no obvious sign of mathematical talent. So the school's new program must not be confined to children with the allele; such a requirement would exclude most of the mathematically talented students from the classes.

You try to explain these difficulties without dampening your audience's enthusiasm for the new program. You point out that they are quite right to be excited about the study's results. Like many other studies in recent years, it shows that there are no important allelic differences between racial groups in the genes that really matter, such as those associated with talent and intelligence. Taken together, these studies show that there are so many alleles at so many gene loci involved in human talents, all widely distributed through the population, that nobody has been left out. The bottom line, you conclude, is that there are untapped riches in everyone's genome. Schools, parents, and children must join together to find and exploit them. The audience applauds doubtfully, not sure whether to believe you or not.

Afterward, you chat with a school counselor who is running a program for children with behavioral problems. He wants to know whether genes have been discovered that explain their behavior. You tell him that, just as with the allele associated with mathematical ability, alleles have been found that are more common among criminals than among the general population. The pattern is the same as with the mathematics "gene," however. Although certain alleles are found more often among criminals, most criminals do not carry them, and most people who carry the alleles are not criminals.

Scientists do not yet understand the function of most of the "criminal" alleles, although they know that some are involved in controlling endocrine hormone levels. Denmark is starting a genetic screening program to look for children with an unusually high number of these alleles, with the intention of putting these children in special classes where they will receive intensive counseling. The Danish parliament is currently debating whether this special handling will unfairly stigmatize these children. You tell the counselor that in your opinion too little is known about these alleles as yet to design a sensible care or therapy program. The Danish scheme is flawed because many children with no problems at all will be put into it and perhaps stigmatized as a result. Valuable data may emerge from the program, however, leading to the design of better programs in the future.

When you finally collapse exhausted into bed that night, you wonder what the next day will bring. Whatever it is, it is sure to be exciting.

CHOICES

In the last few pages I have tried to give you a glimpse of the sort of future that may spring from our current exploration of the human genome. Everything I have suggested is a very real possibility, an outgrowth of trends currently in progress. The exciting scientific and medical ferment that has already begun is sure to continue. Things will get more complex, not simpler. Uncertainties will remain, as in the case of the mother worried about the possibility of schizophrenia in her offspring. We will have to make painful choices unimaginable today, and indeed far more numerous than those we now face. But far more information than ever before will be available for people to make rational decisions about their lives and those of their children.

While the Human Genome Project has helped to focus attention on these emerging changes, it is only a part, and in some ways a very small part, of the current explosion of knowledge about human genetics. This is worth remembering: Vast as the goal of this project is, it is only a beginning. If the project is successful, it will have determined the entire store of information in our chromosomes by the year 2005. In that year, if all goes according to plan, it should be possible for any research worker to call up a computerized data base and locate the sequence of any region of the three billion bases of DNA that constitute our genome.

This information will give us the capacity to design experiments in gene therapy and gene manipulation that are much more sophisticated and far-reaching than anything we can conceive of today. The outstanding problems in human biology—the genesis of cancer, the routes by which the human embryo and fetus develop, the mechanisms of aging, the origin of mutations—will all be illuminated in a strong and steady light by the results of this undertaking.

Still, it must always be remembered that simply determining the sequence of all this DNA will not mean we have learned everything there is to know about human beings, any more than looking up the sequence of notes in a Beethoven sonata gives us the capacity to play it. In the future, the true virtuosos of the genome will be those who can put this information to work, and who can appreciate the subtle interactions of genes with each other and with the environment.

A nightmare future I did not explore, because I think it far less likely, would be a totalitarian one in which knowledge of the human genome might be used to pigeonhole people—to determine their jobs, whom they should marry, whether or not they should have children: in short, a world in which our genes have been shown to be the ultimate determiners of our destinies. This book will show why exploration of the human genome should never lead to such a conclusion. Anyone who claims otherwise is a liar.

To see that this is so, we must examine the new human genetics in detail, looking at just what it means to map and sequence the DNA, and what relationship all this information has to the humans who carry it. We will see how some genes that contribute to a disease or some other characteristic are relatively easy to track down, while others are much harder to find and still others—the great majority—are effectively impossible to detect. We will also see how they will remain impossible to detect even when we know the sequence of every one of the hundred thousand or so genes in our genome. This may seem paradoxical, but by the end of the book you will understand why it is true.

I also hope to show by the end of the book that the flawed but optimistic future I have just painted is not an unlikely one. This introduction has offered a possible future filled with exciting discoveries and challenging problems. In the next few chapters I will trace some of the many lines of research that have led us to the point where we can imagine such a future. I will begin with the ideas and discoveries that have made it possible to conceive, for the first time, of the incredible idea of reading the entire sequence of human genes. This idea occurred to many scientists simultaneously and it is, with remarkable swiftness, becoming translated into reality through the Human Genome Project. The project's birth was not easy; its current shape is as confusing and amorphous as the field of human genetics itself, and it is not at all clear how and when it will reach its goals. Yet it has seized the imagination of scientists and the public like few other ideas in the history of modern biology.

Part One

ORIGINS

CHAPTER 1

Cancer

Eventually the techniques of nucleic acid chemistry should allow us to itemize all the differences in nucleotide sequence and gene expression that distinguish a cancer cell from its normal counterpart, and perhaps at that point the steps in carcinogenesis will cease to be in doubt. —John Cairns (1981)

The Human Genome Project is an idea whose time has come. To understand why, we must look at the many intellectual, technical, and political currents that came together at this moment in the history of science—none more important than those springing from cancer research. This most intractable of human diseases cries out for draconian measures, and just such an approach was proposed by the Nobel Prize–winning virologist Renato Dulbecco in an editorial in the journal *Science* in 1986. His was the first public call for a Human Genome Initiative:

> We have two options: either to try to discover the genes involved in malignancy by a piecemeal approach, or to sequence the whole genome of a selected animal species. . . . I think it will be far more useful to begin by sequencing the cellular genome. The sequence will make it possible to prepare probes for all the genes. . . . Classification of the genes will facilitate the identification of those involved in [cancer's] progression.

Why is cancer so difficult to deal with, and why did Dulbecco feel he had to call for something so sweeping as a complete sequencing of every gene in the human genome? To understand this we must look at cancer in a broad perspective—the one taken by molecular geneticists and cell biologists, not that of the doctor who treats individual patients. Scientists working with cancer at these fundamental levels have discovered that the disease has many forms, which seem to grow in number the more it is studied. It can be likened to the ancient Greek myth of the Hydra.

The killing of the nine-headed Hydra was the second of the twelve labors demanded of Hercules by his cousin King Eurystheus of Tiryns. Like all of Hercules' labors, this task turned out to be much more difficult than it first appeared. When Hercules finally tracked down and confronted the monster, he discovered that as quickly as he lopped one head off, two more grew in its place. Hercules finally slew the Hydra with the aid of fire, scorching the stumps that were left after he cut off each head. The middle head, however, was immortal, and went on hissing and biting even after it was severed.

Hercules dealt with this problem resourcefully by burying the head under a huge rock. He was able to return triumphantly to Eurystheus with the news that he had killed the monster, at least to a first approximation.

Cancers, too, can be killed to a first approximation, through the application of radiation and drugs that destroy the continually dividing cancer cells while doing less damage (though sometimes only marginally less) to normal cells. Often, however, like the immortal head of the Hydra, cancer cells have the capacity to come back. In spite of President Richard Nixon's proclamation of a war on cancer in 1971 and the expenditure of many billions of dollars on research since then, there is still no certain cure or vaccine for any of the more than 200 types of cancer that affect humans. A small number of cancers, including testicular teratocarcinoma and some childhood leukemias, have a very high cure rate through radiation and chemicals, but these are the exceptions.

Other organisms are luckier. A cancer vaccine has been developed for, of all creatures, chickens. Most of the chickens we so thoughtlessly consume are raised under conditions of extreme overcrowding. Feathers, dust from dried excreta, and dander—literally, dandruff—fill the air of a henhouse, producing optimal conditions for the rapid spread of bacteria and viruses. Outbreaks of disease are common. Control of Newcastle disease often requires that millions of chickens be slaughtered and cremated.

So when in 1906 Dr. Jószef Marek was petitioned by a distraught breeder to investigate a mysterious plague that was paralyzing his chickens, he was not surprised. Treating chickens formed a large part of his practice as head of the Department of Veterinary Medicine in Budapest's Royal Hungarian Veterinary School. Marek examined four of the breeder's paralyzed cocks and found that the nerve roots of their spines were unusually thickened, due to an infiltration by cells that, whatever they were, were obviously not nerve cells. This infiltration had led to the paralysis, but beyond that, Marek could tell the farmer nothing. The only thing to do was to destroy the infected chickens, and hope that this would check the spread of the disease. Now that Marek and other veterinarians knew what to look for, however, they soon found other outbreaks, in Hungary and elsewhere. For decades afterward, scientists dealing with what came to be known as Marek's disease argued about whether it was a true cancer or simply some kind of cellular response to a bacterial or viral infection.

During the 1950s, when the introduction of antibiotics allowed chicken farmers to raise their animals under even more crowded conditions, new and more virulent strains of Marek's disease appeared. Luckily, methods for studying and growing viruses in the laboratory were also improving. We now know that Marek's disease is indeed a cancer, and that it is directly infectious; indeed, it is the most infectious cancer that has been found to afflict any organism. It spreads through the dust and dander in chicken houses, as virus-loaded cells are sloughed off the skins of infected chickens in immense numbers. The herpes-like virus responsible was first successfully characterized in 1967. Exploitation of the discovery was remarkably swift: An attenuated live virus vaccine became available in 1970. The disease is now largely under control, at least in developed countries, though there are continual small outbreaks as resistant strains of the virus evolve.

Other infectious cancer viruses of chickens have been studied intensively, particularly the famous sarcoma (soft tumor) virus found in 1910 by Peyton Rous of the Rockefeller Institute. Rous showed that filtrates of solid chicken tumors, from which all cells and bacteria had been removed, could produce new tumors when injected into unaffected chickens. His sarcoma virus was one of the first viruses to be discovered, and was the first to be shown responsible for a cancer.

Rous' finding gave an enormous impetus to the study of cancer viruses. The hope of repeating in humans the success of the treatment for Marek's disease in chickens dominated cancer research in the middle part of this century. The story quickly became complicated, however, even in chickens. Each virus that was discovered turned out to have its own set of properties. For example, it has proved very difficult to develop vaccines against the Rous sarcoma virus, and the vaccines that have been created remain far less effective than the vaccine against Marek's disease. Yet in some strains of chickens the tumors produced by the Rous virus tend to regress and disappear spontaneously. This does not happen with Marek's disease, or with most other cancers.

In spite of the great difficulties, much has been learned about human cancer viruses since Rous's discovery. Some strains of papilloma virus have been implicated in human cervical cancers. An early indication of the infectious nature of this cancer came from the observation that women whose husbands were uncircumcised (and therefore harbored the virus under their foreskins) were more likely to develop cervical cancer than those whose husbands had been circumcised. The Epstein-Barr virus, responsible for mononucleosis, has been linked to the development of a lymphoma common in tropical Africa. And another virus, similar in many ways to the one responsible for AIDS, has been associated with, and is probably responsible for, a form of leukemia that is common in Japan.

Perhaps the most striking association between a virus and a human cancer is found in Asia, where the hepatitis B virus is very widespread. In some parts of China, such as southern Guangxi province, as many as half the deaths in males result from an incurable liver cancer. Most of the men who contract the cancer also test positive for antibodies to the virus, but many who are cancer-free harbor the virus as well. The deaths are concentrated in areas where the food contains high levels of certain toxins produced by fungi. The virus and the toxins seem to act together to increase the incidence of cancer, although liver cancer can still appear in the absence of one or both of these factors.

Cancer can be catching in humans, although this seems to be very rare. In one dramatic treatment for childhood leukemia, the child's bone marrow is destroyed with radiation and chemicals, then replaced with cells from a donor, usually a brother or sister. The leukemia can return, and in a few cases it appears that the resurgent cancer may have come not from a residual cell of the original cancer but from a cell of the donor. Something in the recipient's system appears to be able to cause a cancerous change in the cells received from the donor, who has remained healthy and cancer-free. Is that something a virus? Nobody knows.

In humans, unlike chickens, a virus cannot usually produce cancer by itself. It needs help. The help can come in the form of widespread genetic damage to the target cells, such as that produced by food toxins. In fact, genetic damage alone can produce cancer, apparently without the aid of viruses. The damage needed to trigger cancers can come from a great variety of environmental insults to living cells, provided they are severe enough.

One such environmental insult plagued nineteenth-century chimney sweeps. You may recall the jolly chimney sweeps from the film *Mary Poppins,* but these men and children were anything but jolly in reality. A Dickensian underclass, they were responsible for keeping the millions of coal-burning fireplaces in England drawing well. This meant freeing the chimneys of the partially burned coal residue that blocked them and sometimes burst into unwanted flame. Small boys, the smaller the better, were the cheapest way to clean soot from these chimneys. They crawled or were pushed up into them and knocked the soot free. Unsurprisingly, the boys often died. If asphyxiation, tuberculosis, or starvation did not carry them off, a crusty black skin cancer, usually appearing first on the scrotum, would. Percivall Pott, a physician and surgeon at St. Bartholomew's Hospital in London, showed in 1779 that this "soot-wort" was a true cancer, which could spread to the lymph nodes and other tissues. Because the disease was confined to chimney sweeps and others who

Figure 1–1. Two young chimney sweeps are overcome by fumes, and their tragedy suddenly intrudes into the drawing room of a middle-class family.

handled soot, Pott was quite sure that soot was to blame, but he had no idea of the mechanism involved.

We now know that the black oily soot from incompletely burned coal contains many powerful carcinogenic agents. When simply applied to the skin, the soot can produce cancers in organisms ranging from mice to humans. Similar agents are found in other incompletely burned substances of biological origin, notably tobacco, which is by far the most powerful carcinogen in our present environment.

Usually a cancer appears only after years of exposure to a carcinogen. Some carcinogenic agents, such as coal tar, interact with the cells chemically, while others appear to do so physically. The air of many buildings and factories is filled with microscopic fibers of asbestos, some forms of which can cause cancer. While these fibers are chemically quite unreactive, they can literally pierce cells from one side to the other like fowl on a spit, causing chromosome damage and opening up the cells' interiors to chemical influences from the outside.

The immune system, the body's primary defense, is often surprisingly ineffective in defending against cancers. Some viruses alter the surfaces of cells when they invade them, enabling the immune system to "see" which cells have become infected and destroy them. Unfortunately, most cancer cells, including the commonest kinds, show few external signs of the changes that have turned them into potential killers. As a result they remain virtually invisible to our immune systems.

Scientists are now trying to make the immune systems of cancer victims more effective by growing components of their systems to high levels outside the body and then reintroducing them. This approach can bring about remarkable cures, but more often it does not, and the side effects are often severe. The limited success of these efforts suggests that trying to boost the immune systems of people who already have advanced cancers may be like closing the barn door after the horse's departure.

Scientists once thought of the immune system as a kind of policeman, which kept an eye on all parts of the body and destroyed most cancers before they got a chance to start. This concept of *immune surveillance* had to be modified when it was discovered that mutant mice with no functioning immune systems did not seem to develop most kinds of cancers any more often than normal mice. It is possible that other protective systems are at work in these cases. Recently, mice with severely disabled immune systems were found to develop cancers at high rates when their cages were neglected and dirt was allowed to accumulate.

The recent and widely publicized observation that people in good physical health show low incidences of cancer presents a challenging puzzle to researchers: Which aspects of "good health" are actually protective? Diet may play a role—not only the kind of food, which everybody worries about, but the amount. Experimental animals given unlimited food develop cancers far more often than those on a more austere regimen. One possible explanation is that this Lucullan fare results in ill health, which in turn damages the body's defensive patrols against cancer. In contrast, malnutrition during youth actually decreases the lifetime risk of cancer (though it increases the risk of other diseases).

THE CANCER TWILIGHT ZONE

It is clear, then, that ill health during adulthood causes a breach in the body's defenses against cancer. What remains uncertain is the nature of these defenses. Because in humans most cancers do not seem to run in families, the idea that there could be a genetic component to cancer was rather slow to take hold. Once again, lower animals showed the way. For years it has been a simple matter to breed strains of mice that are highly susceptible to certain cancers. Strains of chickens resistant to Marek's disease were actually produced by selection as early as the 1930s. Individual poultry breeders could accomplish this simply by removing diseased animals from their flocks. This was the chief way flocks were protected before the introduction of a vaccine.

It took a long time for the discoveries of genetics to penetrate the world of cancer research. One reason is that there is no single gene for cancer. Whenever a gene is found that seems to "cause" cancer, the story quickly turns out to be much more complicated. The rare skin disease xeroderma pigmentosum (dry pigmented skin) is a good example. Children with this disease show no symptoms for the first year or two. Then, gradually, their skin becomes more and more susceptible to damage from ultraviolet (UV) light, until even the light from fluorescent lamps is enough to cause severe freckling. The precancerous skin lesions so common among older people appear in these patients when they are young, and many develop rapidly into cancers.*

*It may be that people with fair skin and freckles, who also have a high incidence of skin cancer (though nowhere near as high as that of the victims of xeroderma), will be found to have milder problems with the same repair systems. This would be very interesting to look at, and if true would raise the problem of why the genes for fair, freckled skin are so common in some human groups—far commoner than the genes for xeroderma.

It is now known that homozygosity for mutant alleles at any one of a number of genetic loci will lead to xeroderma. The genes at these loci control the repair system that normally heals damage caused to DNA by UV light. If any one of the steps in the repair system is missing, the entire system cannot function, leading to uncontrolled damage in cells exposed to UV.

Because of the early onset of skin cancer, people with the disease usually only live to around forty, but this is still long enough for various health statistics to be collected on them. One odd pattern in these statistics was first noticed by John Cairns, an epidemiologist at Harvard. He found that while the incidence of skin cancers in xeroderma victims eventually approaches 100 percent, the incidence of cancers that affect other tissues and organs is no higher than in the general population.

Clearly, then, a defect in the UV repair system is not sufficient to trigger the appearance of a cancer; it has to be coupled somehow with DNA damage by UV light. Cells deep inside the body of a xeroderma patient are unaffected because, while they all carry xeroderma genes, they are protected from UV light. It is not a xeroderma allele itself that "causes" cancer, but rather the genetic havoc that ensues when the xeroderma allele is present, preventing any UV-damaged DNA in the cell from being repaired.

Such genetic disruption is typical of all cancers. Cancer cells characteristically possess broken chromosomes, or too many or too few, and they have all kinds of things wrong with their physiology. A very serious genetic condition called Bloom's syndrome causes chromosomes to break at a very high rate in all the cells of the body, not just those exposed to UV light. Patients with this syndrome show a high incidence of many kinds of cancer. Indeed, cells that are going to become cancerous are typically marked by an ever-increasing number of genetic problems. For some cancers, a few of the steps leading to *carcinogenesis,* the onset of cancer, have been worked out, but estimates for the number of events that must occur inside a cell for it to turn cancerous range as high as ten.

In large part through the work of Michael Bishop and Harold Varmus of the University of California Medical School in San Francisco, it is now known that the genome contains several dozen genes called *proto-oncogenes.* All of these are involved in various aspects of the regulation of cell growth, and they can be classified into two groups. For a cell to become cancerous, one or more proto-oncogenes from each group must be converted into *oncogenes* (cancer genes) by mutation. Such a conversion is a rare event unless it is triggered by something specific, like the invasion of a virus or the breakage of a chromosome at just the point where the proto-oncogene is located. The chance of two such events, each involving a proto-oncogene from one of the two groups, happening in the same cell should be very small indeed. And even this unlikely happenstance is not enough to trigger a cancer, for there are other genes, called *anti-oncogenes,* that protect against the effects of mutant oncogenes. For the next steps toward cancer to be taken, one or more of these anti-oncogenes must be destroyed, or else be converted by mutation into new forms that are unable to carry out the protective function.

Genetic mayhem at the gene and chromosome level can continue even after the cell becomes cancerous. This confuses the picture, covering up the traces of the original cancer-causing events. Different cells taken from the same tumor can exhibit very different genetic changes. The job of cancer biologists resembles that of investigators from the Federal Aviation Association, who must comb the fragments of a plane wreck to try to pin down one of hundreds of possible causes for the accident. FAA investigators have a much simpler time of it, however, for a plane accident may be caused by as few as one or two human or mechanical failures. Cancers, or at least most human cancers, seem to be caused by many more failures than that.

"THERE IS NO LIMIT TO OUR KNOWLEDGE"

The possibility of finally unscrambling this multitude of influences led Renato Dulbecco to his dramatic proposal in *Science* magazine. A quiet, courtly man, Dulbecco is currently the president of the Salk Institute in La Jolla, California, and his office looks west over a dazzling view of the Pacific. Sitting under an old framed map of Europe that dominated one wall, he talked with me about the reasoning that led him to write the *Science* editorial.

I began by trotting out my comparison of cancer researchers to FAA investigators. Dulbecco swiftly extended the analogy even further, saying that cancer research was more like investigating the scattered fragments of a plane of a completely unknown type, so that it would be almost impossible to determine whether a part was normal or damaged—much less whether it was the one responsible for the accident. Dulbecco's work on the effects of tumor viruses on cells had shown him how difficult it was to track down the causes of cancer. Advances in molecular biology and gene sequencing led him to think about the possibility of sequencing the entire genome. In this way the unknown aircraft could be converted into one in which the structure was completely understood.

Dulbecco felt certain from the first that the human genome would be sequenced. He pointed out that there is no limit to what we can find out about ourselves—the limits are all man-made and not imposed by nature. Before writing the *Science* editorial, which appeared in March 1986, he gave two talks in which he proposed the idea of sequencing the genome. The first took place in May 1985, at the dedication of a new laboratory at Cold Spring Harbor on New York's Long Island, and the second in January 1986 at a tumor virus meeting in California. Each time, the audience reacted with disbelief.

Now, Dulbecco can look back with pleasure on how the idea has grown. His current involvement is substantial; he is helping to coordinate the efforts of twenty-nine different groups scattered throughout Italy, each involved in mapping a particularly puzzling portion of the X chromosome. In some people this region is fragile and breaks easily, and the breakage leads to mental retardation and often to autism. The connection between the fragility of the chromosome and mental retardation is an even greater mystery than the connection between chromosome fragility and cancer.

Yet the connection between cancer and the genesis of the Human Genome Project is clear. Finding out about every gene that could possibly be involved might be the most cost-effective way to slay the Hydra. Once all the sequences of the genes of a normal cell are known, it should be possible to track down all the changes that take place in a cell as it becomes cancerous. There are certain to be many such changes, and a good fraction of them will turn out to be extraneous to the generation of the cancer. If the genesis of many cases of a particular type of cancer can be followed in this kind of detail, certain changes and sequences of changes will be found to be common to all of them. Finding this common set of changes will be a difficult and painful task. But at least the work will not be hopelessly confused, as it is at present, by the smokescreen of irrelevant genetic alterations that seem to be an unavoidable by-product of the generation of cancers. Gradually, a far more complete understanding of the changes that are important will begin to emerge.

We know already that the anti-oncogenes and other genes that protect against cancer form part of an interlocking set of defenses that has developed over the course of evolution. These defenses must be breached by outside forces before *oncogenesis,* the transformation to cancer, can occur. Much evidence suggests that defenses against cancer are more complex, numerous, and subtle in humans than they are in other organisms. Indeed, our increased resistance both to mutational change and to cancer is part of what enables us to live so much longer than chickens or mice. It is not a coincidence that chickens are so much more prone to cancer than we are, and that they can catch some cancers as easily as we catch a common cold.

A major goal of the Human Genome Project as Dulbecco envisions it will be to discover the nature of this enhanced protection. The project is open-ended, for in order to find out why our defensive fortress is less easy to breach than those of other animals, it may be necessary to sequence not only all the human genes but all the genes of mice and perhaps those of chickens, in order to find out how their defenses differ from ours. The final goal will be not just to understand our defenses, but to strengthen them, in large part by pinpointing and dealing with the environmental factors that are most likely to breach them. As Cairns and Dulbecco perceptively realized, a profound understanding of the nature of our genes—all of them—will be essential to this enterprise.

CHAPTER 2

DNA

When Jean François Champollion (1790–1832) was a boy of eleven, he met the famous mathematician Jean-Baptiste Fourier, who showed him some papyri and stone tablets bearing hieroglyphs. "Can anyone read them?" he asked. When he was assured that no one could, the boy promptly replied: "I will do it."
—F. Gladstone Bratton, *A History of Egyptian Archaeology*

Fred Sanger, upon retiring from his chemistry post at Cambridge University, moved from a typical faculty house on Hills Road to a much quieter home in the nearby village of Swaffam Bulbeck. Sharing lunch with me at the nearby Red Lion Pub, he waxed enthusiastic about the fun he was having taking care of his acre of garden. Men like Sanger, quiet, cheerful, and bronzed by the sun and wind, can be seen working in gardens throughout the English countryside. Yet he is surely the only one of these legions who has received two Nobel Prizes.

The first, awarded in 1958, recognized his work in determining the complete structure of a protein. The second, twenty-two years later, was for finding a simple and elegant way to determine the sequence of bases, the subunits making up DNA. His invention opened up the possibility, now rapidly being realized, of reading the genes of any organism like a book. Thus, perhaps more than anyone else, Sanger made the Human Genome Project and the current ferment in human genetics possible. His two great achievements are related to each other in fundamental ways; to understand one we need to understand them both. Sanger encountered difficulties in his earliest work that strongly resemble those confronting the scientists trying to sequence the human genome today. His problems were on a much smaller scale, but at the time they seemed to loom very large indeed.

His career began, like most people's, through a series of accidents. Biochemistry was a very new discipline at Cambridge when the young Sanger arrived there as an undergraduate in 1936. Frederick Gowland Hopkins had become Cambridge's first Professor of Biochemistry in 1914, but it was not until 1925 that the Institute of Biochemistry officially opened. The Institute was housed in an ugly building on Tennis Court Road in the very center of town, and in Sanger's time it was filled mostly with teaching laboratories and anxious medical students. So new was biochemistry, both in Cambridge and the world at large, that Sanger had never heard of the field until he saw it in a list of potential courses for his first year. The idea of explaining the complex world of biology in chemical terms immediately appealed to his logical mind. The supervisor of the subject for undergraduates was Ernest Baldwin, an enthusiastic comparative biochemist and teacher. He aroused Sanger's interest so much that Sanger stayed on for an extra year in 1939, taking advanced courses in biochemistry and ending up (somewhat to his surprise) with a first-class degree.

When war broke out, the newly graduated Sanger applied on religious grounds for conscientious objector status, which was granted without difficulty because of his Quaker upbringing. A career in medicine was a real possibility. His father was a physician, and indeed had written a thesis on methods of distinguishing human from animal blood while in medical school at Cambridge at the turn of the century. Yet Sanger found the idea of a medical career rather unappealing. He particularly disliked the fact that in medicine one somehow never seemed to be able to carry a problem to completion. The possibility of doing exactly that was what attracted him to science. His parents were well off, and had no objection to his taking up a scientific career. So, on the strength of his excellent degree, he applied to do a Ph.D. in biochemistry at Cambridge.

His disappointment was intense when he received no reply, not even a rejection. A less determined student might have given up, but Sanger knew the department well enough to realize that his letter had probably been lost in a pile of papers. He went up to Cambridge in person, and it turned out that plenty of people in biochemistry were anxious to have a bright young Ph.D. student. After some false starts, he ended up working with Albert Neuberger, a very junior member of the department.

Between the wars, biochemical research had focused primarily on understanding the pathways followed by small molecules in the cell—molecules such as the sugars and the relatively simple compounds that could be built from them. The steps used by living organisms to make these compounds were slowly becoming understood. It was known that protein molecules called *enzymes* catalyzed the steps transforming one of these small molecules into another. Each step was catalyzed by a specific enzyme. A given metabolic pathway might involve a dozen or more steps, each of which had to follow others in a defined sequence. Enzyme molecules are huge and extremely complicated—much harder to study than the sugars and simple organic acids that make up the metabolic pathways. The chemistry of what enzymes did to their *substrates,* the small molecules on which they acted, was well understood. But in place of an understanding of how the enzymes themselves worked there was instead a series of black boxes.

Sanger was drawn to this difficult challenge. After getting his Ph.D., he obtained a Beit Fellowship and began working with A.C. Chibnall, who had just arrived to take over the department from Hopkins and was interested in the small protein *insulin.* Insulin is not really an enzyme at all, but rather a hormone, the one that is missing in many diabetics. Yet its structure was as unknown as those of the enzymes. Luckily, it turned out to be easy to study.

Biologists and chemists are now so familiar with proteins and their structures that it is hard to realize how little was understood about them in those days. They were known to consist of smaller building blocks called *amino acids.* These small molecules have the properties of both an acid and a base. The acidic and basic parts of different amino acids can hook together to form chains, which in turn make up proteins. The mystery lay in the order in which the amino acids of a particular protein were hooked together. It was not clear at the time whether the chains were linear or branched—both were possible. It was not even clear whether a purified sample of a protein such as insulin consisted of many copies of just one kind of chain or was made up of a mixture of chains.

About twenty different kinds of amino acids are commonly found in proteins. Sanger knew that only seventeen of these were found in insulin, and indeed, it was becoming clear that the insulin molecule consisted of nothing but these seventeen kinds of amino acids. If he could determine the order in which they were hooked together, he would have discovered essentially all the basic information needed to understand the insulin molecule. The problem was a tough one, since the amino acids form a kind of molecular alphabet. The number of letters in our alphabet, twenty-six, is in fact very close to the number of amino acids. Just as there are an enormous number of ways to string letters together to form a sentence, there are an enormous number of different ways to string amino acids together to make a protein. Even if Sanger could find out the exact number of each amino acid in the protein (and when he started out, he didn't even know that), there were still a huge number of possible arrangements.

At the end of the war Sanger was given some laboratory space in the basement of the Tennis Court Road building. His approach was very different from that of more timid scientists, who tend to try one technique at a time. He attacked the problem the way General Patton charged across Europe, using every technique he could think of to break up the protein and then using every available technique to separate the resulting pieces.

He knew that boiling a protein in a strong mineral acid would break it down into its component amino acids, which could then be analyzed as a mixture. Unfortunately, this would be the equivalent of using a hammer to determine the structure of a Ming vase. He had to find a less destructive way to break the protein apart, in order to obtain larger fragments. He also wanted to be able to identify these fragments quickly and easily. It was not obvious how to do this, because both a solution of the protein and a solution of protein fragments appeared transparent to the naked eye. He had to find a way of labeling the pieces so that he could keep track of them.

It turned out to be possible to attach a molecule called FDNB to an amino acid at one end of a chain, which had the effect of turning the chain bright yellow. When he used strong mineral acid to break the chain apart, most of the amino acids in the chain fell away and effectively disappeared, but the FDNB was so firmly attached to its amino acid that it did not come free from it. He could determine what kind of amino acid this was by pouring the mixture through a glass column packed with a separatory material. Different amino acids traveled through this material at different speeds, and since the amino acid attached to FDNB appeared as a yellow band, he could identify it easily and measure its speed. With this method, he was able to determine which amino acids were at the end of the insulin chains. There turned out to be just two, glycine and phenylalanine, which told him that insulin is made up of only two chains. Sanger then learned, after many false starts, how to split the chains apart, giving pure samples of each chain.

The arrangement of the amino acids *inside* the chains remained inaccessible to his methods, however. Perhaps, he reasoned, if one amino acid could be separated from its companions so easily, then groups of two or more might be separable as well. This idea led Sanger to a second discovery, one with far-reaching consequences. He found that by reacting the protein with FDNB, then treating it with acid gently so that the protein molecules did not break apart completely, he could obtain a mixture of fragments consisting of single amino acids, pairs, trios, and so on. This mixture could be separated in the glass column. The fragments without FDNB attached were of course invisible to his eye as they traveled through the column, but those with FDNB separated themselves out into a series of yellow bands. These were the fragments in which the FDNB was attached to one, two, three, or more amino acids. Such fragments, containing more than one amino acid, are called *peptides.* By modifying the peptides chemically, he had made them visible.

Sanger then analyzed the amino acids in each FDNB-labeled band. In peptides from the chain with glycine at the end, the fastest band contained only glycine; the second fastest contained glycine and isoleucine; the next contained glycine, isoleucine, and valine; and the one after contained glycine, isoleucine, valine, and glutamic acid. Real sequence at last!

A diagram of Sanger's method is shown in figure 2–1. The method was crude, and he could not determine the sequence of more than four or five amino acids in a row, but it was the first time that anyone had penetrated into the unknown interior of a protein. And just as importantly, it was the first time anyone had produced a series of *nested fragments* of a molecule, each fragment one amino acid larger than the one before it, a technique that would later prove to be absolutely basic to the sequencing of DNA.

Sanger could only penetrate a few amino acids into the molecule this way, but he had found in the meantime that light acid treatment would break certain parts of the protein more easily than others. With his collaborators Hans Tuppy and Ted Thompson, he spent many hours separating these peptides from the huge mixture of fragments produced by the light acid treatment. Then they made nested sets of fragments using these new peptides from the interior of the molecule. While the work went well at first, it gradually reached a halt. Certain gaps in both chains seemed impossible to fill. For example, they could determine the sequences of five different fragments of the longer chain, but try as they might, they could not join them together, because they were unable to isolate peptides spanning the critical regions that joined these fragments together.

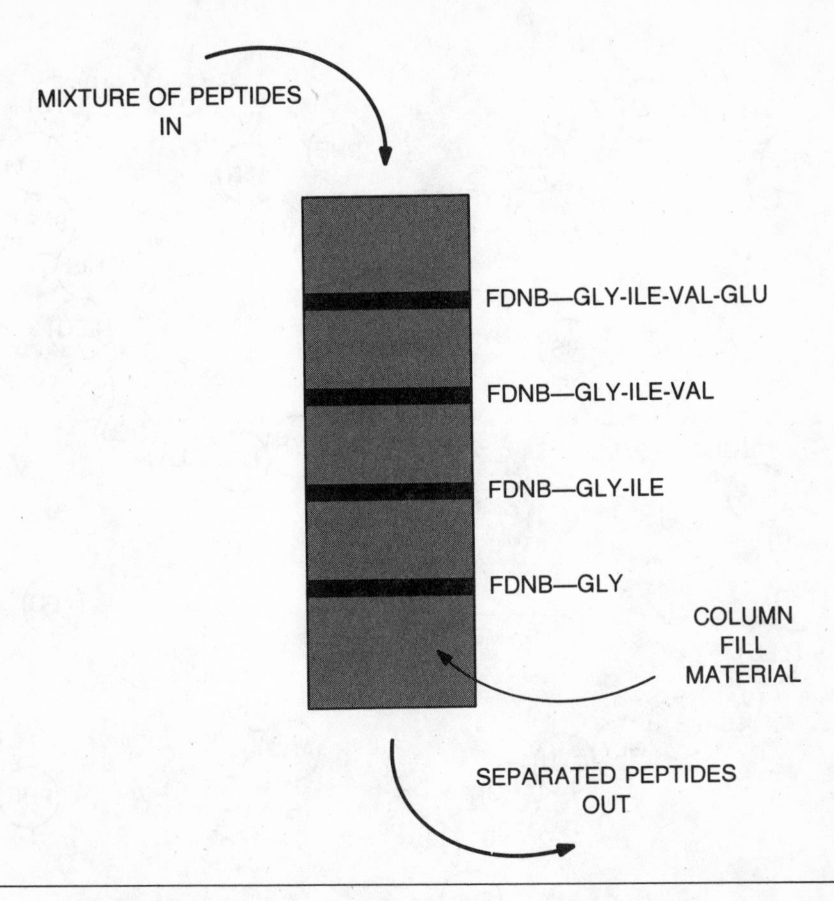

Figure 2–1. How a mixture of nested peptides was separated according to size by being passed through one of Fred Sanger's crude columns.

Even in those early days, a kind of rule of thumb had begun to emerge. The rule says that the first 80 percent of a sequence is relatively easy to determine. The last 20 percent, on the other hand, is always the devil of a job, and takes at least as long as the first 80 percent. Indeed, exactly the same scenario that Sanger faced is being played out now, some forty years later, on a much vaster scale, as many different groups of scientists try to determine the sequences of DNA fragments that make up entire human chromosomes. Relatively long stretches of the fragments can be fitted together, but many gaps remain, and it is proving very difficult to find the critical fragments that bridge them.

Sanger found a new way to penetrate further into the insulin molecule

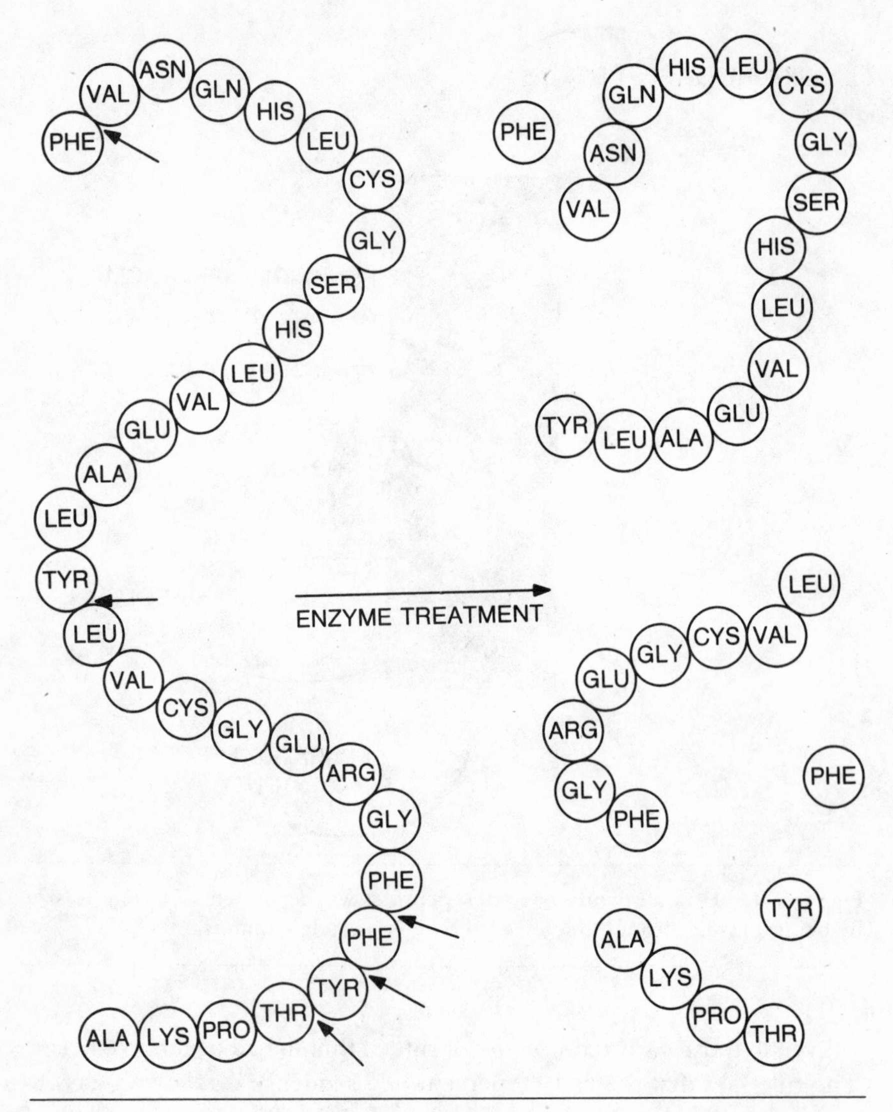

Figure 2–2. The digestion of one of the chains of insulin with the enzyme chymotrypsin. The enzyme cleaves the peptide only between certain pairs of amino acids.

in the form of enzymes capable of digesting protein. An organism's digestive enzymes act very specifically on the proteins taken in during feeding. Typically, a digestive enzyme will only attack certain parts of a protein molecule. When insulin is digested with one such enzyme, chymotrypsin, it splits into nine pieces (see figure 2–2). Other digestive enzymes split insulin in different but equally specific ways. Sanger was able to use the peptides produced by digestion with various enzymes to solve most of the rest of the insulin sequence. A few holdout bits finally succumbed to a battery of other techniques. The sequences of the two chains, one twenty-one amino acids long and the other thirty, had finally been solved.

Ironically, the last problem turned out to be the worst. The two chains of insulin are connected by bridges made up of pairs of sulfur atoms, and a third bridge is formed between two points on the shorter chain. Determining the location of these bridges turned out to be a much harder task than determining the amino acid sequences, and it took at least as long. Not until 1955, twelve years after he began his work on insulin, did Sanger finally complete his analysis of the structure of this tiny protein.

DNA MAKES ITS BOW

Determining the sequence of insulin was a chemical tour de force, but Sanger had done far more than simply solve a difficult structural problem. At the time, there was a general feeling among geneticists that since proteins were so complicated, the genes carrying the information needed to make them must be at least as complicated. Thus, geneticists tended to dismiss the growing evidence that the DNA found so plentifully in the nucleus of the cell—"that stupid molecule," as the virus geneticist Max Delbrück called it—was the genetic material that somehow coded the amino acid sequence of proteins. DNA was much simpler than proteins, with only four building blocks rather than twenty. How could such a simple molecule code for such a complicated one? Yet the evidence was growing, chiefly from the newly understood genetics of bacteria and viruses, that DNA really was the genetic material.

Sanger had shown, and others were soon to verify, that proteins were unique *linear* orders of amino acids; even though the linear chains were often formidably knotted, they did not branch, as many had worried. If proteins were arranged in a linear fashion, then presumably so was the information needed to code a protein. It might not be beyond the capacity of a seemingly simple-minded molecule like DNA to manage the job of coding this information. Perhaps different combinations of its building blocks could be used to specify different amino acids, combinations that would form a series of code words.

When James Watson and Francis Crick unraveled the structure of DNA during the years 1952 and 1953, they did it at Cambridge's Cavendish Laboratory, a few hundred yards from where Sanger was wrestling with the final problems of insulin structure. Thanks to Sanger's work, Watson and Crick were well aware that proteins had linear structures, and this important insight helped lead them to the structure of DNA. Chemically, the DNA molecule is very stable. It consists of two backbone strands that wind around each other, each a long chain made up of alternating molecules of phosphoric acid and deoxyribose, a sugar. Attached to each deoxyribose molecule and pointing toward the interior of the molecule is one of four nitrogenous *bases*—adenine, thymine, guanine, or cytosine. (It is standard shorthand to call these A, T, G, and C.) Each of these bases is paired up with another base that points inward from the other strand.

Watson and Crick made many false starts before they realized that the molecule must be two-stranded, with the phosphoric acids on the outside and the bases on the inside. Once this realization came to them, it fell to Watson (with Crick looking constantly over his shoulder) to try to figure out how the bases fitted together inside the molecule. Adenine and guanine are close relatives of a fairly large and complicated molecule called purine. Over the years they have come simply to be called purines rather than the more chemically accurate but cumbersome "purine derivatives." (There are many other purines in the natural world, including caffeine.) Cytosine and thymine, derivatives of a smaller molecule called pyrimidine, have by a like process of language evolution come to be called pyrimidines.

At first Watson tried to make adenine pair with adenine, thymine with thymine, and so on. This would make a molecule with two identical halves, a perfect candidate for the genetic material, because it could split apart and make more identical molecules. He quickly discovered that because purines and pyrimidines are different sizes, the backbones would have to be pushed further apart at some points and pulled closer together at others, resulting in a lumpy-looking molecule with a highly irregular helix. He knew this could not be correct because X-ray diffraction studies showed that the backbones were always the same distance from each other. Further, there was no obvious chemical means by which identical bases could pair with each other.

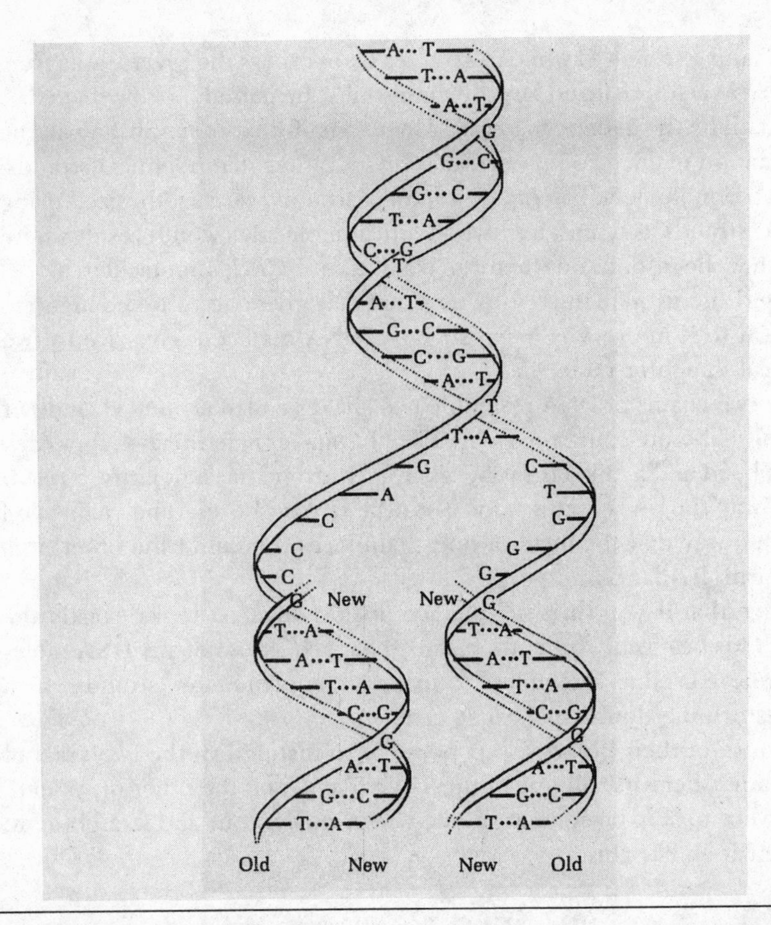

Figure 2–3. The DNA double helix, showing how A pairs with T and G with C. In this simplified picture the helices are coming apart and new daughter strands are being synthesized. The process is actually more complicated than this, but the end result is the same.

Finally, with the assistance of the crystallographer Jerry Donohue, Watson was able to determine that adenine would only pair with thymine, and guanine with cytosine. This kept the backbones a constant distance apart, since a large molecule was always paired with a small one. These simple pairing rules also explained a puzzling observation made a few years earlier by Erwin Chargaff of Columbia University. Chargaff had found that in any sample of DNA, the number of molecules of adenine always equaled the number of molecules of cytosine. These relationships always held, even though in samples from some organisms there was more

A + T and less G + C while in samples from others the reverse was true.

When Watson realized how the bases must be paired, he saw that DNA could still be the genetic material. The two half-helices simply had all the information needed to make *complementary* rather than identical strands. When a double helix split apart and new strands were synthesized using the old strands as templates, two identical molecules would result where there had been only one before. Watson and Crick's molecular model reflected life in miniature: Just as a cell can give rise to two daughters identical to it in every respect, so can a DNA molecule give rise to two identical daughter molecules.

The two chains of DNA resemble two snakes coiled around each other in some elaborate courtship ritual. Like the snakes in figure 2–4, they each have a head and a tail. This is because each strand has a polarity, a result of the way the phosphates hook onto the deoxyriboses. The chains coil in such a way that the head of one chain faces the tail of the other, in a kind of molecular *soixante-neuf.*

Notice that if you turn the picture upside down, it looks exactly the same. This beautiful symmetry means that each strand of any DNA molecule can be used as a template to make a complementary strand, which will restore the double-helical structure.

Because of their polarity, it is possible to distinguish the two ends of the chain. Chemists call one of these the 5′ end and the other the 3′ end. If we were to snip a segment of one chain of DNA out and straighten it, it might look like this:

```
5′ -AAAACGTATA- 3′
```

Or this:

```
5′ -GTCCCAGGGT- 3′
```

The sequences of the other strands that pair with these are determined by the rules of base pairing. In the first case, the two paired strands *must* look like this:

```
5′ -AAAACGTATA- 3′
3′ -TTTTGCATAT- 5′
```

And in the second, they must look like this:

```
5′  -GTCCCAGGGT-  3′
3′  -CAGGGTCCCA-  5′
```

Notice that although the first segment happens to be rich in A + T and the second in G + C, both segments obey the pairing rules. In both molecules A always pairs with T and G with C. Because the pairing rules apply across strands, this constrains the pairs of bases that can be formed as the complementary strand is synthesized. There is no such constraint

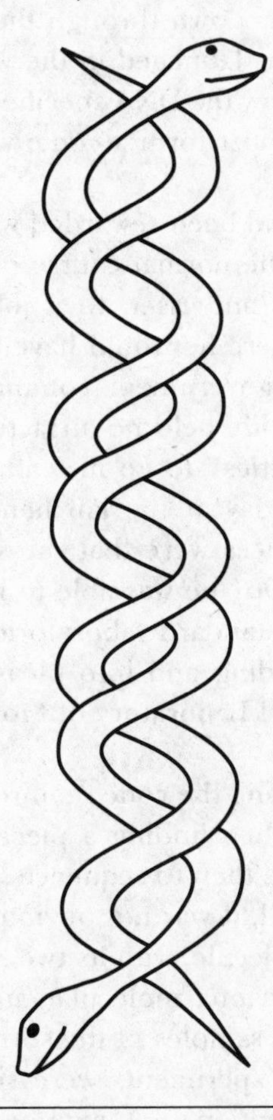

Figure 2–4. The DNA double helix represented as a pair of snakes. By turning the picture upside down, you can see that the molecule is completely symmetrical—each half of the double helix can serve as a *template* for the synthesis of its complementary half.

on the sequence of bases that runs along one strand; they can be in any order.

Ordinarily, one strand of the DNA is the "sense" strand, while its complementary strand is the "antisense" strand.* The sense strand carries the information; a sequence of bases along its length might specify a sequence of amino acids or other genetic instructions. The specific information on any given sense strand has evolved over the three and a half billion years that the ancestors of a particular piece of DNA have been handing their instructions down through the generations.

A worldwide search was launched in the wake of Watson and Crick's discovery to determine how the DNA specified its particular information. The sequence of bases must form a code; what was the nature of that code?

By this time, Sanger had been rewarded with his first Nobel Prize for his insulin research. In the normal course of events he would have acquired a carpeted office and risen to a job administering a vast research establishment, where he would have been far too busy shuffling papers and serving on government committees to do any more research. But because this life held no attraction for Fred Sanger, he ignored all the "opportunities" to go into administration and continued to come to work every day at the lab bench as if nothing had happened. The only differences were that the size and fame of his group had grown and that in 1961 he was able to move away from the last of the series of rather substandard laboratories he had occupied in the Tennis Court Road building and into pleasanter quarters at the new Medical Research Council Laboratory of Molecular Biology on the outskirts of town.

Sanger knew that cracking the code required two things: learning how to sequence DNA, and then finding a piece of DNA that coded for a known protein. If he were then to sequence *that* piece of DNA, the code should become clear. Still, it was not obvious how to begin sequencing the complicated DNA molecule, with its two strands running in opposite directions and its impervious molecular structure. To make matters worse, pure and plentiful samples of the sort of small pieces of DNA he needed for sequencing experiments were simply not available at that time. So he turned his attention to *ribonucleic acid,* or RNA, a molecule that

*Remarkably, more and more genes are being discovered in which both strands can carry information. And there are many other parts of the genome in which neither strand of the DNA seems to be a sense strand, since neither of them carries a message that the organism can read.

Sidney Brenner and Francois Jacob had just shown to be a bridge between DNA and protein.

RNA is very like DNA in its chemical composition, but consists of only a single chain. It is formed as a complementary strand to one of the strands of DNA, then is peeled off the DNA to carry its message elsewhere in the cell. If you think of DNA as being like a lithographer's stone, then the RNA is like a print produced from it. Just as the print carries a negative image of the information on the stone, the RNA carries a complementary image of the information on the DNA. In some cases, thousands of RNA "prints" are made from a DNA gene over the lifetime of a cell. Because of this multiplication process, it is easy to obtain pure samples of certain small pieces of RNA.

Sanger's scheme was straightforward: First, find a method of sequencing RNA, then find a piece of RNA that carries the information from the DNA gene for an already-sequenced protein, next sequence *that* RNA, and finally, crack the code. The project was quite feasible, but he had not counted on the accelerating pace of molecular biology. In one stroke, Marshall Nirenberg at the U.S. National Institutes of Health found a beautifully simple way to crack the code without sequencing any nucleic acids at all. Some brilliant genetic experiments by Crick, Brenner, and others had already demonstrated that the bases must be read in groups of three. Nirenberg was able to use artificial groups of three to work out the code words or *codons* for all twenty of the amino acids.* Just as Sanger was beginning his sequencing work, the entire genetic code was determined almost overnight. This sudden advance left a number of other people flat-footed as well, notably Charles Yanofsky at Stanford, who had begun to use elegant but slow genetic methods to crack the code.

Rather than discouraging Sanger, this discovery spurred him on to greater efforts. Now that the code was known, the need was even more pressing to find ways to sequence RNA or DNA easily, for now any of the information that might be carried by a region of one of these molecules could in principle be read.

Sanger soon realized that the nucleic acids RNA and DNA actually had several advantages over proteins when it came to sequencing, because their structure is much simpler. All four of the subunits are chemically very similar to each other—purines have chemical properties very like

*Seymour Benzer of Cal Tech started the trend of naming various genetic units with words ending in "on." Some of the terms he introduced, like muton and recon, have fallen into disuse, but the tradition has been carried on by others who have coined words we will meet many times in this book, such as codon, replicon, intron, and exon.

those of pyrimidines, and indeed they are much more similar to each other than amino acids are. At first, this might seem to be a disadvantage; it is surely easier to determine a sequence of apples, oranges, and bananas than it is to determine a sequence of Granny Smiths, Macintoshes, and Gravensteins. In fact, though, some amino acids are *so* different from each other chemically that peptides made up of them are hard for chemists to handle.

In contrast, all pieces of RNA and all pieces of DNA have very similar properties. It does not matter whether they code for the amino acids of a protein or do not, and it does not matter whether they come from a sperm whale or a redwood tree. They can be treated physically and chemically in exactly the same way. Further, RNA and DNA contain phosphorus atoms, found in the phosphoric acids in the backbone of the molecule. At the same time that Sanger began to work on RNA, a radioactive isotope of phosphorus, ^{32}P, was just becoming available in quantity for biological work. It emitted far more energetic gamma rays than the carbon or hydrogen isotopes used to label proteins. As a consequence, it was possible to detect vanishingly small amounts of nucleic acids by labeling them with ^{32}P. In this way it was easy to detect a ten-billionth of a gram of RNA or DNA, or even less.

By making nested fragments of RNA in the same way that he had earlier made nested fragments of insulin, and then using clever ways of separating the resulting fragments, Sanger and his students were able to decipher the sequence of an RNA that was 120 bases long—small, but at that time a world record. The RNA happened not to code for a protein, so Sanger was unable to confirm Nirenberg's code, but it was a remarkable beginning. The year was 1968.

THE YIN AND YANG OF DNA SEQUENCING

In traditional Chinese cosmology, yin is the female, passive, or constructive force, while yang is its opposite—male, active, destructive. Up to this point Sanger, and the others who were beginning to follow in his footsteps, had always taken what might be called the yang approach to the sequencing of macromolecules. They tore them down into small component parts and separated the resulting nested sets of fragments. There were problems with this method, however. The chemical or enzymatic treatments often used in breaking down the molecules could give poor results. Further, large amounts of material were needed, and if the treatment went on for too long or too short a time, they risked ending up with too many little pieces or too many big ones.

Sanger knew that if he were ever to sequence longer stretches of nucleic acids, he would have to come up with much more specific methods. The yang approach, he thought, could only be taken so far before it would start to fail. He decided to sequence DNA by yin rather than yang—building it up rather than breaking it down. Having been most impressed with the precise, specific way that digestive enzymes broke down the insulin molecule, he began to look into an enzyme called *DNA polymerase.* This enzyme is the most prominent of the many different enzymes that are required to duplicate or *replicate* DNA. DNA cannot make copies of itself unassisted; that would be like asking a book to make copies of itself without the help of a printing press, or a copy machine, or at the very least a typewriter. The genome is like a book that contains, among many other things, detailed instructions on how to build a machine that can make copies of it—and also instructions on how to build the tools needed to make the machine! Thus, genes in the DNA itself code for both the enzymes that perform the replication and the enzymes that build *those* enzymes.

For DNA to duplicate, the double helix must be unwound into two separate chains. These separated chains can then be used as templates to construct two new chains complementary to them. Through this process, which a friend of mine refers to as Watson-Crickery, the information in the old molecule is accurately transferred to its two daughter molecules. A whole cluster of enzymes, together called a *replisome,* is required to unwind the DNA, position the DNA polymerase molecules correctly, and break and join the growing DNA chains in a variety of ways. The discovery of how the replisome works took lifetimes of effort and has been one of the triumphs of molecular biology.

The daunting complexity of this replication process has actually turned out to benefit the explorers of the human genome. Living organisms need a great variety of enzymes to duplicate, repair, nibble at, cleave, and shape their DNA. Many of these enzymes have now been isolated and purified from living cells, providing laboratory workers with a versatile set of tools.

The first enzyme to be purified and made to perform on its own was one of the DNA polymerases. The enzyme was purified by Arthur Kornberg at Stanford, and used shortly afterwards by Ray Wu and Dale Kaiser of the same university to fill in and sequence a short piece of single-stranded DNA found at the end of an otherwise double-stranded viral chromosome. This little stretch of twelve bases was the first piece of DNA ever to be sequenced, and the result was published in the same year that Sanger published his much longer RNA sequence.

Wu and Kaiser knew enough about the enzyme to realize that it would always begin its work just at the point where the double-stranded viral DNA became single-stranded. By supplying the DNA polymerase with various combinations of radioactive and nonradioactive A, T, G, and C, they were able to make molecules of various lengths that all began at that same point but that ended at different places. Rather than breaking the DNA down to make nested fragments, they were using the enzyme to produce *new* nested pieces, each one base longer than the last. Sanger immediately became interested in scaling up the yin approach of Wu and Kaiser in order to sequence much longer pieces of DNA. To do this he needed a long single-stranded piece of DNA to which small pieces of the other strand could be bound. These short double-helical stretches would act as *primers,* allowing the DNA polymerase to begin its work. Then he could use DNA polymerase and radioactive bases to fill in the long single chain, making it into a double strand. (Figure 2–5 shows what he had in mind.) But where could he find such single-stranded DNA in the large quantities that he needed?

In the early 1970s, Sanger turned to DNA not from the chromosomes of higher organisms but from bacterial viruses. These pieces of DNA were uncomfortably large, but they could be made in great quantities and were very pure. The one he finally settled on was the chromosome of a bacterial virus called ϕX174, which is about 5,000 bases long. It had the great advantage of being normally single-stranded, though as the virus makes copies of itself it does go through a brief double-stranded phase. Details of the structure of the viral chromosome and recipes for making it in quantity were supplied to Sanger by the virologist Robert Sinsheimer, then at Cal Tech, who would later become one of the founding fathers of the Human Genome Project.

Sanger was able to isolate the brief double-stranded phase of the viral chromosome and break it into short fragments with the help of a collection of newly discovered enzymes called *restriction enzymes.* These enzymes each recognize a specific sequence of bases on DNA, and cause a break at that point each time they find it. Sanger purified the resultant fragments, then turned the fragments into the single-stranded form and mixed them with single-stranded viral chromosomes. When the little single-stranded fragments found their complementary regions in the chromosome, they bound to it. The final result was a set of single-

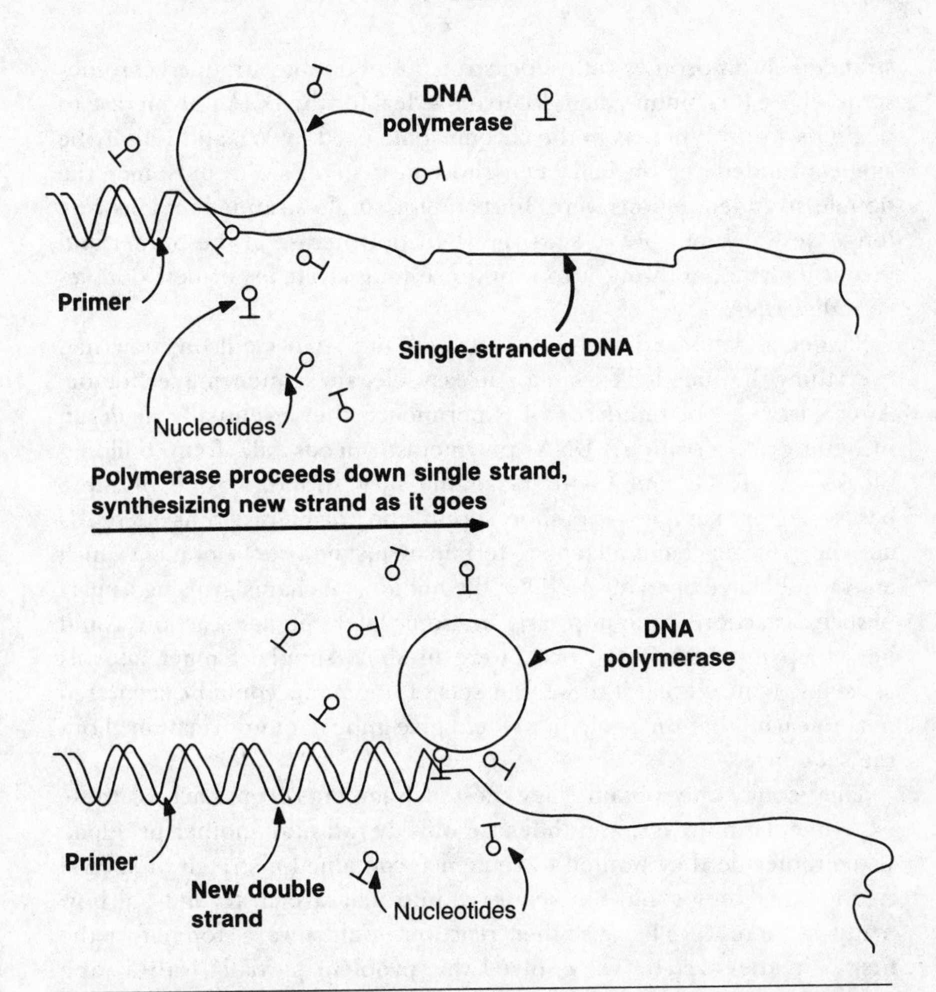

Figure 2–5. A simple diagram showing how DNA polymerase, represented as a ball, synthesizes a new double helix from a primed single DNA strand. For the polymerase to begin its task, part of the molecule on which it is working must be double-stranded. In the laboratory, this is often accomplished by adding a short piece of DNA that is *complementary* to a region of the single-stranded DNA that is to be duplicated. This acts as a *primer* for the reaction. The experimenter must also supply DNA building blocks in the form of nucleotides. The enzyme then can proceed down the single strand, synthesizing a new complementary strand as it goes.

stranded chromosomes with short stretches of double-stranded chromosome where this binding had occurred—ideal for the DNA polymerase to begin its work. Whereas in the chromosome used by Wu and Kaiser the single-stranded region had been short, in Sanger's chromosomes the double-stranded regions were short and the single-stranded regions very long. He was thus able to start the DNA polymerase at the primer and give it its head, allowing it to synthesize long stretches of new double-stranded DNA.

Sanger and his assistants, Bart Barrell and Alan Coulson, now had everything they needed except for an easy, elegant sequencing technique. After carrying out hundreds of experiments, they eventually made an intriguing observation. DNA polymerase needs all four building blocks—A, T, G, and C—to make the new strand. If one of these bases—A, for example—is in short supply, the polymerase runs out of A's and the growing chains all tend to terminate just before the point at which an A would have been inserted. Of the millions of chains growing as part of such a reaction, some stop early and some later. Similar reactions could be set up in which T, G, or C were in short supply. Sanger and his assistants realized that if these four sets of fragments could be separated next to each other on a gel, they would give important information about the sequence.

They could only obtain a few nested fragments from each of these experiments, however, and they also quickly ran into another problem. If the molecule they wanted to sequence contained a stretch of several As in a row, they could not sequence into that stretch to find out how many As there were because their reaction would always stop before the first A in the stretch. They solved this problem partially with a yang method: chewing with a different enzyme at the DNA until the enzyme was stopped by a specific base. Using the two methods, which they called "minus" and "plus," they could attack difficult sequences from both ends. This was complicated and laborious, but it worked. Sanger considered this "plus-minus" combination of methods to be his major conceptual breakthrough. The only distressing feature was that it didn't work awfully well.

Meanwhile, over the years Sanger had made great advances on his first clumsy methods for separating the DNA fragments. The technique he eventually settled on had been developed by others, but he improved it by pushing it to its limits. He constructed a device in which two large plates of glass were separated by about a millimeter of space. Into this tiny gap he poured a chemical solution that set into a gel, rather like a thin layer of Jell-O. When a powerful electric current was passed through this gel, any charged molecules, including pieces of DNA, were forced through it. The smaller the DNA fragments, the more quickly they could move through the tiny pores in the gel.

Sanger found that turning up the current brought better results—up to a point. The heat of the current kept the DNA fragments from curling back on themselves, making it more likely that they would move smoothly and evenly at a rate determined only by their sizes. But if the current was increased too much the glass plates could crack or the gel could melt, spoiling the experiment. Others facing this problem had tried to cool the gel down, but Sanger wondered what would happen if he let it stay as hot as he dared. He found that at the point where the glass plates were just toasty warm to the touch, everything worked perfectly. Fragments of radioactive DNA were separated out into sharp bands that could easily be distinguished from each other when their images were formed on X-ray film. Fragments could be separated in this way if they differed in size by as little as one base out of several hundred.

With much effort, Sanger's group was able to sequence quite long stretches of DNA using the "plus-minus" reactions. In 1974 they embarked on the heroic task of sequencing all 5,000 bases of the ϕX174 chromosome—a huge task, since any one set of reactions produced only a narrow range of sizes of nested fragments. Hundreds of separate sets of reactions had to be carried out, and Sanger soon realized that they were not utilizing the full separatory power of their long gels for sorting out fragments. This recognition led him to his final breakthrough, one that utilized both the specificity of the DNA polymerase and the fact that it could be fooled.

In the meantime, however, another group was coming up fast on the outside.

THE ULTIMATE YANG APPROACH

Sanger thought that the yang approach of tearing down the DNA chemically had reached its limits. He was wrong; it could be taken much further. On the other side of the Atlantic, at Harvard, Walter Gilbert and his assistant Allan Maxam were sequencing small stretches of RNA using Sanger's techniques, and were growing dissatisfied with the slowness and difficulty of the method.

By the time Maxam and Gilbert began their sequencing work in the late

1970s, it had become possible to cleave any piece of DNA into short defined fragments by means of restriction enzymes, and to *clone,* or make multiple copies of, these fragments. These clones were made using *plasmids*—small, semiautonomous pieces of DNA that grow inside bacterial cells. All the plasmids available at the time were double-stranded, so Maxam and Gilbert sought an effective way to sequence double-stranded DNA. They began experimenting with ways of breaking up the cloned plasmid DNA to give nested fragments. First they labeled one end of their DNA with ^{32}P so as to keep track of it. They then divided the sample up and subjected it to various ingenious chemical treatments. One of these preferentially destroyed A's. The demolition was so thorough that whenever an A was destroyed the backbone of the DNA molecule was also destroyed, breaking the molecule into numerous fragments. Following in Sanger's footsteps, Maxam and Gilbert use a very light chemical treatment, so that only a small fraction of the A's were destroyed. The resultant set of nested fragments might have looked like this:

Original sequence:

^{32}P-G-A-C-C-A-T-A-G-G-T-C-A-T . . .

Fragments after treatment:

^{32}P-G-A-C-C-A-T-A-G-G-T-C + T . . .
^{32}P-G-A-C-C-A-T + G-G-T-C-A-T . . .
^{32}P-G-A-C-C + T-A-G-G-T-C-A-T . . .
^{32}P-G + C-C-A-T-A-G-G-T-C-A-T . . .

I have listed the fragments labeled with radioactive phosphate on the left and some of the fragments cleaved off of them on the right. Of course, some of these unlabeled fragments would in turn be partially cleaved; the G-G-T-C-A-T . . . fragment, for instance, would be partly cleaved by the chemical treatment into G-G-T-C and T . . . , and so on. But none of these unlabeled fragments, including all those from the other DNA chain, were visible; only the radioactively labeled fragments could be detected.

After a great deal of work, Maxam and Gilbert came up with several different chemical treatments that preferentially destroyed one or more of the four bases. By running these reactions in separate test tubes, they could produce different nested sets of fragments. When separated next to each other in a gel, these sets of fragments gave the complete sequence of the DNA.

The paper setting out the Maxam-Gilbert technique was published in February 1977. The results of the technique were dazzling. The sequence of a piece of DNA hundreds of bases long could now be read from a single gel. The new method was much easier to use than Sanger's "plus-minus" approach, and it generated the immediate acceptance elegant techniques always gain in science. Many labs around the world began to use it.

THE TRIUMPH OF YIN

Sanger, however, continued to plug away. By early 1977 he had finished nearly the entire sequence of the ϕX174 chromosome; only a few remaining bits worried him. His plus-minus technique could indeed be used to sequence a very large molecule—in the same way that a toy spade could be used to dig the Panama Canal. Other labs were not exactly stampeding to adopt the Sanger method.

About a year earlier, in fact, Sanger himself had begun to experiment with a new approach. He knew that the DNA polymerase enzyme could be fooled into adding *analogues* of the four DNA building blocks onto the growing chain. These molecules were similar enough to regular A, T, G, and C to hoodwink the enzyme, but once incorporated they would often bring the synthesis of the growing chain to a jarring halt. The analogues were chemically so different that the regular building blocks could not be attached to them in order to continue the chain. Sanger used this observation to invent a beautiful and elegant way of making large sets of nested fragments that would always end at a given base.

The idea is simple to grasp in principle. Imagine a child making a chain of snap-on beads. These chains, which you might remember from your own childhood, consist of plastic beads that can be joined together by little ball-and-socket arrangements. The ball on one bead snaps into the socket on the next. Occasionally, a child might pick up a bead from the collection that has a functioning socket but from which the ball has been broken off. Once this bead is plugged into the end of the child's chain, no more beads can be attached.

Imagine further that the beads come in four colors, and that the child is trying to duplicate a chain that somebody else has made. Suppose some defective beads of one of the colors are mixed in with functional beads of all four colors. If the child is too young to realize the nature of the defect, she might add a defective bead to a partially completed chain. Once she finds she can proceed no further, she might drop the chain in annoyance and start to make another. But if the child is not easily bored, she might end up making a series of chains, all following the same pattern, and all ending at a particular color. The chains will be of different lengths, depending on how many beads are added before the child accidentally adds a defective bead and is forced to start making a new chain.

The molecules that Sanger used were exactly like the snap-on beads I have just described. Up to now, for simplicity's sake we have called the building blocks of DNA A, T, G, and C, for the initials of their names (adenine, thymine, guanine, and cytosine), but chemically this is not quite correct. Each base is actually part of a larger subunit called a *deoxynucleotide,* which includes not only the base but also the *deoxyribose sugar* and the *phosphate* that complete a single segment of the linked DNA backbone. In synthesizing DNA, the DNA polymerase builds the chain by joining a deoxynucleotide onto a waterlike structure called a *hydroxyl group* that forms part of the sugar of the previous deoxynucleotide.

The analogues used by Sanger were *dideoxynucleotides,* in which the hydroxyl group had been replaced by a hydrogen. This is the molecular equivalent of snapping off the ball of the snap-on bead. Once a dideoxynucleotide has been added, the chain stops, for it has no hydroxyl group to which the next deoxynucleotide can attach itself. The DNA polymerase drops away, and must find a new chain to work on.

The only dideoxynucleotide Sanger was able to obtain at the time was dideoxy-T, the counterpart of the normal T deoxynucleotide. It worked beautifully, giving sharp bands the whole length of the gel, but the method was useless without the other three dideoxynucleotides. A drug company promised to supply him with them, but after a year it finally reneged on its promise. Sanger and Alan Coulson were forced to set out to make the other three themselves, learning the organic chemistry techniques that they needed as they went.

Once isolated, the other three dideoxynucleotides turned out to work as well as dideoxy-T had. The paper giving the details of Sanger's technique appeared in December 1977, ten months after Maxam and Gilbert's publication. Figure 2–6 shows the final brilliant result of all Sanger's years of painstaking work. To make the fragments that you can see on the gel, four different DNA polymerase reactions were carried out, each one partially "poisoned" by the addition of small amounts of one of the dideoxynucleotides along with the four regular deoxynucleotides. Then the resulting DNA fragments were separated on four adjacent lanes

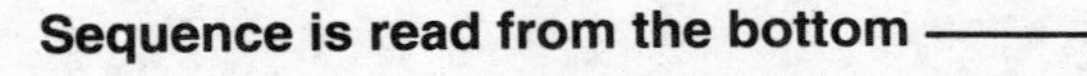

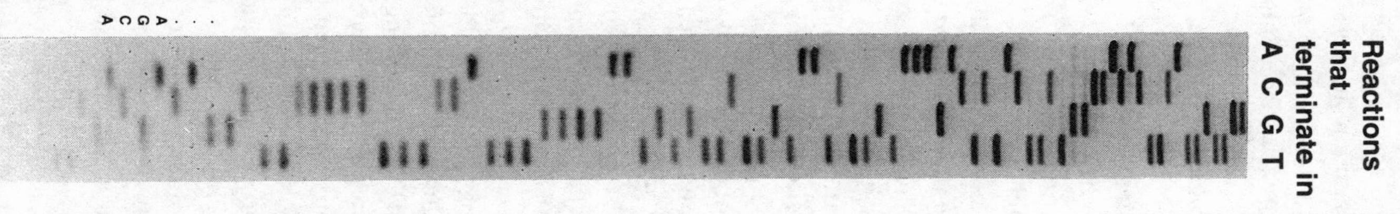

Figure 2–6. Autoradiograph of a typical sequencing gel produced by Sanger's dideoxy method. When X-ray film was laid on the gel, each radioactive piece of DNA formed a picture of itself on the film, producing the series of bands that you see here. The four lanes on the gel represent reactions terminated by small amounts of dideoxy-A, -C, -G, and -T respectively. You can read the ladder formed by these fragments quite easily by starting at the bottom (where the smallest fragments are) and working upward. The sequence of some of the DNA is given on the left. Note that the bands get closer together as you move upward, making the gel progressively harder to read. Eventually, after 300 bases or so, the experimenter can read no further.

of a single gel. In the first lane are the results of the reaction to which dideoxy-A was added. All the nested fragments in this lane end in A, while in the second lane, all the fragments end in C, and so on. An examination of the full length of the gel reveals that each succeeding band is one base longer than the last, and the lane in which each band appears indicates whether the base at the end of the fragment is A, C, G, or T. One can find out the sequence of bases in the DNA simply by "reading" the bands in order, starting from the small fragments that traveled furthest and are nearest to the bottom of the gel and proceeding upward to larger and larger fragments. The sequence, starting from the 5′ end of the DNA molecule, is listed to the left of the gel.

Sanger immediately applied his new method to the job of resequencing those parts of the ϕX174 chromosome that continued to worry him. But even this wonderful new technique was not enough to resolve the last few puzzles. Displaying a total absence of hubris, Sanger asked Ted Friedmann, then working in the lab, to use the Maxam and Gilbert method to do the last little bits.* The final sequence, numbering 5,386 bases, was published in 1978. It differed from the preliminary sequence published a year earlier in about thirty places. Sanger and his coworkers had discovered every bit of the genetic information carried by a virus. It was the first time that an entire genome had been sequenced.

The impact on biology of the DNA sequencing techniques developed by Sanger and by Maxam and Gilbert has been incalculable. In high-energy physics, teams of hundreds of scientists are needed to discover one new particle. If a similar effort were required to sequence a gene, molecular biology would be hamstrung. There would be no way to explore the thousands of genetic problems currently being investigated worldwide. The analytical power that DNA sequencing has provided for molecular biology might be matched in physics if physicists could all set up particle accelerators in their own labs. In molecular biology, the techniques are simple, cheap, and virtually foolproof. With an investment of a few thousand dollars, any lab can be set up to sequence DNA.

In 1980 Sanger and Gilbert shared the Nobel Prize in chemistry with Paul Berg of Stanford, who was the first person to use DNA-cutting enzymes to link together genes from two different organisms. The next year, Sanger's group published the entire 16,569-base-long sequence of the chromosome of human *mitochondria,* which produce energy for all our

*Labs today tend to use the dideoxy method for most sequencing, and to fall back on the Maxam-Gilbert method when everything else fails. In Walter Gilbert's lab, however, the Maxam-Gilbert technique is still used almost exclusively.

cells. After that, the sky became the limit. As I write this in 1990, the Maxam-Gilbert and Sanger techniques have been used to sequence over 60 million bases. About 50 million of these bases are stored in GenBank, the major DNA data bank, and the rest are in the pipeline waiting to be entered. Although these sequences are drawn from organisms ranging from viruses all the way up to mammals, they already represent as much DNA as is found in a small human chromosome. Never before has any biological technique led to the gathering of so much information in so short a time.

In 1988, Fred Sanger made the very public announcement that he would retire at age 70 and leave the burgeoning field to younger investigators. And that is where I found him, happily harvesting vegetables rather than bacteria. He told me he periodically felt the urge to get back into the lab; the habits of a lifetime are hard to break. He realized, though, how difficult it would be to get back up to speed after an absence of years. A graceful exit was surely better than a protracted and embarrassing dotage. And what better time to slip away than at the end of his most magnificent performance, with the applause still ringing in his ears?

CHAPTER 3

Therapy

I wish Your Lordship wolde make trial of my OYLE OF STAGS BLUD, for I am strongly persuaded of the rare and great vertu thereof. I know it to be a most safe thynge, yet some offence there is in the smell thereof.

—The Earl of Shrewsbury to Lord Burghley, Lord High Treasurer of England under Elizabeth I, recommending his cure for gout

In the summer of 1847, Alfred Baring Garrod, a Harley Street physician, was called in as a consultant to examine a middle-aged man who had been admitted to University College Hospital with a severe attack of gout. The man's right hand in particular was greatly inflamed and swollen, and he was in considerable pain. The physician in charge could do no more than bandage the hand to keep it immobile, and put the patient on a simple diet. Garrod, who had been studying gout intensively for some years, noted with excitement that the case was both very severe and of recent onset. This was just the kind of situation he had been looking for. He asked the attending physician to supply him with a small amount of the man's blood.

Back in his primitive laboratory, Garrod poured the blood into a little beaker and let it clot. As the dark red clot shrank away from the walls of the beaker, it left behind clear yellow serum, which Garrod carefully collected. He weighed a small container in a balance, then added a thousand grains of the serum, or about 60 milliliters. Since the metric system had not yet penetrated the world of English clinical research, Garrod measured his samples and chemicals in scruples, grains, minims, and drachms, and felt perfectly comfortable with these holdovers from the alchemy laboratories of medieval times. He slowly dried the serum over a boiling water bath, then crushed the yellow deposit that remained into a powder. When he added pure alcohol to this powder a cloudy suspension resulted, showing that most of the deposit was insoluble in alcohol. He boiled the suspension briefly, letting it settle. The alcohol turned yellow, and the material that settled to the bottom was now white.

When the white material had settled completely, Garrod poured off the alcohol with its yellow pigment. The white material formed a crust on the bottom of the container as the last of the alcohol evaporated away. He then added boiling distilled water. The white crusty substance, although it had not dissolved in the alcohol, did dissolve readily in the hot water. He carefully concentrated a few drops of this solution by gently heating it in a watch glass, and placed the glass under a bell jar with a open bottle of ammonia. As he watched with growing excitement, the clear solution in the watch glass gradually turned a beautiful purple color. The color was a sign that this procedure, which was meant to form the ammonium salts corresponding to any acids present in the original solution, had worked. The fact that the color was purple indicated that the ammonium salt of a particular acid had been formed.

To confirm his discovery, Garrod concentrated down the remaining solution, added a few minims (otherwise known as drops) of hydrochloric acid, and set it aside for a few hours. When he returned, the little dish was filled with tiny rhombic crystals that fluoresced brightly in ultraviolet light. Garrod knew that blood serum from a normal person would not produce these crystals. They were diagnostic for a human waste product called *uric acid.* With this straightforward assay he had succeeded for the first time in detecting uric acid in the blood of a person with gout. Over fifty years before, in 1793, a physician named Murray Forbes had also tried to find the acid, but he did not purify the material from the serum sufficiently and his experiments failed. Now Garrod had not only succeeded, but had shown that the blood from a person suffering a severe gouty attack could contain enormous amounts of uric acid. Using this simple diagnostic procedure, Garrod became the first scientist to peek successfully inside the biochemical workings of a living human being and find that a certain disease was associated with a specific biochemical abnormality.

A century and a half later, we may look back at Garrod's primitive

Figure 3–1. Alfred Baring Garrod.

laboratory and crude experiments with a hint of condescension. After all, we now know a great deal about gout; recently, many cases of the disease have even been traced to the DNA level. In spite of all this knowledge, however, the reasons for 95 percent of gout cases remain undetermined.

Figure 3–2. "I want to know Doctor how I am to get rid of this infernal Gout; can you recommend any thing?"
"Certainly, live on sixpence a day, and Earn it!"

Why? To answer that question we must travel not just through the last century and a half, but through the last sixty million years of our evolutionary history. On the way, we will see how current explorations of the human genome have illuminated very old puzzles, and opened the way to new therapies for devastating diseases. We will also see that our hunt for therapies can only proceed so far before our footing becomes treacherous and maps nonexistent.

EXPLAINING GOUT'S UNIVERSALITY

Even the ancient Greeks and Romans made the connection between gout and overindulgence in food and drink. The disease was first described, with remarkable accuracy, by Hippocrates in the fourth century B.C. He noted that eunuchs did not suffer from it, nor did women until they reached menopause. Today, the picture lingers of the gout sufferer as a fat, bad-tempered old gentleman resting his bandaged foot on a hassock. This is not inaccurate; the disease often affected the feet, and fat, rich old men were frequently its victims. The swelling, pain, and inflammation of the joints led many to assume that gout was a kind of arthritis. Yet unlike other types of arthritis, the agonizing symptoms of gout could vanish as suddenly as they appeared.

Physicians before Garrod, having no idea of the causes of gout, resorted to bafflegab. They often attributed the disease to "vascular plethora" or congestion of the digestive organs with blood. One authority contributed the penetrating assessment that it was due to a "peculiar disposition of the system, combined with weakness of the affected parts."

We now know that the pain of gout arises from the deposit of uric acid crystals in the thin film of fluid that lubricates the joints, so that the slightest movement causes exquisite agony. The bandaging recommended by doctors in earlier centuries served to reduce the swelling and prevent movement, but only when the crystals redissolved did the pain vanish. Severe cases of gout that continue over a long period often lead to grotesque malformations. Huge chalklike concretions can accumulate around the joints, distorting them and rendering them immobile. This actually provides some relief from the pain, but leaves the sufferer crippled.

Based on his observations, Garrod made a remarkable and prescient suggestion. In albuminuria, another disease he had studied, patients seemed unable to excrete another waste product, *urea,* properly, with the result that the substance accumulated in their tissues. Might there be separate excretory functions for uric acid and urea, the former being defective in patients with gout and the latter in patients with albuminuria? Such defects might be hereditary, which would explain why gout sometimes appeared to run in families—although Garrod was quick to point out that he had seen a number of cases of severe gout in which none of the immediate family members were affected. As it turned out, Garrod's hypothesis was remarkably prescient. Some fifty years later, his son Archibald Garrod was the first to suggest that diseases such as gout could be caused by inherited deficiencies of certain enzymes, conditions he called "inborn errors of metabolism." The younger Garrod is now considered to be the father of biochemical genetics.

Gout remains very much with us. The crowded office of Jay Seegmiller, recently retired as the head of the Institute of Aging at the University of California at San Diego, contains a row of rather gruesome exhibits. They are casts of the hands and feet of people who suffered from advanced cases of gout, all of whom he has treated or met at one time or another.

One of the most remarkable pieces in the collection is a cast of a particularly distorted pair of feet, with huge lumps on the soles. The feet belonged to a retired Yugoslav army officer, who worked for the UN in Geneva and suffered so much from gout that he was forced to wear sandals even in the winter. Seegmiller wanted to treat the officer with a then-experimental drug called allopurinol, but the National Institutes of Health bridled at the idea of bringing over a foreigner (and a communist one to boot) for free treatment at American taxpayers' expense. Private funds saved the day, and after a year under Seegmiller's care the difference in the officer's feet was astonishing. Casts taken after treatment showed that the lumps had practically disappeared.

Allopurinol routinely works such miracles now, by inhibiting the enzyme responsible for production of uric acid. (It was in part for her work on this compound that Gertrude Elion of Burroughs Wellcome Research Labs received the Nobel Prize in 1988.) The treatment still has some problems, but when carried out properly it can virtually eliminate the clinical consequences of gout. As a result, Western physicians can now prevent the monstrous deformities and exquisite agonies that used to accompany full-blown cases of the disease. Its causes, however, have turned out to be far more complex than Garrod originally thought.

In fact, if you are a normal human male, you are actually teetering on the brink of gout. Normal males have between 2 and 10 milligrams of uric acid per 100 milliliters of blood serum. In gout sufferers, the range is from 6 to 13. This means that among those men with uric acid levels of, say, 9 milligrams, some will develop gout and others will be quite healthy. Simple chemical experiments show that a uric acid concentration higher than about 7 milligrams should be enough to cause crystallization, triggering the first symptoms of gout. Yet levels often exceed this limit in people without any sign of the disease, and are lower in people who do develop gout, suggesting that we are dealing with a more complicated process.

If you are female, the amount of uric acid in your blood is on average about a milligram less than it is in males, although it rises after menopause to male levels. The difference is so slight that simple chemistry would suggest many women should suffer from gout too. Yet as it turns out, gout is one of the many illnesses women largely avoid through their balance of hormones. Only about 5 percent of all gout cases occur in females. The hormonal protection diminishes after menopause, as well as during the first few months of pregnancy, when uric acid rises, occasionally to dangerous levels.

Essentially, then, we are all living on the edge of a kind of biochemical disaster. Because normal uric acid levels are so high, it is not surprising that so many people in past centuries suffered from gout. The roll call is remarkable: In addition to the Earl of Shrewsbury and Lord Burghley, it included most of the French and English kings, Queen Anne, Cosimo de' Medici, the Holy Roman Emperor Charles V, Robert Boyle, John Milton, the elder and the younger Pitt, Benjamin Franklin, Samuel Johnson, David Garrick, Lord Nelson . . . on and on. Battles remained unfought and fleets swung at anchor as generals and admirals sweated through attacks. Indeed, gout may well have helped form Milton's uncompromising view of hell.

Clearly, gout was not a disease of the masses. Why did it afflict so many highly distinguished people? The answer, it seems, lay in diet—the culprits being "rich" foods such as red meat, brains, sweetbreads, and patés, as well as rivers of heavy port, Malmsey, and Madeira. The high uric acid levels that resulted were enough to push many men over the edge. Some actually took a perverse pride in their suffering; after all, it meant they had arrived.

Fascinating as gout's sociology is, it leaves the main question unanswered. Why are our systems loaded with potentially dangerous levels of uric acid? This biochemical brinkmanship is all the more remarkable because we have no obvious need for all this uric acid. Most other mammals have far lower levels.

Uric acid is derived from the purines, those important components of DNA and RNA that we met in the last chapter. Eating a lot of red meat and other foods rich in these compounds will cause us to make more uric acid. The only warm-blooded organisms with whom we seem to share this excess uric acid production are apes and monkeys, birds, and—oddly enough—Dalmatian dogs.

Other animals make uric acid from the purines in their diets too, but for them this is just a way station. The uric acid is immediately broken down by an enzyme called *uricase,* and then degraded still further to other compounds that can be excreted harmlessly. As part of the gradual process of human evolution, we lost our ability to produce uricase.* The gene for the enzyme seems to have started to lose its function about sixty million years ago, in a group of primitive prosimians from which monkeys, apes, and ourselves all descended. We know it must have started to happen about then, because only we and our immediate primate relatives have been affected. We can trace the gradual loss of uricase activity in our ancestral lineage by comparing uric acid levels among our closer evolutionary relatives, the primates and monkeys. Apes have the same high levels that we do, monkeys have somewhat less, and our more distant prosimian relatives, such as the galago and the loris, have the low levels characteristic of most mammals. The prosimians also show high levels of uricase activity.

The biochemist Bruce Ames at Berkeley has suggested that our large amounts of uric acid stem from what may have been a genetic accident: We had to compensate for the loss of our ability to make vitamin C, an ability our ancestors lost at about the same time that their uric acid levels began to increase. Both uric acid and vitamin C help protect the cell from damage by outside forces such as radiation, but uric acid is actually more effective. Perhaps, Ames suggests, all this uric acid is one reason why we live so long. If so, we may have made a Faustian pact in exchange for this longevity, gout being the price our species pays for being able to repair damage to our cells so efficiently.

*Intriguingly, it has recently been shown by groups in Texas and Japan that we still have the gene for uricase, or at least part of the gene. In chickens, the enzyme is functional for the first two weeks of life, then disappears. Does this happen to us too? Current government prohibitions on the use of fetal tissue for research will make it difficult to investigate this fascinating possibility.

Yet if gout is an unavoidable cost, why is it one some people must pay while others do not?

THE FAR END OF THE CONTINUUM

Jay Seegmiller began working on the problem of gout in the early 1960s in the lab of De Witt Stetten at Johns Hopkins. Between them they showed that most cases of gout were indeed associated with abnormal breakdown of purines. Still, patients with gout differed from each other in so many ways that it was not obvious what biochemical problem they might have in common.

The first real break came through a chance contact with Rodney Howell, a pediatrician at the Johns Hopkins teaching hospital. Howell's wife, a pediatric neurologist, was working with a child who suffered from a remarkable set of symptoms. This particular collection of symptoms, or *syndrome,* had first been described by a medical student, Michael Lesch, and his mentor, William Nyhan. The boy was spastic and mildly retarded, and exhibited strange involuntary behaviors. Unless prevented, he would bite and chew uncontrollably at his fingers, and when his hands were restrained, he would bite at his lips until they were largely destroyed. He was fully aware of what was happening, and equally unable to control it. His younger brother showed exactly the same set of symptoms.

One thing that struck the Howells was that the boy's urine, if left standing, would spontaneously grow crystals. They thought of uric acid and did a simple test for it, which showed positive. The boy appeared to be producing at least six times as much uric acid as a normal person. For help, the Howells turned to Seegmiller, who confirmed their test using a precise enzymatic assay he had developed. Seegmiller was fascinated by the obvious connection of Lesch-Nyhan syndrome with purine metabolism and gout. During a sabbatical in England soon afterward, he came across several other cases of the syndrome. In one young boy the self-mutilating impulse was so extreme that his despairing doctors finally extracted all his teeth. Yet even this drastic step was not enough to prevent the self-destructive drive, for the boy then began poking at his eyes, so that his arms had to be strapped to his sides.

In contrast, another boy Seegmiller visited only bit his hands when he was upset and worried. When Seegmiller saw him for the first time, he was in an asylum and his arms were restrained. He was later transferred to a hospital, where he was given much more care and attention. When

Seegmiller visited him there, the restraints had been removed; the more supportive environment had lessened his drive toward the self-destructive activity. Sadly, the boy subsequently had to be put back in restraints, for his mother became ill, and his worry was enough to trigger the worst symptoms of the syndrome. Apparently, a range of symptoms was possible in this appalling disease.

On his return to the States, Seegmiller began to search for other sufferers from the disease, and quickly found several. All were boys, and all showed enormously elevated concentrations of uric acid in both their urine and their blood. With such a clear-cut symptom, one far more extreme than the variable levels of uric acid Seegmiller had found in the cases of gout he had been working with up to that point, he felt that it should be possible to track down the cause. He began to investigate purine metabolism in these children.

Only a small fraction of the huge amounts of purines our bodies need come from diet, he knew. He also knew that while large numbers of old cells are continually being destroyed in the body, only about a tenth of their purines are broken down into uric acid. Were the rest somehow being recycled? As it turned out, they were. When DNA, RNA, and other purine-containing molecules are digested by enzymes as a part of normal metabolic turnover, their component purines are also broken down. Before the breakdown proceeds very far, however, most of them are "salvaged" in two different ways. In one of these, an enzyme called HPRT helps turn purines back into the building blocks of DNA and RNA. Indeed, HPRT can be thought of as the body's equivalent of a newspaper recycling program, building new genes from old ones just as new newspapers are printed from old.

Just as in a recycling program, however, this salvaging is not 100 percent effective. Currently, the city of New York cannot sell all of its recycled newspaper, because there is not enough of a market for old newspapers in the New York area. The excess papers must be dumped either into landfills or into the ocean. Similarly, some of the purines escape the ministrations of HPRT and continue down the breakdown pathway to uric acid.

Once Seegmiller realized that this process occurred in normal people, he understood the dynamics of the system. If something were wrong with a salvage pathway, the body would "dump" higher levels of uric acid. So the salvage pathways were the obvious places to look for defects. Biochemical studies of Lesch-Nyhan patients and normal volunteers suggested to Seegmiller that the most likely candidate was the enzyme

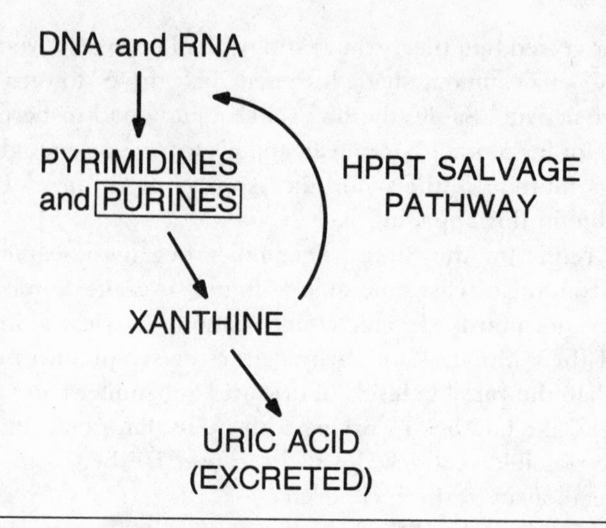

Figure 3–3. A sketch of the salvage pathway for purines, in which purines can be recycled back into DNA and RNA without having to be built completely from scratch. This saves the cell a great deal of energy, just as recycling programs save energy in our macroscopic world.

HPRT. And in fact, when he examined blood samples from Lesch-Nyhan patients, he found they all had undetectably low levels of HPRT. This discovery not only explained the biochemistry of the disease, but also explained why it showed up so much more often in boys.

Geneticists are always hoping for a simple pattern of inheritance in the characters they work with, because a simple pattern usually means that the genetic explanation is simple as well. Lesch-Nyhan disease showed such a pattern, that of "sex-linked" (or, more accurately, *X-linked*) inheritance. Males have only one X chromosome, while females have two; the male Y chromosome takes the place of the second X. Most of the genes on the X do not match the few on the Y, with the result that males have only one copy of the majority of their X-linked genes. A recessive gene carried on the X is therefore far more likely to reveal itself in males than in females, where it could be masked by the normal dominant gene on the other X. Females who carried the Lesch-Nyhan mutation on one X could still manufacture HPRT with their other, functional allele. Thus, it was easy to conclude that the gene for Lesch-Nyhan syndrome had to be on the X chromosome.

Seegmiller had been able to add Lesch-Nyhan syndrome to the small but growing number of human metabolic defects that were understood at the biochemical level. More importantly, he had shown that metabolic pathways in the body were connected in subtle ways to each other, like pipes in a complex plumbing system. Levels of various biochemical compounds could be regulated by turning enzymatic "taps" at various points in the system.

Unfortunately, Seegmiller's discovery did little for patients. Putting them on a purine-free diet, for example, would not help because their synthesis pathway was intact and they made most of their own purines anyway. In fact, they actually had an excess of purines in their blood that would normally have been salvaged by HPRT and turned back into DNA and RNA. This purine excess, not the excess of uric acid, led to the uncontrollable tendency to self-mutilation in Lesch-Nyhan children.* So it was no surprise that drugs capable of alleviating the symptoms of gout were ineffective for Lesch-Nyhan patients. Various other treatments sometimes gave transient relief, but without permanent cure. The problems that caused Lesch-Nyhan lay deep within the metabolism of the patients' cells. Thus, although both Lesch-Nyhan and some cases of gout are caused by missing enzymes, gout can be cured and Lesch-Nyhan cannot.

Today we can be optimistic, however, because of the accelerating pace of research. A short version of the gene for HPRT was one of the first human genes to be sequenced, in 1983. The complete HPRT gene has now been cloned and sequenced by Thomas Caskey's group at Baylor University, drawing on new and rapid DNA sequencing methods invented by Wilhelm Ansorge of the University of Heidelberg. So intense is interest in HPRT that it is one of the small number of human genes to have been sequenced in their entirety. It is quite large, stretching about 50,000 bases along the X chromosome. Only small parts of it code for the HPRT protein, and these coding regions are broken up by long stretches of seemingly meaningless DNA, called *introns,* that do not code for anything (although in a moment we will see that these are not so meaningless after all.)

By comparing the sequence of the normal gene with that of one with the Lesch-Nyhan mutation, it is possible to determine the exact alterations in the DNA that cause the disease. Many different DNA changes have been found in different patients, changes that have occurred independently in different families. You can see why this would be the case with an X-linked disease like Lesch-Nyhan. Because of their severe

*Rats given high doses of another purine, caffeine, will bite and injure their paws.

effects, Lesch-Nyhan mutations do not last long in the population. A new Lesch-Nyhan mutation may arise in a female X and produce no immediate effect, but within a generation or two it will probably match up with a Y chromosome in a male. The lack of a normal allele to mask the mutant will cause the male to suffer from the disease. Since its effects will make him unable to reproduce, his copy of the mutant allele will be lost. New cases of the disease continue to appear only because new Lesch-Nyhan alleles continue to arise through the process of mutation.

A new mutant allele will turn up at this locus in about 1 in every 50,000 people. Some of these changes will be very slight, perhaps an alteration in a single base. In one such mutation, a G is changed to a C. This change specifies that the seventy-first amino acid from the end of the HPRT protein will be an arginine rather than a glycine. Arginine is large and negatively charged, while glycine is small and electrochemically neutral. This tiny change, involving 1 out of 217 amino acids in the HPRT protein, is enough to destroy the function of the enzyme completely.* In other mutants, parts of the gene have been removed, and in still others small bits of extra DNA have been inserted.

You might think one could just supply Lesch-Nyhan children with the missing HPRT itself. Unfortunately, there is no way to get this huge protein molecule past the body's defenses to the place where it can do its job. It cannot be taken by mouth, because the digestive enzymes would pounce on it and break it down just as they do all other proteins. Nor can it be injected directly into the bloodstream, for the patient's immune system would destroy it there. Repeated attempts would surely cause a violent and dangerous immune reaction.

Genetic studies have shown, however, that enormous amounts of HPRT are not needed; even tiny amounts of the enzyme can have a huge impact on the severity of the disease. Genes can be changed by mutation in many ways, giving rise to a spectrum of mutant enzymes ranging from hopelessly crippled to infinitesimally altered. Since enzymes are robust molecules, they can often suffer quite severe damage and still go on working after a fashion. Some Lesch-Nyhan children show many typical symptoms but lack the mental retardation and the uncontrollable urge to mutilate themselves. One percent or so of normal HPRT activity is found in the cells of these children. Such "leaky" activity among mutant enzymes is very common.

*Note how much larger the HPRT molecule is than the tiny insulin protein Fred Sanger needed twelve years to sequence! Yet even HPRT is dwarfed by some proteins we will meet later.

Clearly, in some Lesch-Nyhan children a tiny amount of HPRT activity is enough to reduce the severity of the disease dramatically. A working gene inserted into as few as 1 percent of the body's cells ought to be enough to alleviate the worst symptoms of the Lesch-Nyhan syndrome. Indeed, Lesch-Nyhan sufferers are excellent candidates for the replacement of the malfunctioning gene with a normal allele—a so-called *transgene.* This process, christened *gene therapy,* has already been accomplished in tissue culture using cells extracted from humans and mice. The cells were treated with a virus carrying a version of the HPRT gene from which the introns had been removed, and a few were found to take up the virus and *express* the transgene (meaning that they proved capable of generating HPRT). In such experiments, the researcher has no difficulty seeking out those cells that are able to make the most enzyme with the use of the transgene, even if these true successes number as few as one or two out of every million. Yet if a gene transplantation is to work in a living person or an experimental animal, a much larger fraction of the cells receiving the new transgene must be able both to express the gene and to keep on expressing it. This has not yet been achieved with any consistency in any animal in which it has been tried.

Oliver Smithies of the University of Wisconsin is among those who have come closest. After infecting mouse cells with a virus carrying an altered HPRT gene, he selected cells that expressed the altered gene and injected them into tiny mouse embryos. The embryos were then implanted into the uterus of a female mouse and brought to term. Most of the resulting mouse pups were *chimeras,* mixtures of the original embryonic tissue and the introduced cells. Those male chimeras whose testis tissue happened to have been made from the introduced cells were able to pass the modified gene on to the next generation.

Of course, such an experiment is a far cry from real gene therapy, which must be carried out not on early embryos but on babies or adults who show the signs of the disease. Human embryos cannot be manipulated in the ways that Smithies manipulated his mouse embryos. Even in these relatively simple experiments, Smithies encountered difficulty in selecting cells that expressed the new gene. Recently he may have found one answer to this problem. His group discovered that the shortened versions of the HPRT gene commonly being used in such experiments worked very badly in cells recently taken from mouse embryos, although they did work well in cells taken from mice years before and grown in culture. To make the gene work properly in the fresh cells, Smithies' group had to leave in two of those "meaningless" stretches of DNA, the introns. Thus, it turns out to be important to sequence and examine introns, in order to determine which parts of them aid in gene expression and why.

One criticism of the Human Genome Project, leveled particularly by Martin Rechsteiner of the University of Utah, has been that it is a foolish waste of resources to sequence these long stretches of DNA that do not code for proteins. This criticism assumes that we already know what parts of the DNA are important and what parts are not. Yet the recent findings of Smithies and others suggest that the apparently useless introns contribute somehow to the success of transgenes. Would a transgene work even better if even more of its introns were left intact? This experiment cannot be done at the moment, because leaving in too many introns can make the gene so long and unwieldy that it cannot be carried by a virus. Both genome sequencing and advanced virology are needed to understand the significance of introns. Only when this can be accomplished will gene therapy be able to move beyond its infancy into the medical mainstream.

"THEY BROUGHT IT ON THEMSELVES, MISS TWIDDLE!"

At the time of the early work on Lesch-Nyhan, gene therapy was no more than a glimmer in the minds of a few farsighted scientists. Seegmiller had been frustrated in his search for a cure for that devastating disease. Still, he now knew far more than he did before about the dynamics of purine metabolism. Armed with this new knowledge, he turned his attention to gout.

One obvious possible explanation for gout was that its victims suffered from the effects of a leaky HPRT enzyme, capable of supplying some intermediate amount of HPRT that would stave off the severe symptoms of Lesch-Nyhan syndrome but still slow down the salvage pathway enough to allow more purines than normal to be broken down into uric acid. Since such an error would run in families, Seegmiller searched for families in which there were a number of people with gouty arthritis. One very promising group consisted of three middle-aged brothers, each of whom had been suffering from severe gout since his twenties. The brothers' symptoms were so severe that they overlapped some of those seen in Lesch-Nyhan. Yet Seegmiller, already confident that he knew the key to gout, was able to tell the men with great excitement that they would soon know the reason for their disease.

Late one evening in 1966, Seegmiller's assistant, William Kelley (now dean of the University of Pennsylvania Medical School), began to test the blood of all three brothers using new and sensitive assays. By three in the morning, Kelley had found that all of them had a mere half of 1 percent of the usual levels of HPRT. He stayed up until Seegmiller arrived at six and elatedly told him the news. Was it the answer to gout? No, they quickly discovered—only a small piece of it. Many familial cases of gout showed lowered levels of HPRT, but other familial cases, as well as most isolated cases, showed normal levels. Reduced HPRT was actually only found in 2 or 3 percent of gout cases.

A few years after Kelley's observation, a group of Israeli scientists showed that some victims of gout have elevated levels of an enzyme called PRPP synthetase, one of the enzymes normally responsible for making purines. Purines flood the systems of these patients, making it impossible for even normal amounts of HPRT to salvage them all before they are broken down into uric acid. It is the equivalent of a glut of newspapers overwhelming the recycling facilities. More recently, a group at Guy's Hospital in London has found that some other rare cases of familial gout result from an inherited inability of the kidney to clear uric acid from the blood in normal quantities. This is exactly what Alfred Baring Garrod had predicted 150 years earlier.

The fact remains, however, that most victims of gout have no family history of the disease. Their HPRT, PRPP, and kidneys appear to function perfectly normally. The great majority of these cases have been associated with factors in the environment.

A small fraction of you may remember a 1940s comic strip called "The Katzenjammer Kids." At the end of each strip these mischievous but endearing kids would be seen embroiled in the dreadful consequences of their latest escapade. It was the job of another character to rub salt in their wounds. This was the virtuous little prig Rollo Rhubarb, who always appeared dressed in a spotless Lord Fauntleroy suit. He would point to them and exclaim to the governess, "They brought it on themselves, Miss Twiddle!" Is Rollo right? Have the great majority of gout victims brought the disease entirely on themselves, or have they been given an assist by their genes?

Clearly, uric acid excretion can be affected by intake of rich food and alcohol. If kidney function is damaged—either through alcohol abuse or through the effects of illness—then the excess uric acid stays in the blood without being excreted, helping to trigger gout. It has been known since ancient times that moderation in food and drink can help prevent or alleviate many attacks of gout. People who happen to have mild defects in HPRT or PRPP synthetase can lead perfectly normal lives if their diets are not overrich in purines and if they refrain from abusing their kidneys. Conversely, people with quite normal enzymes can give themselves gout if they persist in imitating the lifestyle of Henry VIII.

Evidence suggests that certain gene variants in the human population predispose their carriers towards alcoholism, and these would certainly contribute to gout as well.* Might genetic variation also affect our behavior toward food, leading some confronted with rich fare to overindulge and others to be more abstemious? The effects of alleles capable of influencing such behavior would be subtle, complex, and difficult to trace. Such alleles are probably far more widespread in the human species than the drastic mutations that lead to Lesch-Nyhan.

Yet why on earth should there be alleles in our species that would contribute to such apparently self-destructive behavior? There could be a subtle evolutionary reason. If Bruce Ames is right that there are advantages to having high levels of uric acid in our bodies, then a great deal of genetic adjustment must have taken place in the course of our evolution to compensate for this drastic physiological alteration. Many alleles that ameliorate the harmful effects of uric acid must have arisen at many different genetic loci in the course of the sixty million years during which this compound was gradually increasing in the bloodstreams of our ancestors. This enormous span of time gave these further evolutionary changes plenty of time to occur.

Like most other evolutionary processes, this genetic "fix" was imperfect. After all, the old alleles being replaced by new ones must have had their own valuable characteristics, or our ancestors would not have had them in the first place. The replacement of old alleles by new ones often leads to a balance of competing evolutionary advantages, in which both alleles eventually coexist in the species. Some people will have one allele and some another, giving rise to a great variety of genetic possibilities.

Consider then, as only one of dozens of possible interactions between genes and environment, our behavior toward food. Alleles contributing to abstemious behavior in the face of a sudden bonanza of rich food would not be likely to present an advantage in a world in which food is

*A recent, widely publicized study by researchers in Texas and Los Angeles reported that a particular allelic variant for a receptor protein found in the brain is significantly more common in alcoholics than in nonalcoholics. But this variant is certainly not the only cause of alcoholism. Even though the association appears to be significant, at least among severe alcoholics, the majority of alcoholics do not have the variant—and the great majority of people who carry the variant in the population are not alcoholics.

in short supply; there, any alleles contributing to greedy behavior would confer an immediate advantage. This situation would be expected to reverse itself if food became more consistently available. Still, even in the new world of food abundance that has existed for most people since the agricultural revolution, the long-term advantages of moderation are probably balanced against the immediate advantages of greed, with neither side coming out ahead. We can begin to glimpse a complex evolutionary history of relationships among various alleles of dozens or perhaps hundreds of genes, their selective advantages and disadvantages altering and interacting over time as the environment changes.

Rollo is partly right. Gout reached its apex during the gastronomic and bibulous excesses of the eighteenth century. We know that too little time has passed since then for the human gene pool to change appreciably, meaning that we have more or less exactly the same genes, with their various alleles, that our gout-ridden ancestors had. Now, however, increased knowledge of the disease and changing societal patterns have enabled most of us to avoid bringing it upon ourselves.

Other genetically based diseases that we will meet later in this book will prove far more difficult to deal with than gout has been. They will need all the understanding that medicine, biochemistry, education, sociology, and human genome research can bring to bear.

CHAPTER 4

Politics

All that I mean to say is this: permanent inequality of conditions leads men to confine themselves to the arrogant and sterile research of abstract truths; whilst the social condition and the institutions of democracy prepare them to seek the immediate and useful practical results of the sciences. . . . [M]en living in democratic ages cannot fail to improve the industrial part of science.

—Alexis de Tocqueville, *Democracy in America* (1840)

As a result of the astounding technical breakthroughs of Sanger and of Maxam and Gilbert, DNA sequencing became faster and cheaper—though not, as we will see, as fast and cheap as it needed to be to tackle a really big job such as sequencing the entire human genome. While Renato Dulbecco was the first to suggest publicly the idea of a human genome project, an idea he had arrived at on his own, similar proposals were already being discussed elsewhere, particularly at the U.S. Department of Energy (DOE).

In retrospect, it seems astonishing that the DOE should have turned out to be the most important shaper of the Human Genome Project. The department is the lineal offspring of the Atomic Energy Commission, a civilian agency established immediately after World War II to continue the work started by the Manhattan Project and to oversee the development of atomic energy as a peaceful resource. From the beginning, the agency's mandate also included support of research in medicine, biology, and engineering. Innovative research has been carried out with this support, both on university campuses and at the remarkable collection of national laboratories founded at the time of the Manhattan Project.

Over the decades, the DOE's mandate has grown to include not only atomic energy but all forms of energy production in the United States. Its problems have grown as well. It is beleaguered on every side, with challenges posed by unsafe nuclear power plants, radioactive waste disposal, acid rain, groundwater pollution, environmentally caused cancers, and so on. It was these societal pressures on the agency that led to the development of the Human Genome Project.

By 1984, legal clouds were gathering around a number of government agencies. Some of these were a legacy of the thoughtless early days of the atomic era, when the panicky transition from hot to cold war had inspired some unnecessary experiments later found to cause harm. In the forties and fifties, groups of servicemen had been exposed to radiation from atomic blasts, both in the United States and at South Pacific testing areas such as Bikini Atoll. Civilians living near the U.S. testing grounds in Nevada and New Mexico had been exposed as well. Growing awareness of the dangers of radiation in the intervening decades brought the prospect of endless litigation on the part of these people and their families. And this was only the beginning. Lawsuits were pending from Vietnam veterans who had been exposed to Agent Orange, and others were likely to come from people living near Three Mile Island and the far more dangerous nuclear plants at Hanford, Washington, and Savannah River, South Carolina. More distant but still worrisome was the possibility of class action lawsuits against both the government and private polluters from people living in parts of the country like the Mississippi delta, where the levels of potential carcinogens in the groundwater are very high.

All these suits, both real and potential, are complicated by both epidemiology and politics. While most of the large-scale fluctuations in cancer incidence across the country can be traced to regional differences in the level of cigarette smoking, environmental influences have been linked to cancer in asbestos workers, uranium miners, and other people in high-risk occupations. Just how resilient are our genomes to various kinds of environmental damage? If future lawsuits are not to be decided simply (and unfairly) on the basis of who has the most resources and the cleverest lawyers, clear-cut and unequivocal data about the effects of this damage will be needed. If, for example, Agent Orange has indeed damaged the genomes of thousands of servicemen, leading to cancers in them and to congenital and genetic malformations in their offspring, then it would be a gross injustice not to compensate them. But if it has had no effect, and the courts still force the government to pay out lavish compensation, then society suffers.

At a 1984 conference sponsored by the International Commission for Protection Against Environmental Mutagens and Carcinogens, held appropriately enough in Hiroshima, a possible way to cast light on these problems emerged. Several speakers pointed out that it was essential to preserve the DNA of the survivors of the explosions at Hiroshima and Nagasaki, against the day when more advanced technology could finally determine the true extent of the genetic damage caused by the bombs. In the same year, the U.S. Congress mandated a report on the feasibility of detecting low-level damage to the genome from environmental hazards. The job of figuring out how to do this fell to David Smith of the DOE and Mortimer Mendelsohn of Lawrence Livermore National Laboratory.

Smith and Mendelsohn reasoned that in order to detect small amounts of genetic damage, one would have to be able to look at and compare a great deal of damaged and undamaged DNA. Why not get together a group of people at the forefront of DNA sequencing technology, to find a cost-effective means of carrying out such tests? Ray White of the University of Utah was called in on the planning of this meeting, and he eventually took over most of the work. The plan was to try to get as expert a set of participants as possible to attend, even though the general feeling was that the goal of the meeting was unlikely to be realized.

James Neel, a population geneticist from the University of Michigan, was invited, as were various luminaries of the DNA sequencing world. Prominent among these were Charles Cantor, then at Columbia and the co-inventor of a way to separate very large pieces of DNA on a gel; Maynard Olson of Washington University at St. Louis, who was looking for ways to put large pieces of DNA into fast-reproducing yeast in order to make many copies of them; and David Botstein of MIT, who had suggested a way to map DNA in humans by using the genetic variation that already existed in natural populations. The meeting took place in December 1984 at Alta, a ski area in the Saguache Mountains south of Salt Lake City. The participants were able to concentrate fully on their thankless task, as they were snowed in for the entire five days.

Neel began the conference by presenting the best data yet acquired on how the environment can damage the human genome. This information came from his own long-term study, carried out under the auspices of the Atomic Bomb Casualty Commission, of people who had been exposed to radiation from the explosions at Hiroshima and Nagasaki. These people and their children had been subjected over a period of more than forty years to the most intensively detailed genetic investigation ever carried out on a large human group.

Yet were it not for the political sensitivity of the situation, the study would certainly have been discontinued years earlier for lack of results. The survivors, and their children who had been *in utero* at the time of the blasts, had undeniably been affected by their exposure to the radiation. Birth deformities and cancers were common, and cancer rates continued to be elevated decades after the end of the war. But these effects did not extend to those children who had been conceived by the survivors *after* the explosions had taken place. Their levels of cancer and birth deformities were the same as those of the control group. The major question addressed by Neel and his coworker, William Schull, along with their many Japanese and American collaborators, was whether genetic damage done to the survivors had been passed on to the next generation.

The investigation received a boost when, a few years after the war, new techniques made it possible to examine human chromosomes in detail. The chromosomes of the survivors did indeed show injury, even years after the event, but no damage could be detected in the chromosomes of those children of the survivors who had not themselves been exposed to the radiation. Most of the damaged chromosomes in the parents appeared to have been lost before they could be passed on. Yet these results did not rule out the possibility that other, less visible, changes might have been transmitted to the children. Moving one step closer to the actual genome, Neel and Schull took blood samples from four groups: one that had been exposed to radiation, an otherwise similar group that had not been exposed, and children of both groups who had been conceived after the explosions. Since DNA sequencing methods were not yet in use, Neel and Schull looked at proteins made by cells in the blood. A small army of technicians in their lab began measuring the rates at which blood proteins moved through gels. These gels resembled Sanger's sequencing gels, but were designed to separate whole proteins rather than fragments of DNA.

The availability of proteins from both parents and children made it possible to trace the genes responsible for these proteins from one generation to the next. If a child's protein moved through a gel at a rate different from that of the same protein from its parent, a change must have occurred in the gene. Only about a third of all the possible genetic changes could be picked up in this way, however.

By the time of the Alta meeting, about half a million genes had been scored in the irradiated group, and three new mutants had been found. A similar number of genes had been scored for the unirradiated group, and three new mutants had also been found. Exactly as many mutations had been detected in the unirradiated group as in the irradiated one. Even at this level, then—only one step removed from the DNA sequence itself—no long-term genetic effect from the explosions could be detected. These negative results represented an enormous amount of work and a great deal of money. The bottom line was that whatever long-term genetic damage had been done at Hiroshima and Nagasaki, it was too slight to be detectable, even through massive study.

Hans Mohrenweiser, now at Lawrence Livermore National Laboratory, was a leader in the group that collected the data on those million genes. He described to me recently the mind-numbing boredom of the task. Of course, if an apparent difference had been detected, he and his group would have been forced to go on to collect perhaps ten times as much data to see whether the difference could be considered significant. He was personally very glad that this was not necessary. But the possibility remained that the mutation rate really *had* increased markedly, and that they had missed the fact by a statistical quirk. Alternatively, the radiation might have increased the mutation rate above the spontaneous rate by a mere 10 percent or so. The amount of labor that would have been needed to be fairly confident of the existence of such a slight increase boggles the mind.

One striking fact did emerge from the research. In earlier experiments, mice had been exposed to the same amount of radiation received by the survivors of Hiroshima and Nagasaki. Unlike these survivors, the mice did pass on detectable numbers of damaged genes to the next generation. The huge study of Neel and Schull showed that the human genome is much more resistant to radiation damage than the genome of mice—at least four times as resistant, according to their calculations. Furthermore, nothing in their data ruled out the possibility that it might be even more resistant than that.

As the Alta conference proceeded, it became apparent that if DNA technology were to be applied to this problem it would have to be used to scan many genomes' worth of DNA to pick up what were sure to be small alterations in the mutation rate. The conclusion reached by the participants was that it was simply not yet possible to do this.

Frustrated in their main goal, they nonetheless found to their pleasure and surprise that DNA mapping and sequencing techniques were advancing further than any of them individually had realized. Indeed, new ideas about how to sequence DNA rapidly actually emerged in the course of the conference itself. These and later ideas were incorporated in a report on the conference prepared for the Congressional Office of Technology Assessment under the direction of Michael Gough. This report made exciting reading—so much so that when Charles DeLisi, then director of the DOE's health and environmental research programs, read a draft of it in October 1985, it occurred to him that all this new technology could be applied to a more manageable goal, that of sequencing the entire human genome once rather than many times. At the time he knew nothing of the fact that Renato Dulbecco had come to the same conclusion for different reasons a few months earlier.

DeLisi realized that a project of this scope was tailor-made for the DOE. It would require the cooperation of molecular biologists, physicists, chemists, computer experts, and engineers, exactly the kinds of groups already working together in the national labs. As an added incentive, successful completion of the genome project might actually salvage the department's original goal of measuring damage to the genome.

He immediately called Mortimer Mendelsohn, who was then chairman of the DOE's Health and Environmental Research Advisory Committee (HERAC). Mendelsohn's initial reaction was negative, but DeLisi was unwilling to let his exciting idea go; he was sure that the more people thought about it, the more it would appeal. To try to harness that enthusiasm, DeLisi asked Mark Bitensky, a senior fellow at Los Alamos, to organize a workshop at Santa Fe, New Mexico. This workshop, unlike the one at Alta, would be specifically geared to address the possibility of a human genome project.

BIG SCIENCE VERSUS SMALL

DeLisi's feeling that general enthusiasm about a human genome project would grow was reinforced when he learned about the experiences of Robert Sinsheimer. About a month before the Alta conference, Sinsheimer, then chancellor of the University of California at Santa Cruz, had—again independently—come to the same conclusion about the feasibility of a human sequencing project that DeLisi had later reached. Sinsheimer had been involved in the development of sequencing technology from the very beginning. His lab had performed the detailed studies of the single-stranded DNA of the virus ϕX174, which Fred Sanger had then used to invent his elegant sequencing methods.

Yet like DeLisi's, Sinsheimer's first attempt to gather support for the project failed. He wrote a letter to David Gardner, the president of the nine-campus University of California, suggesting that a bequest the University had recently received be used to establish an Institute to Sequence the Human Genome at Santa Cruz. Gardner replied that the bequest money would not be available, but that Sinsheimer was perfectly free to seek other support for such an institute—if he could find it. Sinsheimer promptly began gathering support in time-honored fashion by bringing together many stars of the sequencing world at a workshop the following May, five months after Alta (and just before Dulbecco's talk at Cold Spring Harbor). This meeting has the distinction of being the first to address directly the possibility of sequencing the entire genome. Initially many were skeptical, but, just as at Alta, a feeling of excitement about the new developments in mapping and sequencing gradually emerged. Sinsheimer was encouraged enough to begin searching for funding for his institute.

Meanwhile, an agenda was quickly crystallizing at the DOE. By March 1986, when DeLisi's Santa Fe workshop was held, Mortimer Mendelsohn had become a convert. DeLisi realized from the outset that a project of this magnitude could not belong to the DOE alone; its resources were dwarfed by those of the National Institutes of Health (NIH), the immense agency based in Bethesda, Maryland, that has dominated the biomedical research landscape since World War II. He contacted Donald Fredrickson, a former director of NIH who was then head of the Howard Hughes Medical Institute. Fredrickson expressed some interest, but of the NIH bigwigs invited to the Santa Fe meeting, only one—Vincent deVita, then head of the National Cancer Institute—responded, and in the end even he could not make it.

The forty-three scientists who did participate in the workshop helped outline a general plan to sequence the human genome, an outline that has survived more or less unchanged through innumerable meetings since. Two phases of research were envisioned. The first would concentrate on creating a detailed physical and genetic map of the genome, on developing the appropriate computer tools, and above all on finding a way to make DNA sequencing less expensive and less slow. Only when these goals had been achieved would the second phase, the actual sequencing, begin.

The cost of all this was not discussed at the meeting, and DeLisi, who is currently dean of the College of Engineering at Boston University, now speaks with some bemusement about the $3 billion price tag—a dollar a base—that has somehow become inseparably linked to the project. In a 1988 paper in the *American Scientist,* he pointed out that if the cost of sequencing could be reduced a hundredfold and the rate increased by the same amount, the entire human genome could be sequenced in a few months for a piffling $30 million.

When Bitensky's enthusiastic summary of the Santa Fe workshop reached him, DeLisi made a number of politically shrewd moves. He used the report to enlist as an ally Alvin Trivelpiece, then the director of the DOE's Office of Energy Research. Trivelpiece in turn asked HERAC to produce a more extensive report. A subcommittee of HERAC was promptly established, including Dulbecco and Sinsheimer among others. DeLisi and Mendelsohn also made contact with the White House Office of Management and Budget, and with a number of members of Congress who might have a direct interest in such a project were it spearheaded by the DOE. These included former senator (now governor) Lawton Chiles (D-Fla.), Senator Pete Domenici (R-N.M., where the largest DOE installation, at Los Alamos, is located), and the powerful congressman John Dingell (D-Mich.), the head of the House subcommittee in charge of the DOE budget. Congressional staffers, particularly on the House side, were very supportive, and momentum gathered for a series of Congressional hearings to be held in conjunction with the normal DOE appropriations hearings.

At every stage in the process, the Department of Energy forced the issue. In early 1987 the HERAC report appeared, endorsing the sequencing project in the strongest terms. It also recommended that the DOE be in charge of the overall project, noting its experience in running large multidisciplinary programs. Mortimer Mendelsohn, author of the report, said at the end of it, "It may seem audacious to ask DOE to spearhead such a biological revolution, but scientists of many persuasions on the subcommittee . . . agree that DOE alone has the background, structure and style necessary to coordinate this enormous, highly technical task." On the strength of that very favorable report, Secretary of Energy John S. Herrington directed that centers to study the human genome be set up at Los Alamos, Livermore, and Berkeley, three major DOE weapons and research laboratories. Eleven million dollars were secured to begin the process.

Other things were happening as well. In August 1986, the Board of Basic Biology of the National Academy of Sciences (NAS) established a blue-ribbon committee to study the feasibility of a sequencing project. (The NAS, a distinguished group of some 1700 senior scientists drawn from every field, frequently provides expert advice to the government on matters involving science and technology.) The new committee was chaired by the distinguished cell biologist Bruce Alberts, and its members included James Watson. A mixture of proponents and doubters about the project, they embarked on a series of deliberations that finally led to the production of a favorable report fourteen months later. And in addition, some months after the establishment of this committee, congressional pressure (particularly from the New Mexico delegation) led the Congressional Office of Technology Assessment to begin putting together yet *another* report on the project's feasibility.

This left NIH uncomfortably out in the cold. The agency did not have its first meeting to discuss the feasibility of the project until November 1986, a year and a half after Sinsheimer's workshop. At an earlier meeting at Cold Spring Harbor in 1986 and at a subsequent series of NIH-sponsored meetings through 1987, participants repeatedly stressed that if the project was to consist of "good" science, NIH should be the one to run it. The general perception was that the quality of science supported by NIH was higher than that supported by the DOE, because of NIH's stringent peer-review system.

The great majority of NIH funds are dispensed through a system whereby dozens of panels of respected scientists from each institute examine and rank all the applications pertaining to their areas of expertise. Of course, the machinery can be toyed with—program directors can overrule a panel's recommendations in a few cases, and there is an appeals process—but for the most part NIH follows the recommendations of its panels. This system has resulted in the funding of excellent science, but there have been repeated complaints that really innovative research is not encouraged because of the necessity of obtaining a consensus within the review panels.

At the DOE, a large fraction of the funds for research are dispensed through a system of contracts, or funneled directly to the national laboratories. For the kind of highly directed, goal-oriented research normally carried out by the DOE, such a system makes sense. Tony Carrano of Lawrence Livermore National Laboratory remarked proudly to me of one particularly imaginative DNA sequencing project that it had been developed primarily using in-house money. It had received NIH funding, but only when it was well on its way to success.

The DOE is understandably defensive about its procedures. Ron Walters, the program director for biological and environmental research at the Los Alamos labs, states that peer review is indeed alive and well at the DOE, but that this is not widely recognized. Indeed, the majority of the money the DOE has given to human genome projects so far has been dispensed using a peer-review system that was set up by Dr. Irene Eckstrand, an experienced administrator lent by NIH for the purpose. DeLisi has also pointed out that proponents of the "NIH style" conveniently forget that NIH also dispenses an appreciable fraction of its research budget in-house at its Bethesda campus, and that intramural directors and lab chiefs have the final say on allocation of these funds.

Still, the feeling among participants at the NIH conferences on the human genome project was that in this sort of endeavor, the very shape of which had yet to be clearly formed, goal-oriented contracts were not the way to go. While applicants for NIH funds must state the goals of their research, these goals rarely stay fixed during the course of a project. As long as an investigator does exciting and innovative research with the money, no one ever objects: "But in 1982 you said that you were going to do such-and-such, and now it turns out that you have done so-and-so! Explain yourself!" Indeed, a much more exciting goal often beckons partway through the project. Knowing that he or she will not be held accountable for failing to reach some outmoded set of goals gives the investigator an unparalleled sense of freedom and adventure.

The debate about styles of funding for the genome project first surfaced in February 1988, when the National Academy of Sciences report appeared. In the course of the year of discussions that led to the report, the NAS committee had undergone the same transformation from skepticism to enthusiasm that had marked the earlier meetings and workshops. The committee ended by drawing up a thorough set of guidelines very similar to those proposed at Santa Fe. Their report concluded that there should be a strong push toward a detailed genetic map of the genome and another toward faster and more efficient methods of sequencing DNA. While perhaps ten multidisciplinary research centers should be set up around the country, the paper stated, the majority of funds for the project should be dispensed to individual investigators through an NIH-style peer-review program.

The committee was split, however, over who the agency in charge should be. Some members thought that the DOE should be rewarded for having originated the project and pushed it so effectively. Others, including the powerful voice of James Watson, insisted that since the DOE's funding methods often led to second-class science, NIH should be the lead agency. Yet even Watson was forced to admit that NIH seemed less than wildly enthusiastic about the project.

It was a bad time for NIH, one marked by difficulty in funding new research proposals because of falling revenues and many simmering political controversies. The most widely publicized of these was an attempt to recruit Washington University chancellor William Danforth to be NIH director, a plan that failed as a result of the Reagan administration's insistence on using attitude toward abortion as a litmus test for the appointment. Such incidents had thrown the NIH bureaucracy into disarray.

At the time the NAS report appeared, two bills were introduced in the Senate to establish a National Advisory Panel on the Human Genome. One, sponsored by Pete Domenici, specified a national policy board on the human genome to be headed by the Secretary of Energy. The second, sponsored by Lawton Chiles, would give NIH and the DOE rotating control over the panel and the project. The possibility of having either of these schemes imposed on them had the ultimate effect of spurring the two agencies to greater efforts at cooperation.

The public was exposed to this debate in hearings before the Subcommittee on Oversight and Investigations of the House Energy and Commerce Committee in April 1988. The hearings were held at that time to coincide with the release of the detailed, readable, and highly favorable report on the Human Genome Project from the Office of Technology Assessment. At the hearings a number of scientists were quizzed about the details of the project. The testimony of James Watson, with his dazzling credentials, particularly impressed the subcommittee members.

Later in the hearings, James Wyngaarden, then the head of NIH, and David Nelson, the executive director of the Office of Energy Research, representing the DOE, were grilled much more severely. Nelson had taken the place of DeLisi, who had left the DOE to take a position at the Mount Sinai School of Medicine. (DeLisi's DOE position, associate director of health and environmental research, was not filled until May 1990.) Representative Ron Wyden (D-Ore.) asked whether, in view of the wrangling about which of the two agencies should head up the project, a memorandum of understanding should be drawn up between them on how to divide up responsibilities. Nelson replied that they had no plans to do so.

But plans were quickly developed, in spite of the agencies' stance. In subsequent private discussions, House committee members expressed concern that so much depended on a few enthusiastic and competent people at the two agencies. A few personnel changes along the lines of

DeLisi's departure could easily break down whatever cooperation existed between NIH and the DOE.

Pushed by all these forces, the two agencies swiftly drew up a short memorandum in which they agreed to cooperate on the project and formalize mechanisms for obtaining outside advice. This had the effect of defusing the clumsy dual-leadership provision of the Chiles bill, which later died in subcommittee. The human genome provisions of the Domenici bill were then dropped during a House-Senate conference, leaving the agencies free to seek funding for the project through regular budget hearings and in-house arrangements.

In September 1988, James Watson agreed to head up NIH's part of the genome project. Initially, he was reluctant to shoulder the burden in addition to his heavy responsibilities as director of the Cold Spring Harbor Laboratories. But the temptation proved too great. "Only once," he wrote later, "would I have the opportunity to let my scientific life encompass the path from [the] double helix to the 3 billion steps of the human genome." He set up shop within the NIH director's office with the avowed goal of ensuring that any funds for the project would be "new" money, not simply a diversion from NIH's missions on behalf of basic science. Watson's appointment enormously enhanced the project's credibility.

After all its birth pangs, the project received $41 million at the DOE and $60 million at NIH for 1991. Yet even these numbers were small potatoes in the world of big science, and only about half of what was originally envisioned as necessary. Funding was the most obvious problem facing the project, which had the bad luck to come along just at a time when many other large research projects were competing strongly for scarce federal dollars. (Robert Sinsheimer never did get the money for his institute at Santa Cruz.) So intense is current competition that even a human genome proposal approved by an NIH review panel has only about a 20 percent chance of being funded, or roughly the same chance as that of a proposal submitted to other parts of NIH.

The arrival of the Human Genome Project at a time of budgetary stress is not a coincidence, though it has greatly added to its problems. The more successful science becomes, the greater the number of wonderful and eminently fundable ideas that are likely to appear. And science has certainly been successful. Since the great increase in federal science funding began in the late 1940s, the population of the United States has increased from about 150 million to 250 million. Were this rate of increase to continue, the population would double every fifty-five years. But during this same postwar period the number of scientists and engineers has increased tenfold, from 550,000 to over five million. This means that the number of scientists and engineers has doubled every eleven years. At that growth rate, by the year 2070 everyone in the country, including babes in arms, will be a scientist or engineer!

Since the time of Malthus, it has been known that such explosive growth is a recipe for disaster, or at the very least for retrenchment. In a Malthusian world where resources are perceived to be leveling off, it is not surprising that the first impulse of most scientists will be to object strenuously when a new, large, and greedy kid appears on the block.

A second problem is that the biomedical community has had no experience with anything quite like this. The 1991 funding for AIDS research, currently budgeted at $800 million, is eight times that for the entire Human Genome Project. AIDS money, however, is distributed over thousands of separate research efforts, sociological studies, and testing programs; it does not have the monolithic feel to it that the Human Genome Project does. The distinction may appear to be slight, but perceptions often drive reality. The AIDS program is seen as a huge agglomeration of Small Science. The Human Genome Project is perceived to be an extremely modest example of Big Science.

The reluctance of NIH to participate in the project can be traced in part to the fact that scientists themselves are primarily responsible for the allocation of resources within the agency. They are vividly aware of just how finite the resources for biomedical research are. By contrast, the initial enthusiasm of the DOE could be traced in part to the fact that it was able to obtain funds from outside the biomedical community. Indeed, with the geopolitical scene rapidly changing, it was the DOE's hope that funding, resources, and people that might have been devoted to defense could be diverted to the Human Genome Project rather than being lost to the agency. Diatribes against the quality of their science notwithstanding, the DOE staff feels strongly that they are able to do things NIH cannot.

Some other objections to the project are now surfacing, but they are oddly muted. Martin Rechsteiner, a researcher at the University of Utah Medical School, has had perhaps the largest impact, with an open letter to the biomedical community in early 1990. In it he complained about the diversion of money from exciting and intellectually stimulating research to a nightmare world of robotlike technicians endlessly sequencing meaningless stretches of DNA.

Shrewd politicking has been carried out in this area as well. Watson and

Charles Cantor were the co-organizers of the second Human Genome Conference, which took place in San Diego in October 1990. They invited two articulate opponents of the project to speak: Donald Brown from Washington's Carnegie Institute and Bernard Davis of Harvard. Given the thankless task of attacking the project before an unimpressed audience made up largely of people already committed to it, they focused on budget problems and the threat of starvation of small science. The general feeling among many audience members was impatience and a desire to get on with the scientific part of the program.

Davis's major point was originally to have been that a heavy concentration of gene mapping should precede any actual sequencing. Watson, who spoke earlier in the program, defanged Davis rather effectively by a preemptive strike. Such a ban on sequencing, said Watson, is like saying, "Let's get married, but no sex!"

The other masterful preemptive move on Watson's part was to deliberately earmark, from the beginning, 3 percent of the NIH human genome budget for the consideration of ethical matters arising from the project. Nancy Wexler, a psychologist from Columbia University heavily involved in the search for the gene for Huntington's disease, was put in charge of this effort. A public television series is being planned, and tweed-clad ethicists have already congregated for meetings in Tokyo and in Monte Picayo, Spain.

There is no sign that researchers are rallying to the side of the dissenters in any numbers. In spite of their own misgivings, most are fascinated by the prospect of finding out about the entire human genome. As Watson said at the time of the NAS report, "It has got to go ahead, it is so obvious."

Part Two

CONNECTIONS

CHAPTER 5

Through the Genome with Gun and Camera

Man is only man at the surface. Remove his skin, dissect, and immediately you come to machinery. —Paul Valéry

On the evening of April 8, 1663, the Royal Society, a mixed group of dedicated natural philosophers and dilettante aristocrats chartered by Charles II, gathered at Gresham College on Broad Street in the old city of London. There, some primitive rooms had been set aside for the Society's weekly meeting. Though the group had only been in existence for a year, its meetings were already becoming crowded. All attention focused on the stooped and wizened figure of Robert Hooke, the newly appointed curator of experiments. It was his demanding job to come up with a remarkable new scientific demonstration for each meeting.

This time he had cut a thin shaving of cork with a razor and placed it on the pin of a simple single-lensed microscope. The Society's members eagerly lined up to hold the microscope up to a lamp and peer through the lens. They saw clearly that the slice of cork was made up of an apparently endless array of tiny, thick-walled rectangular chambers. Each chamber trapped a minuscule pocket of air, giving the cork its lightness.

When Hooke had first seen them, the chambers had reminded him of rooms in a medieval dungeon. Because of this, in his 1665 book *Micrographia,* he named them *cells.* Although he did not realize it, the walled chambers were really a kind of skeleton. Each had at one time contained a diaphanous bag of protoplasm, and during its lifetime had secreted the thick surrounding wall. When the bag disappeared, the wall remained for

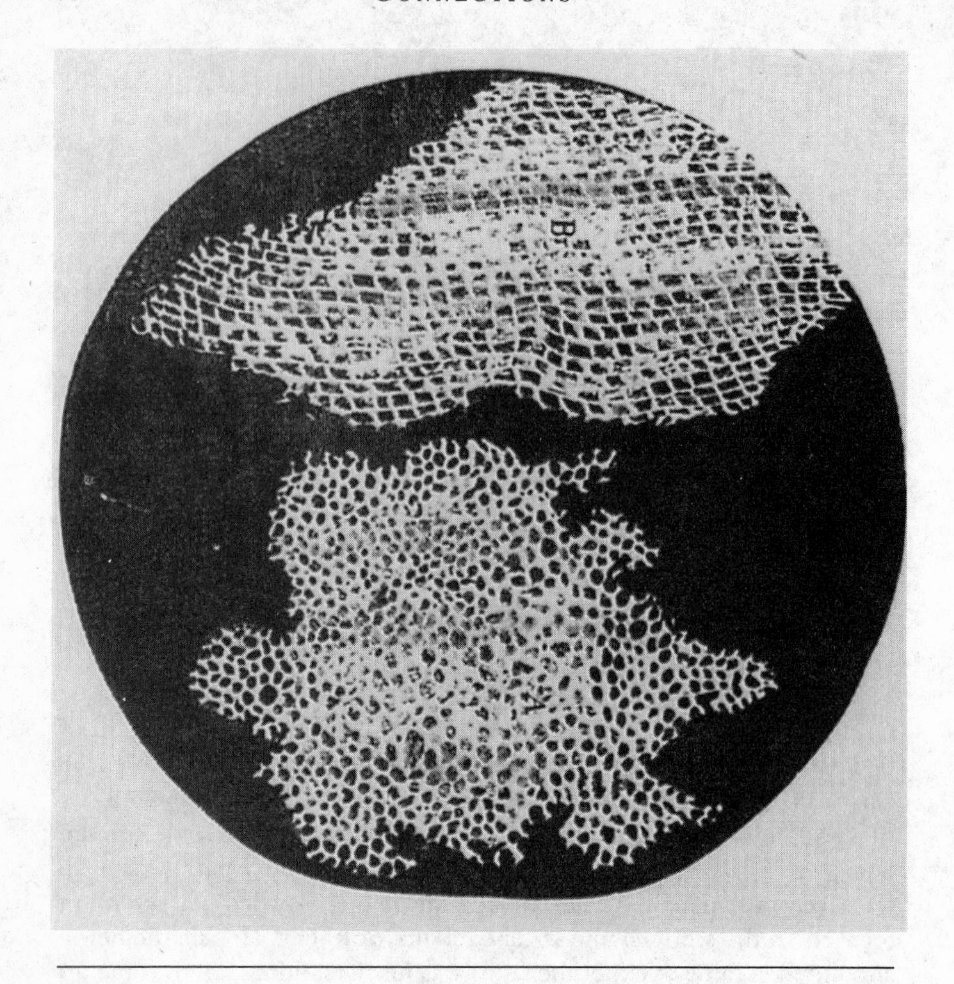

Figure 5–1. Robert Hooke's drawing of thin sections of cork seen through his simple microscope.

Hooke to see with his crude microscope. As scientists came to understand this process, they transferred Hooke's name for the wall to the living structure that made it. In the process, the word "cell" lost its grim medieval connotations.

Despite their diverse functions, the ten trillion cells that make up the human body are all fragile bags of protoplasm very similar to the transient inmates of Hooke's chambers. A few are large enough to hover on the edge of visibility. The human egg cell, if illuminated like a dust mote in a sunbeam, would be just barely detectable to the naked eye. While the egg measures about a tenth of a millimeter across, the average cell in the body has only a fifth of that diameter, and thus lies quite beyond the reach of our unaided vision.

Under an ordinary microscope an unpigmented cell such as a white blood cell is virtually invisible. It consists mostly of cytoplasm, where metabolism is carried out and the majority of the proteins are made. The nucleus is a small transparent blob within the transparent cell, taking up about a hundredth of its volume. Along with the chromosomes it contains, the nucleus can only be made visible by staining or by certain optical manipulations. The sex chromosomes are named X and Y, and the remaining twenty-two chromosomes, the *autosomes,* are numbered roughly in decreasing order of size. Chromosome 1, shown in Figure 5–2, is about six times the size of our smallest chromosome, 21. While chromosome 1 contains almost three hundred million bases of DNA, about a tenth of the genome, chromosome 21 contains only about fifty million.*

The figure shows a representative of human chromosome 1 alongside the corresponding chromosome of our three nearest relatives, the chimpanzee, gorilla, and orangutan. Scientists had their first glimpse of the human genome in the form of such chromosomes. The chromosomes in this figure would not be recognizable to nineteenth-century microscopists, however, for the art of chromosome staining has come a long way in the last hundred years. The beautiful examples in the figure were caught by Jorge Yunis of the University of Minnesota, just at the point in the cell's career when they were shortening and thickening in preparation for cell division. Then they were subjected to drastic chemical treatment to emphasize slight differences in different regions. Finally, they were elaborately stained to bring out those differences. The genes, of course, remain quite invisible at this level. Chromosome 1 carries about 10,000 genes, but even superb preparations such as the one in this picture show fewer than seventy fuzzy bands (not all of which are visible here). Each band contains many genes.

Invisible though the genes ordinarily are, it is now possible to locate them directly on the chromosomes—at least in a rough sort of way. Molecular biologists can isolate one strand of a DNA double helix, attach

*The fact that chromosome 21, not 22, is the smallest was not known at the time that the chromosomes were given their numbers. The ability of Down syndrome children to survive the genetic disturbance caused by an extra copy of chromosome 21 may be due in part to the fact that the chromosome is so small. People carrying extra copies of the other human chromosomes usually suffer much more severe effects.

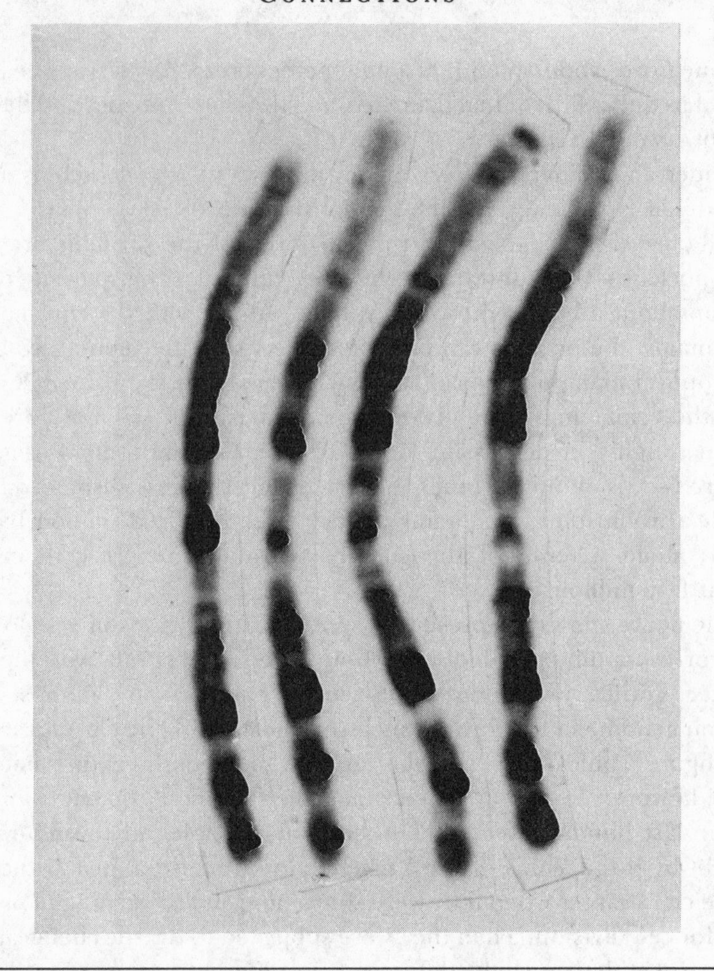

Figure 5–2. Chromosome 1, the largest chromosome in, from left to right, humans, chimpanzees, gorillas, and orangutans.

molecules of a brilliantly fluorescing compound to it, and then use the strand as a probe, sending it to seek its complementary partner strand in the set of chromosomes. When the probe matches up with its partner, the two strands are said to *hybridize.* The chromosome band the probe settles on fluoresces brightly, so that it can be detected with a UV microscope. Yet each visible band contains so much DNA that it is not possible to tell which of the hundreds or thousands of genes in a band is hybridizing to the probe. The process is like sending a letter using only a zip code as an address. The letter will arrive at the right post office, but can go no further.

If one chooses two copies of chromosome 1 from two different people, or indeed picks the two copies that are found in any one cell, and examines them at the level of magnification in figure 5–2, they will appear exactly alike. In fact, they are not. Even if both happened to descend from the same chromosome 1 just a few generations back, they will each have acquired different mutations in the interim, tiny changes in the DNA that are usually far too small to be visible under the microscope. Further, each time sex cells are produced, the two copies of chromosome 1 exchange pieces with each other, a process called *crossing over.* Over many generations this process distributes the ever-accumulating mutations among many different copies of chromosome 1, producing almost infinite numbers of new combinations of genes each generation.

The differences between chromosomes can build up quite quickly. You will probably recall being taught that all the cells in your body are genetically identical, since they are all descended from a single zygote formed by the fusion of a sperm and an egg. In fact, however, the chromosomes in your body are constantly becoming more and more different from each other, as new mutations appear throughout your lifetime. If you were to pick two presumably identical number 1 chromosomes, inherited from one of your parents, from different parts of your body and then sequence all their DNA, you would find them to differ by an average of four or five hundred mutations. One DNA base in every half million or so would be different as a result of the changes that have accumulated over your lifetime.

When these mutational differences are summed up over many generations, they become significant. An examination of the DNA of two unrelated human beings (unrelated in the sense that their common ancestors lie far back beyond the range of any genealogical sleuth) would reveal them to be different in about one out of every thousand bases. Many of these differences seem to have no effect, but a fraction are responsible for all the obvious and not-so-obvious genetic variation among human beings.

The chimpanzees, our nearest relatives, differ from us even more, by about one base in every hundred. This is ten times the average difference between two human beings, and reflects the fact that our ancestors parted genetic company from those of the chimpanzees about five million years ago. Yet in spite of this long separate evolutionary history, the human and chimpanzee chromosomes in figure 5–2 appear very similar. In fact, about three million differences have accumulated on this pair of chromosomes alone, but most of these differences are not visible until one gets down to the level of DNA sequence. Many of the changes are indeed tiny, such as the substitution of one base for another. But other differences are more dramatic—stretches of DNA deleted, duplicated, or shifted from place to place. Even most of these larger changes are quite invisible under the microscope. One must look to more distant branches of the evolutionary tree, beyond our closest relatives, the great apes, in order to see really obvious chromosomal differences.

The finding that humans and chimpanzees differ at the DNA level by only 1 percent is often greeted with astonishment. Many people respond to this news by gathering the proud mantle of their humanity around themselves and pointing to all the obvious differences that separate them from these lower beasts. Can a mere 1 percent really account for what seems to be such a huge gulf?

Actually, the cold evolutionary fact is that the differences that really matter are even fewer than that. Most mutational changes in DNA have no overt effect on either the person or the chimpanzee who carries them. It is extraordinary, and rather humbling, to realize that the essential genetic differences between humans and chimpanzees probably amount to something between a tenth and a hundredth of one percent of our respective genomes. And it is even more extraordinary to think that we will soon be able to track down just what these differences are. The trick will be to separate the differences that matter from those that don't. As we continue our exploration of the genome, we will find some rules that may help us to make that distinction.

IMAGINING THE GENOME

I have just, with some difficulty, hefted Webster's Third New International Dictionary onto my lap. This is the one that you see in libraries, sitting proudly on its own little lectern. I find that there are about 60 characters (not words, but individual letters or spaces) to a line, 150 lines to a column, and 3 columns to a page. This works out to 27,000 characters to a page. The dictionary has roughly 2,600 pages, adding up to 70 million characters in all. Since there are about three billion bases in one set of human chromosomes, it would take forty-three volumes the size of this enormous dictionary simply to list the information that they carry. Let us call each of these volumes a Webster.

Each Webster is three and a half inches thick, so a human genome's worth of information in the form of Websters would fill a shelf twelve feet long. This may not sound very impressive when compared to the information in a whole library of books. There are about seventy million books and periodicals in the Library of Congress, one for every character in a Webster, filling some 400 miles of shelf. But it must be remembered that the information in the library is different from the information in a sequence. The books in the library only had to be written. If the Human Genome Project is to be completed successfully, each piece of DNA that will eventually be listed in all forty-three of the Websters will have to be cloned and sequenced in a complex series of steps. Further, to be reasonably sure of the information's accuracy, both strands of the DNA will have to be sequenced; Sanger's rule is that you should sequence both strands twice. And there will always be ambiguities requiring a region to be sequenced several times using different techniques. The actual sequence that will need to be generated in order to be fairly sure of the genome will probably be, at a minimum, five times as long as the final sequence. So our final set of Websters would actually represent a distillation of information that, if printed in full, would fill a shelf of Websters stretching sixty feet or more.

Now the task confronting sequencers of the human genome begins to take on its true proportions. It is very different from writing a book. It is as if the author of each book had to dig up and smelt the lead, antimony, and tin needed to make, mold, and set the type individually.

Suppose you took one of these Websters off the shelf and opened it at random. You would be confronted with a gray expanse of featureless type, with no spaces and no breaks into paragraphs. Examined more closely, each line would look something like this:

TTTTTTTTTTTTTGAGACAGAGTTTTGCTCTCGTCGCCCTAGGCTGGAGTGCAGTGGTGC

There would be, in each volume of this library of forty-three books, over a million such lines of type, all looking at first glance the same. Ideal reading for insomniacs!

The four bases making up the information sequence of DNA are all represented here, but you can see that the sequence itself is not completely random. In fact, this stretch of DNA turns out to be a bit unusual. The long run of Ts on the left end serves as a signpost, indicating that this region is part of a particular kind of sequence. The sequence began as a piece of *messenger RNA* that was reverse-transcribed back into DNA, which was in turn inserted into the chromosome. (The row of Ts is complementary to a row of As that is a prominent feature of messenger RNA.) Such an event is not unusual; more than a million of these pieces of DNA, some quite large, appear to have been pasted into our chromosomes at various points in the fairly recent past. The function of most of them is unknown, but, astonishingly, they account for over 10 percent of our genome.

This kind of insertion is sometimes due to the action of a virus—called a *retrovirus*—and sometimes to the action of other regions of DNA on the chromosome, collectively called retrotransposons or *retroposons.* Retroposons and retroviruses are closely allied to each other, though the retroposons do not have the ability to exist as separate viruses. Later in the book I will look more closely at the properties of retroviruses and retroposons, the dangers they pose, and the opportunities they provide for gene therapy.

Continuing to leaf through the volume, you might discover a stretch of DNA that looks like this:

CACACACACACACACACACACACACACACACACACACACACACACACACACACACACACA

It is estimated that thirty to fifty thousand such short sequences are scattered through the genome. Their utility is doubtful; their monotony, from the standpoint of the poor scientist who has to sequence them, is unquestionable.

Other immense regions of the human genome consist of sets of longer sequences, each a few hundred bases long, repeated over and over again. Some of these repeats occur just a few times; others can occur thousands of times. Some of these repeating stretches would go on for two thousand pages in our set of Websters. They are a kind of desert, in the sense that they are almost empty of genetic information.* Although the function of these grim regions remains obscure, it is strongly suspected that they have something to do with the structure of the chromosomes. Sidney Brenner, the discoverer of messenger RNA, has suggested that scientists caught faking their data should be sentenced to work on the Human Genome Project. It would be a fiendish punishment, worthy of Gilbert

*Of course, all this monotonous DNA *is* genetic information in the sense that it is carefully replicated and passed on to the next generation. But we normally think of genes as stretches of DNA that code for proteins or that act as regulatory elements for other genes. Since we have little idea what these long dreary stretches of repeated DNA are for, we cannot really classify them as genes.

and Sullivan's *Mikado,* to condemn some errant molecular biologist to the task of sequencing a span of DNA like this!

Those who are not insomniacs will probably close the book with a shudder at this point and turn to the latest Danielle Steel. What we desperately need is a plot. There is one, but only with the aid of computer analysis can it be extracted from this gray mass of information. Computers can be used to obtain two major kinds of information: evolutionary and structural. Substantial progress has been made on the first. The second is much more difficult.

NEW KINDS OF COMPUTERS

The largest computerized DNA data bank, called GenBank, currently contains about fifty million bases, a quarter of which come from humans. This number is a bit inflated, containing as it does some multiple entries of the same gene. Because these genes are particularly important, they have been sequenced more than once by different groups. Still, even ignoring these duplications of data, the fact remains that about ten million bases of the human genome have been sequenced during the first thirteen years of DNA sequencing technology. This represents about one three-hundredth of the entire genome. GenBank is also filled with DNA sequences from many other organisms, ranging from those closely allied to ourselves like the chimpanzees, to our more remote relatives like mice and rats, to even more remote relatives like fruit flies, nematode worms, plants, and, most distant of all, bacteria.

As soon as a new human sequence is worked out, the first thing the excited investigator does is search through the data already in GenBank, looking for sequences that resemble it. The data bank is now so rich that all sorts of things are bound to turn up. For example, when a new hemoglobin gene was discovered recently, it joined a family of nine other previously sequenced genes, each of which makes a slightly different kind of hemoglobin. These genes all arose from a very distant common ancestor through a process called *gene duplication.* In this rare occurrence, an extra copy of a gene gets placed elsewhere in the genome, sometimes even on a different chromosome. The two copies are then free to evolve separately. They will accumulate different mutations, and as they diverge they will be shaped by selection and chance to take up different tasks. The end products can be similar or very different, depending on the amount of time that has passed and the selective pressures involved.

Imagine that you have a copy of a very important memo stored in your word processor, and you want everybody in your department to look at it and suggest changes. One way to do this would be to place the memo into each person's word processor in turn. As each person made changes, the memo would alter over time much as a gene does, and at the end you might have a very different memo. This process resembles what happens to a gene over evolutionary time if it does not experience gene duplication.

Suppose, however, that you decide to make two copies of the memo and send each copy around to half the members of the department. While you might save time, the two memos could end up so different from each other that it would be hard to combine them into a finished product. Indeed, by the time the process ended the two memos might address very different questions, and come up with very different solutions! This is what gene duplication has the ability to do—in effect, to make new genes from old. This is especially true if much evolutionary time has elapsed.

The hormones glucagon and secretin, for example, resemble each other very closely in amino acid sequence (and therefore in the DNA sequence of the genes that code for them), but have very different functions. Glucagon increases the level of sugar in the blood, while secretin stimulates the secretion of pancreatic enzymes. Both are important members of a family of hormone regulators, and their genes can be traced back to a common ancestor even further back in time than the ancestor of the hemoglobin genes. Because these genes duplicated such a long time ago, they are like our two very different memos. Their accumulated differences have led them to take up very different tasks in the body. Yet in spite of all these differences, a computer analysis can still easily detect the resemblance between them. Such an analysis is called a *homology search.* (The word *homologous* is from the Greek for agreement. When applied to structures such as genes, it means that they had a common ancestor in the distant past, even though by now their functions might be very different.)

As homology searches range further afield, more things turn up. One recently sequenced gene, when mutated, can give rise to the devastating disease cystic fibrosis (CF). A search of GenBank revealed that part of the CF gene resembles all sorts of other genes. The CF gene was already known to be involved, though perhaps only indirectly, in the transport of small molecules across membranes in the cell. Many of the genes that resemble it do similar tasks. But not all. One relative of the CF gene repairs damage to DNA caused by UV light in bacteria. Another helps to make eye pigments in fruit flies. Yet another is involved in cell division. It may be that all these genes, too, do their varied jobs by transporting molecules across membranes, but it seems likely that at least some of them have acquired totally new functions.

Tim Hunkapiller is one of the many computer experts involved in these homology searches. Tall, rangy, and usually clad in a plaid workshirt and jeans, he has a flowing reddish beard that makes him look like a very large gnome. He is so full of ideas that the words trip over each other as he tries to get them all out. Give him something new to think about, and he visibly chews it over for a moment, then brings out a new and expanded version of your idea that is much better than the one he was given. When I first met him, I thought that Tim looked like an ecologist, and I learned to my surprise that his bachelor's degree is in fact in ecology, from Oklahoma Baptist University. He went to Cal Tech for graduate school, partly because it was free and partly because his brother Mike was already there. When he arrived he was quickly pulled into the orbit of Lee Hood, a molecular biologist who is currently very much involved in the early stages of sequencing the human genome.

Tim proudly showed me his new baby, which had been consuming much of his time for the two years previous to our meeting in early 1990. Together we unrolled part of an enormous sheet of paper, four feet across and twenty-five feet long, on which was drawn the circuit diagram for one of the most complicated special-purpose computer chips ever designed. Tim and a few coworkers had designed the chip themselves; the final refinements and testing of the scheme had then been carried out at the nearby Jet Propulsion Laboratory using computer-aided design. The chip itself, a quarter of an inch across, is being fabricated by Hewlett-Packard. It contains 400,000 transistors and 16 separate microprocessors, all designed to do only one thing and to do it superbly, with incredible speed: to look at evolution in action through homology searches.

As GenBank grows, and at this writing it is doubling about every two years, homology searches become harder and harder. It can now take hours of a supercomputer's valuable time to search all of GenBank for the sequences homologous to various parts of a newly obtained long DNA sequence. When the human DNA data base is completed, it will contain a hundred times as much information as GenBank does now, making such a search prohibitive.

Tim estimates that his chip will be able to do searches a hundred times as quickly as a supercomputer can. He predicts that the chip and the input-output devices needed to use it efficiently could be manufactured for a mere $25,000. He hopes to put it in a stand-alone box that can be hooked directly to a microcomputer. As copies of this and other similarly specialized chips find their way into the scientific community, they will help many different scientists carry out huge homology searches. It will be possible to organize the family relationships of the genes in the human genome (and of course of all the other genomes that are being sequenced along with that of humans) to a degree unimaginable today. Indeed, as the data base grows, it should be possible eventually to trace all our hundred thousand or so genes back to just a few primitive genes, and perhaps to gain a glimpse of the genetic architecture of some of our earliest ancestors.

Tim's approach is only one of the many that are being taken. Parallel processing computers are being developed at Cal Tech and elsewhere that will greatly speed some of the repetitive tasks of homology searches. Several people have suggested that the problem with special-purpose chips such as Tim's is that if new and better search algorithms come along, or new kinds of searches are required, the chips could quickly become superannuated. Because of these disagreements, competition among various kinds of computer architecture and various software approaches to the problem continue to be brisk.

THE SHAPE OF THINGS

Titan, the largest moon of Saturn, was thought at the time of its naming to be the largest moon in the solar system. Recently, this has been shown to be incorrect—Jupiter's Ganymede is slightly larger. In what will probably set off a reprise of this history, the name titin has recently been bestowed on the largest protein yet discovered in the human body. Titin is indeed very large for a protein, but others even larger are sure to be found soon. The choice of what to name them (ultrin? immensin?) is a bridge that has yet to be crossed.

You will remember that the tiny protein insulin has a piffling fifty-one amino acids. The average protein is somewhat larger, made up of perhaps two or three hundred amino acids. Titin, by contrast, is an enormous molecule consisting of some *thirty thousand* amino acids. It is found in striated muscles, where it appears to add needed structural strength to muscle fibers during times when the muscle is relaxed and dangerously few connections hold the filaments together. At two-thousandths of a millimeter long, a titin molecule could stretch halfway across a typical cell nucleus. Were it not so thin, it could easily be seen under an ordinary microscope. Under the electron microscope, it can, and a picture is shown in figure 5–3.

As you can see, it looks very much like a string of tiny beads, with a distinct head at one end of each string. Each bead, or *subunit*, is actually a complex structure of about a hundred amino acids. A complete titin molecule contains three hundred such subunits, each one different. They fall into two general groups, however, with two or three from one group following one from the other in a fairly regular pattern. A homology search of GenBank revealed that one of the two kinds of beads bears a remote resemblance to antibodies, despite having an utterly different function. In the titin gene, this antibody-like region has been duplicated again and again. Like our two duplicate memos, all these duplicated regions that would eventually give rise to parts of titin evolved along very different pathways from the one followed by the antibody gene.

Only about 7 percent of the titin gene has been sequenced at this point.

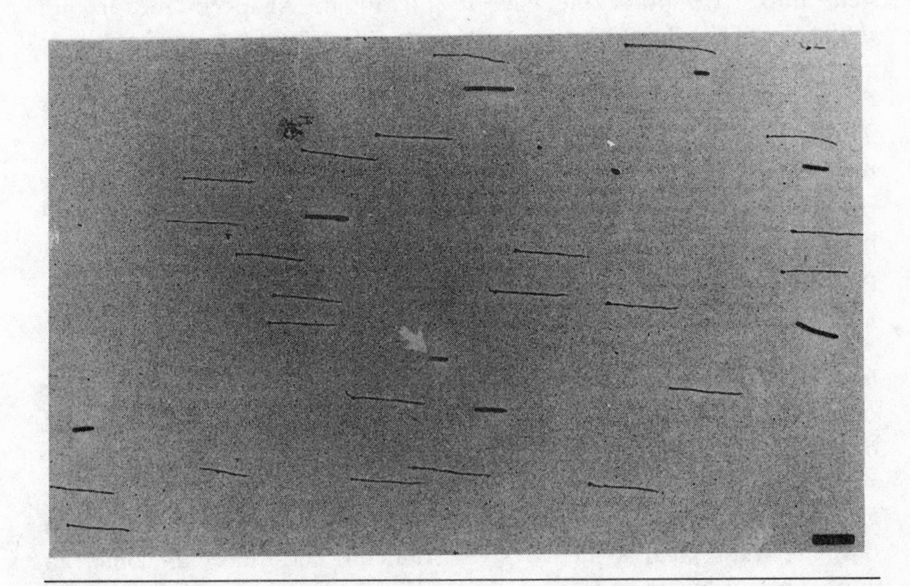

Figure 5–3. The long fibers in this electron micrograph are individual molecules of titin, the largest protein yet to be discovered. The heavy lines (arrow) are tobacco viruses made up of thousands of molecules of a more ordinary-sized protein.

(The sequence is from the rabbit gene, though all indications suggest that the protein is very similar in humans). Eventually, the whole enormous gene for titin will be sequenced, giving new hope of understanding muscle structure and perhaps providing insight into various wasting muscle diseases. To understand the role titin plays in the structure of muscles, however, it will be necessary to know its exact three-dimensional shape.

Proteins are made by the sequential addition of amino acids. Since each of the twenty amino acids is chemically different from all the others, a lengthening chain of amino acids can contort and fold in many ways. One common structure formed by such a chain is helical—a single helix rather than the double one of DNA. Even though there are quite long helical stretches in most proteins, they are all interrupted sooner or later by other regions with quite a different shape. All of these structures are established within a few seconds of the manufacture of the protein. To make things more complicated, helper proteins appropriately called *chaperonins* sometimes play a role in easing a growing protein chain into the right conformation.

How marvelous it would be to be able to feed the sequence of the titin gene into a computer and have it spit out the shape of the protein molecule! This should in theory be possible, for there must be strict chemical rules that determine how any given chain folds up on itself. Otherwise, any protein could take a great variety of shapes. Currently, if a protein can be persuaded to form a crystal, its three-dimensional structure can be worked out by X-ray crystallography or nuclear magnetic resonance. These processes are laborious in the extreme, however, and have been used to determine the structures of only about 300 different proteins. One of the first structures to be determined, that of the hemoglobin molecule, is shown in figure 5–4.

In the course of determining all these structures, a great deal has been learned about the rules that govern how proteins fold. Some of these rules rely on an understanding of the chemistry of amino acids, but others, unfortunately, are entirely empirical. They state that if such-and-such a sequence of amino acids is observed, it will probably form such-and-such a structure. Attempting to predict the course of folding given only the amino acid sequence is very difficult, and quickly becomes an impossible task if the protein is large. Yet this is what must be done if most of the information concealed in the genome is to be understood.

It turns out that many of the hundred thousand genes in the human genome can be translated in more than one way, to produce proteins that

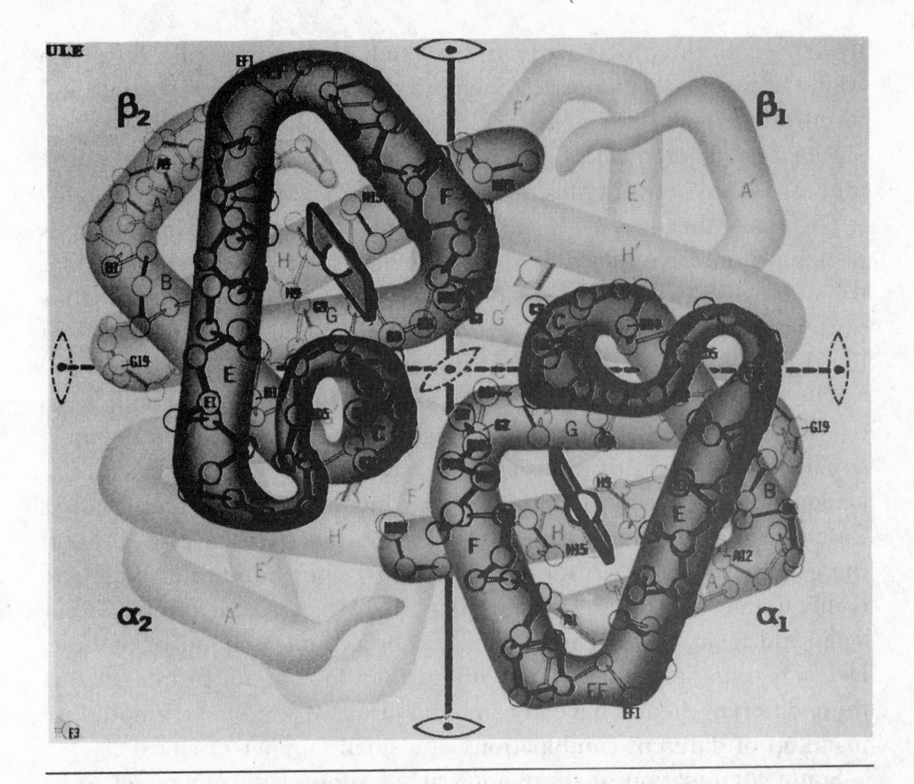

Figure 5–4. The three-dimensional structure of the hemoglobin molecule. Notice how part of the molecule consists of fairly regular helices, the famous alpha helices of Linus Pauling, interspersed with more complicated structures.

are longer or shorter or scrambled in various fashions. So the total number of proteins that the cells in our bodies can make is probably closer to a million—though no single kind of cell makes more than a fraction of this number. There is no way that all of the million proteins of a human being can be crystallized. Solving the structures will have to be done by computer.

Computer programs have been written that can predict the structure of many small proteins with some success. More recently, a new type of programming has been introduced, in which the computer is initially provided with a number of rules and methods. Using these rules, it repeatedly tries to guess the structures of proteins from the amino acid sequence, then compares the results of its guesswork with the real structures. In the process it learns to emphasize certain rules in certain contexts and to deemphasize others, to discard some rules entirely and to introduce new rules that work. This is called *neural networking.*

When confronted with fairly large proteins like hemoglobin, the best of these neural network programs can now guess their structures roughly two-thirds of the time. They still have a long way to go. Two homologous proteins often have very different amino acid sequences but very similar three-dimensional structures. Confronted with such pairs of homologous proteins, the neural networking programs can often guess correctly that the two proteins are similar—but not as similar as they really are.

The exciting feature of these programs is that they can continue to learn and evolve. At their present stage of development, even the most sophisticated programs can only make vague guesses at the structure of a huge molecule like titin. Better rule-making may be the answer. Charles Cantor at Berkeley and Charles DeLisi at Boston University head two of the groups trying to wrestle with the rules. They hope to arrange and codify themes in protein structure that, like Wagnerian leitmotifs, occur again and again. Cantor thinks there may be a couple of hundred of these. DeLisi is trying to reduce things still further, looking for properties that these different themes have in common, just as Wagnerian leitmotifs are made up of different combinations of a small number of chords.

Some combination of these approaches, along with others yet to be invented, will be needed to extract the structural information that is concealed in those dry stretches of DNA sequence in our Websters. If the efforts succeed, even partially, these growing masses of information will leap to vivid three-dimensional life on thousands of computer screens around the world, giving new insights into the structure of living cells. Then the fun will really begin.

A GLIMPSE INSIDE THE NUCLEUS

Devices for storing information and making it accessible to computers are now far more compact than the written word. It turns out that we could easily fit all the information in our forty-three Websters onto six of the small disks that are used in computer optical storage devices, and still have lots of room left over for numbering and notes.

Even with this high density of information, the volume of our set of six disks will still far exceed that of the nucleus of the cell in which the original information was found. The disks, taken out of their cases, would form a stack about twelve centimeters across and half a centimeter thick. We would need two sets of disks to carry all the genetic information in a typical nucleus. This is because, as you will recall, each somatic cell nucleus has two sets of chromosomes, one maternal and one paternal. This gives the set a total volume of 40 cubic centimeters, or 40,000 cubic millimeters. The nucleus of a typical cell is about five-thousandths of a millimeter across, so it would be quite possible to stack 200^3, or eight million, nuclei into a cubic millimeter and still have some space in between them. The volume of our sets of disks, which we thought to be so tightly packed with information, turns out to be three hundred and twenty billion times the size of a nucleus.

Remarkably, the genetic information in the nucleus is even more closely packed than that. Most of the nucleus is water, and the rest is filled with a wide array of macromolecules in addition to the DNA. Indeed, DNA makes up only about 0.3 percent of the volume of the nucleus. So the DNA molecules themselves pack over a hundred trillion times as much information by volume as our most sophisticated information storage devices.

Of course, there is far more going on in the nucleus than simple information storage. If you will allow me, I will shrink you down and take you on a quick tour of the nucleus and its neighborhood in a typical cell. I hasten to point out that this is not an original idea. Christian de Duve wrote an excellent textbook a few years ago in which he used the same scheme to introduce students to all the parts of a cell. You may have come across the late George Gamow's exciting explorations of the subcellular world in his book "Mr. Tompkins Explores the Atom." And late-night movie buffs will recall a wonderful and unintentionally uproarious movie called "Fantastic Voyage," in which a cast of characters, including a scantily clad Raquel Welch, were shrunk down and set adrift in somebody's bloodstream. The scene in which dozens of hungry antibodies attacked her is burned indelibly in my memory, as it may be in yours.*

Ms. Welch and her companions were shrunk down to a size where they could slip handily through a capillary, but we will have to become much smaller in order to examine the activities of a nucleus. If we were to shrink ourselves down to a millionth of our normal height, the nucleus would

*You can take the same ride yourself now at Disney World, I learn, though unfortunately without the company of Ms. Welch.

appear to be a roughly spherical object about fifteen feet across. Yet we really ought to make ourselves ten times smaller still, in order to swim comfortably through the pores of the nuclear membrane and around the crowded collection of molecules within. If we do so, the nucleus will appear to have a diameter half that of a football field, and as we swim closer it will dominate our field of vision like an ocean liner. For those familiar with units on the atomic scale, an angstrom unit, which is normally one ten billionth of a meter, is now a millimeter in our enormously magnified world.

In order to see the fine structure of the nucleus we will have to invoke a little magic, so that we can circumvent some uncomfortable physical laws. We have entered a world in which the interactions of single atoms and molecules are visible, and this world is a very dangerous one.

Our first requirement is a magic shield, to protect us from the molecular tumult caused by thermal motion. We are now so small that without such a shield we would be buffeted unbearably by molecules, chiefly water molecules, hurtling at us from every direction. At our size each water molecule is an eighth of an inch across, and even at body temperature they will be flying at us far faster than bullets. It is this bombardment that results in Brownian motion, causing objects as small as ourselves (and we are now at the very limit of resolution of a light microscope) to bounce around in a random pattern.

As we look about, we can see that the membranes and other large molecules that make up the cell have a kind of magic shield too, consisting of layers of water molecules that surround them, clinging to their electrically charged exteriors. Wherever these large structures are crowded together, their protective layers of surrounding water molecules almost touch, slowing the frantic activity and forcing the thermal motion into narrow channels and restricted passages. But there are still many parts of the cell where the frantic jiggling of water molecules is virtually unhindered.

The nucleus and the rest of the cell depend on this tumultuous motion, for the ceaseless thermal activity means that any small molecule will diffuse rapidly through these tiny spaces. So many collisions occur that even types of molecules that are quite rare in the cell will often encounter each other. This allows enzymes to operate at a high rate even when their substrates are uncommon.

We can get some idea of the power of the diffusion process by imagining the following experiment. If we were to swim to the center of the nucleus and open a box filled with, say, sugar molecules, they would instantly be whipped away. An observer able to follow these molecules as they diffused outward would see half of them reach the membrane surrounding the nucleus, seventy-five feet away, in about two ten-thousandths of a second.

The nucleus and all its structures, including the precious DNA, must be tough enough to withstand this hurricane of forces. As we will see, much of the DNA is usually sequestered in a complex set of folded molecules that entwine it with RNA and proteins and shield it from outside influences. A great deal of the structure and function of the nucleus, and indeed of the cell that surrounds it, can be explained by this protective function.

In addition to our magic shields, we will need two other devices in order to explore the nucleus properly. One is a sort of flashlight that can shine very short-wave radiation on the objects we are looking at. Our eyes are now ten million times smaller than they were, and so to function properly they require radiation with a wavelength ten million times smaller than that of visible light, which at our size has a wavelength the equivalent of two feet. To allow us to see, our special flashlight would have to send out a beam of gamma rays.

If this were a real expedition, we would have a problem. Since the energy of a beam of radiation is inversely proportional to its wavelength, our beam of gamma rays would blast a hole in anything we shone it on, and few if any of the rays would be reflected back to our eyes—which would certainly be blinded if they did. Luckily, this is an imaginary expedition, so we will assume that our flashlights are somehow able to work.

The third device we will need is a gadget to slow time down, so that we can see some of the complicated molecular happenings in detail. In real time, the thermal energy of the molecules is so great that they would dance furiously around, making it impossible to discern much.

Finally, we should take with us a less miraculous device: a guide showing the shapes and sizes of the various molecules we will see. It might resemble the waterproof picture guide to tropical fish that the well-prepared snorkeler takes along.

Armed with our guide and our modest array of magical widgets, we are ready to explore the nucleus. The first thing we find is that the nucleus is actually rather difficult to approach. Extending out from it, and rooted on it like huge fungoid growths, are massive collections of folded membranes that spread through and fill much of the rest of the cell. We must thread our way through the membranes in order to approach the surface of the nucleus, and as we do so we are struck by the fact that all the various parts of the cell are connected to each other by these membranes. The membranes make up the *endoplasmic reticulum,* which acts as an important channel of communication from one part of the cell to another and from the cell to the outside.

We make our way through the endoplasmic reticulum and through a complex scaffolding of tubes and filaments that also help to bind the various parts of the cell together. As we swim toward the nucleus, its looming bulk grows to resemble even more that of an ocean liner, for it is studded with what appear to be portholes. These are actually pores, appearing at our size to be about two feet across, and they are the scene of frenetic activity.

The pores are partially closed by eight large protein molecules, which snap back and forth in a blur like sped-up heart valves. Smaller molecules are entering these pores at a high rate. We can tell from our picture guide that these small visitors are primarily the building blocks for DNA and RNA, along with an enormous excess of adenosine triphosphate (ATP), the energy-rich molecule essential for most metabolic processes. (The ATP is made in the cytoplasm, chiefly in great sausage-shaped structures called *mitochondria* that are distributed through the cell.)

At the same time, different molecules are streaming *out* of the pores in equally large numbers. Most of them are ADP and AMP, the energy-depleted forms of ATP, which will be converted to ATP again by the mitochondria. Also streaming out are large, complicated molecules several inches across. These are pieces of *ribosome,* and their function will become apparent in a moment.

The most remarkable molecules oozing out of the nuclear pores are long ones, inching like worms, that are helped along by the flapping of the molecular valves. These are molecules of *messenger RNA:* single strands of nucleotides that carry information from the DNA inside the nucleus to the cytoplasm outside. The messenger RNAs vary in length, averaging to our eyes about a yard or two when fully stretched out, with a diameter of a third of an inch. As they emerge from the nucleus, these messenger molecules fold back on themselves in a variety of complex shapes—a different shape for each gene.

Most of these messengers have one part of their shape in common, a little fold at one end. As they drift away from the nuclear pore into the cytoplasm, this fold is suddenly seized upon by two molecules that clamp the RNA between them like a piece of string between two jaws. These two jaws are those pieces of ribosome that we saw coming out of the nucleus a few moments ago. Together they form a huge and complicated structure about nine inches across. Large numbers of these ribosome jaws are normally drifting free, ready to latch onto passing messenger RNA molecules. Others stud the surface of the huge folded membranes of the endoplasmic reticulum.

The messenger RNAs do not have to drift far before they are grabbed by a ribosome. As soon as the ribosome clutches the messenger, its jaws begin to munch furiously, moving the RNA sideways like tape through a tickertape machine. After one ribosome has worked its way down the message a little way, another one grabs onto the fold at the end. Many of the RNAs have four or five ribosomes gumming away at them simultaneously.

For energy, the ribosomes gobble ATP molecules from the swarm surrounding them. They also engulf small molecules called transfer RNAs. These little molecules are an essential part of the process of protein-building, for they are responsible for carrying the amino acids that will be added one by one to a growing protein chain. In a human cell there are thirty-two different kinds of transfer RNAs, each carrying one kind of amino acid. (While only twenty different amino acids are used in protein synthesis, some amino acids can attach to more than one kind of transfer RNA.) The amino acid is attached to one end of the transfer RNA; at the other is a set of three bases that can hybridize to a complementary set, or codon, on the messenger RNA. (You will recall that each codon, with its sequence of three bases, codes for a particular amino acid.)

The messenger RNA is moved through the ribosome in increments of one codon at a time. As each codon is moved into position, the ribosome must engulf and reject an assortment of transfer RNAs until it finds one that matches and thus will hybridize to that particular codon. You might think that finding the right transfer RNA would take a long time, but it takes only about a fiftieth of a second, for this part of the cell is crowded with rapidly moving transfer RNAs.

After the right transfer RNA is found, the amino acid it carries is attached to the growing chain of amino acids that were specified by the previous codons of the message. As the ribosome travels along the message this chain lengthens and begins to extrude from a hole in the bottom of the larger of the two ribosomal jaws, knotting and convoluting as it emerges. When the ribosome reaches the end of the message the chain breaks free and becomes a finished protein, already taking up its final three-dimensional form. The sequence of bases on the messenger RNA has now been *translated* into the sequence of amino acids of a protein.

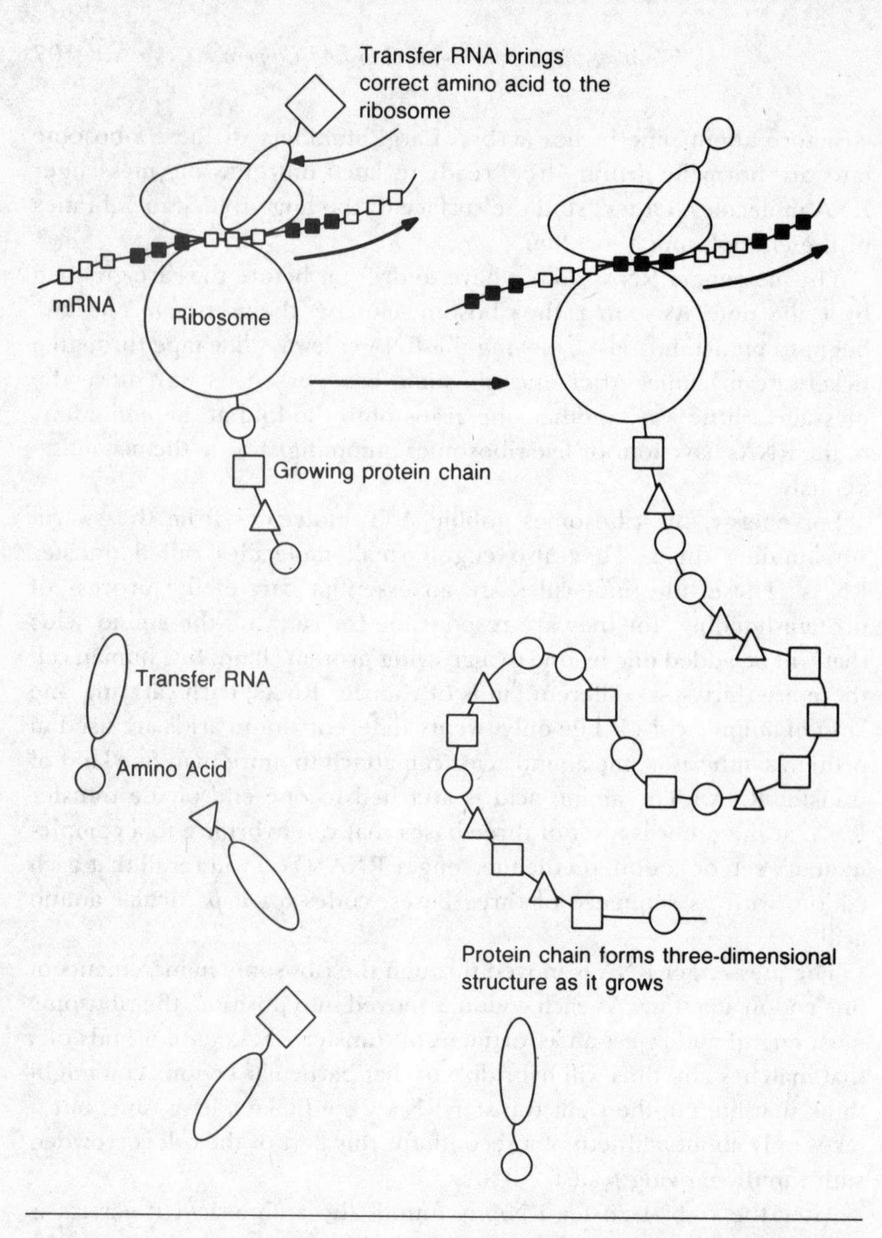

Figure 5–5. A simple diagram showing how the RNA message is read in groups of three bases by the ribosome. Many transfer RNAs carrying different amino acids will be taken up one at a time by the ribosome, but only the one that recognizes the three bases of the messenger RNA will be able to add its amino acid to the growing protein chain. Once this happens, then like the tape in an old-fashioned tickertape machine, the message moves on to the next three bases and the process is repeated. The process occurs rapidly because the ribosome and its messenger RNA are immersed in a sea of transfer RNA molecules.

When the ribosome gets to the end of the message, its jaws release. The two halves fall apart and are swept away in the thermal jostle to find other messages.

Figure 5–5 shows in diagrammatic form how protein synthesis occurs, but a static figure can give no idea of the sheer speed and frenzied activity of the process. A ribosome can manufacture a small protein of a hundred amino acids in about two seconds. You will be forced to use your magic time-dilating device to slow things down to the point where they can be followed.

Much more is going on in the immediate neighborhood of the nucleus than we can begin to comprehend. We could spend days following various fascinating processes, such as the specific targeting of proteins to various parts of the cell as it grows and renews itself. But now it is time to turn our attention to the source of all these messages: the nucleus itself.

ENTERING THE CITADEL

We squeeze through one of the nuclear pores, battered in the process by the vibrating molecules that guard it. Once inside, we find that our surroundings are surprisingly open and peaceful compared with the crowded and frantic activity of the surrounding cytoplasm. Although huge numbers of tiny molecules still swarm about us, we are able to swim around in relative freedom.

The first thing we note is that there is no sign of the chromosomes, those sausage-shaped structures we looked at earlier in the chapter. This is because the nucleus we are visiting is in between cell divisions. The chromosomes are present, but they are in their active state, enormously stretched out. Since they are, we might expect to see long strands of chromosomal DNA coiled everywhere like spaghetti. At our size, the DNA molecule looks as if it were an inch in diameter, like thick rope. The equivalent of more than 20,000 miles of this rope should be packed into the 150-foot sphere of the nucleus. As we look around, our first impulse is to ask where it all is.

It is all there, of course, but coiled, folded, and organized like the miles of rope carried on a sailing ship. As we swim away from the nuclear membrane toward the interior of the nucleus, we pass huge and complex structures that are firmly attached to the membrane at various points and that resemble nothing so much as the wire brushes you once used to clean out test tubes in chemistry class. As we look closely, we see that there are indeed DNA molecules in these structures, but they are corkscrewed and doubled back on themselves, and buried in a mass of proteins and RNA. It is obvious that it would be very difficult to get at this DNA in order to read its message.

The ends of these structures are unraveled, however, and loops and strands of DNA extend out from them into the space in the middle of the nucleus. Swimming out into this central region, we find that even here there is organization. Just as we saw in the cytoplasm, a network of fibers extends through the nucleus, though here it is much sparser. The partially unraveled chromosome strands are attached to these fibers, and loop from one to the next in a complex but regular pattern. The strands themselves consist not of naked DNA but of a much fatter coil, about six inches in diameter, that has been formed by the repeated looping of the DNA around a series of protein cores.

Since the DNA and the protein cores attract each other strongly, this springlike arrangement is obviously stable. Most of the thick supercoils are quite inactive, though in some places they are surrounded by a great deal of commotion. In these regions, a variety of free-floating proteins have clustered like surgeons around a busy operating table to unwind part of the supercoil and remove its core. This has the result of freeing up short stretches of the DNA. A large protein then attaches to this freed DNA and travels briskly along the molecule, spinning as it goes. Peering into the heart of this protein, you see that the spinning motion is actually causing a short section of the double helix of the DNA to unwind. This exposes the bases that ordinarily lie at the center of the molecule.

As the protein travels along, it reels out a long thin strand you recognize as a messenger RNA, very like the ones you saw being extruded through the nuclear pores a few minutes ago. The protein constructs this messenger RNA by using one of the DNA strands as a template. The strand it uses is the sense strand, which you remember is the one that carries the genetic information. The other, antisense strand carries the complementary sequence of bases that will later be needed to make a new sense strand. At this time, though, the protein simply pulls this antisense strand away from the sense strand in order to get at the sense strand's information.

You pause to spend a few moments with your time-slowing device, in order to examine in detail the process by which the messenger RNA is being built. The spinning protein is an enzyme called *RNA polymerase,* and it is performing a central step in the transfer of the information stored in the nucleus to the world outside it. Like the ribosome, the RNA polymerase sucks in small molecules from the surrounding swarm and tries to fit them onto the growing chain. Here, however, the small molecules are not amino acids, the building blocks of protein, but the building blocks of RNA. There are four types of building blocks: two purines and two pyrimidines, very like the building blocks of DNA. Because the enzyme must follow the strict pairing rules, it is only successful a quarter of the time, but it still has an easier task than the ribosomes we saw out in the cytoplasm because the variety of building blocks is smaller. Even so, it moves more slowly than the ribosome, adding about twenty-five bases per second to the growing chain. This is still an incredibly rapid process, and we must slow time down considerably in order to be able to observe it in detail.

As it happens, we have stumbled on the gene that codes for the protein that is defective in patients suffering from cystic fibrosis, a disease we will deal with in chapter 9. What strikes us immediately about the message being made from this gene is its enormous length. The RNA polymerase seems to travel endlessly down the DNA strand, spewing out messenger RNA, which loops and coils behind it in a growing tangle. When the completed message detaches from the RNA polymerase, it consists of a string of 250,000 bases, stretching 700 feet. We are surprised, for when we were outside the nucleus a few moments ago we did not see any molecule so monstrous emerging into the cytoplasm through the nuclear pores.

As we watch, we discover the explanation. In the nucleus, as in the cytoplasm, we are surrounded by a gnat-like swarm of small RNA-protein molecules of seemingly inexhaustible variety. One type of these is particularly common, and several of them attach themselves at various points along the length of the huge new messenger RNA. Deftly, they loop out parts of the messenger molecule and snip them free. Then they join together the remains of the original molecule, now very much shortened. The excised fragments, now in the shape of lariats, drift free to be broken back down into their building blocks by other eager enzymes.

The slang name for the ingenious little molecules that perform this procedure is *snurps,* which stands for *small nuclear ribonucleoproteins.* As you watch, they remove twenty-six pieces of RNA from the messenger molecule, snipping them out in a complex and highly specific order. By the time they are finished, the messenger RNA has been shortened from 700 feet to a mere 10. A shadow of its former impressive self, it moves toward one of the nuclear pores and the cytoplasm beyond, where ribosomes wait to translate its message. Small as it now is, it nonetheless carries the information for a large, complicated, and important protein, thirty times the size of insulin (but still only one-twentieth the size of titin).

The pieces that were removed from the message are the introns, those noncoding stretches of DNA discussed in chapter 3. They make up 98 percent of the length of the gene we have been looking at. The other 2 percent consists of *exons,* the regions that actually carry the genetic information of the gene. It is still not completely clear why most genes of higher organisms should be broken up into short exons separated by these huge stretches of seemingly useless DNA. What makes it particularly puzzling is that the introns are all painfully transcribed into RNA only to be snipped out by snurps and thrown away almost immediately. At first introns were thought to be nothing more than evolutionary relics of life's early days when genes were first being assembled. Now, however, scientists are slowly beginning to understand that these stretches can play important roles in gene regulation. Some introns have recently been found to contain genes of their own, though what the function of these genes might be and why there should be such genes within genes still remain to be discovered.

As you look up and down the chromosome, you see that other genes are being transcribed here and there. Most of them are a good deal shorter than the cystic fibrosis gene, and have fewer (and shorter) introns. There are also long stretches of DNA between the genes where no proteins are clustered and nothing is happening. These regions, some of which are very long indeed, are not even transcribed into RNA, much less translated. Some of these are the mindlessly repeating regions we found as we leafed through the Websters; others are just regions of ordinary-looking DNA with no obvious function.

The nucleus resembles a large, busy library. In a library, a very small fraction of the books are in great demand, and circulate continuously. Others are consulted much less frequently, and most not at all. Indeed, in a typical large library only about 20 percent of the holdings ever circulate; the others never move after they are first placed on the shelves, but simply gather dust. Libraries are now getting so overcrowded that many of these noncirculating holdings are being sequestered in storage areas in remote locations, from which borrowers can only retrieve them after endless paperwork and delay.

This is just the pattern we see in the nucleus. The DNA molecules that surround us in all directions are obviously highly organized, like books in library stacks. Parts of the nucleus are scenes of brisk activity, while other parts seem quite tranquil. You will remember that on our way into the center of the nucleus we passed inaccessible tangled hanks of DNA, RNA, and protein from which it would obviously be very difficult to extract information. These regions are very like the long-term storage areas of libraries.

Before we can become bored with this placid scene, however, it begins to alter, and the metaphor changes from library to printing shop. The nucleus has begun to move from its normal metabolic state into another phase, in which it begins to duplicate its DNA in preparation for cell division. At some unknown signal, vast coils of DNA are liberated from the sequestered regions, and move out into the middle of the nucleus. They unravel as they come, like wound-up rubber bands that have suddenly been released. As we watch, clusters of proteins seize on these released strands and begin to duplicate them.

We drift closer to examine this complicated process in detail. We have just seen how messages are made, and we saw earlier how they escape from the nucleus into the surrounding cytoplasm to be translated into proteins. But what we are seeing now is the ultimate mystery of all: the duplication of the genetic information in the DNA itself. This is the process that ensures that when the cell divides, accurate copies of all the DNA molecules will be passed on to the daughter cells.

The closest human activity to which I can compare this process is the embroidering of the Bayeux Tapestry. Chronicling the Norman conquest of England, this work occupies a single strip of linen 231 feet long. The Norman English ladies who created it had to do it a section at a time, for no castle's great hall would have been large enough for them to work on it all at once. Yet the job of duplicating these twenty thousand miles of DNA is far vaster and more complex. (Of course, the DNA only stretches for twenty thousand miles at our magnified scale. Still, even after we return to our usual size, it is remarkable to realize that there are over three yards of DNA packed into each of our nuclei. All this DNA must be duplicated within the confines of a speck far too small to be seen with the naked eye.)

As you will remember from the second chapter, the nuclear equivalent of the Norman weavers is a cluster of proteins called a replisome. Replisomes work in pairs. As we watch, about 100 pairs of replisomes seize specific places on each of the chromosomes, and each pair begins to work in opposite directions. Since all the chromosomes are being duplicated at once, there are about ten thousand replisomes operating throughout the nucleus. They work at incredible speed, spewing out new DNA strands at the rate of 150 nucleotides per second—far faster than the processes of RNA transcription or protein translation. At full bore, the DNA can be replicated at one and a half million nucleotides per second. Even at this rate, it would still take about half an hour to duplicate all six billion nucleotides.

Actually, the whole process of DNA duplication in a human cell takes much longer—about seven hours. The reason is that the cell must repeatedly unravel endless strands of DNA, duplicate them, and then ravel them up again. Still, it can do in seven hours what the DNA sequencers of the Human Genome Project are likely to take fifteen years to accomplish. And it performs its task with astonishing accuracy. Far more inadvertent errors will surely creep into the DNA sequence resulting from the Human Genome Project than are introduced by the busy replisomes in the actual course of DNA replication.

The reason for this accuracy becomes apparent as you watch the replisomes working along the DNA. Occasionally they back up a notch, cut out a base that they have already joined to the new chain, replace it with another, and then continue on. The DNA polymerases in the replisome have the capability of proofreading as they go, so that even if they do make a mistake (which will happen once in every ten thousand or so nucleotides, for unavoidable chemical reasons) they can back up, cut out the mistake, and continue on. The level of accuracy they can achieve is astonishing. Probably no more than a few dozen mistakes are introduced into the new DNA strands as the whole set of six billion bases is duplicated. But it is these mistakes, as well as other changes caused by external agents such as radiation, chemicals, and viruses, that provide the mutations that ultimately power the process of evolution, and give rise to the accumulated differences that gradually distinguish initially identical chromosomes from each other over the course of evolutionary time.

It is time for us to leave the nucleus and return at last to our normal size. We are dazzled by the array of macromolecular activities we have seen, and questions fill our minds. How does the cell know which parts of the chromosomes to unravel and gather up again as the DNA is duplicated? How do the snurps know which introns to snip out, and when? Why are some parts of the chromosomes attached to the scaffolding of the nucleus, while others float free? It is now known that particular chromosomes tend to inhabit certain parts of the nucleus, and that these preferences differ in different tissues. Why? Why are some parts of the chromosomes gene-rich and others gene-poor? Indeed, what is all that extra DNA doing in the nucleus anyway? Surely we only need the parts that code for genes.

One of the most exciting aspects of the Human Genome Project is that when it is completed it will give us a detailed view of all our DNA. We will be able to look for large-scale patterns in the sequence of the DNA, and use probes to locate where these regions are found in the nucleus. (Simple experiments of this type are already being carried out.) We will be able to delete and modify critical sequences to see what effect this has on cell division, on nuclear structure, and on how the chromosomes move about and pair with each other. Once we can view the entire genome as an entity, we will be able to understand why it is constructed as it is. In ten or fifteen years, it should be possible to take a Cook's tour of the nucleus that is far more detailed and fascinating than the one we have just completed.

I have tried to communicate in this chapter some inkling of how vastly complex the human genome actually is. This complexity makes it even more remarkable that we can actually contemplate taking the genome apart bit by bit, with the eventual goal of understanding it.

CHAPTER 6

Dismantling the Genome

The map is not the territory.

—Alfred Korzybski

One of the great pleasures of motoring or cycling in the French countryside lies in wending your way down the quiet little roads that snake everywhere through the charming groves and fields. You may be lucky enough to encounter a perfect picnic spot under the willows at the edge of a slow-moving river, where you can briefly imagine yourself back in the nineteenth century. Later, around the bend of a road, you might find a quaint gray-stone village with its ancient church, and perhaps a restaurant where you will be served delicious food by happy locals to whom tourists are still an object of curiosity rather than disdain. To make these discoveries, however, you must be prepared to follow the network of tiny roads, drawn in white, that appears on only the smallest-scale Michelin maps. Each of these maps covers a small section of the country, some twenty-five by fifty miles, in loving detail.

These are not the sorts of maps you would need to find the shortest route from Paris to Lyons. That requires a much larger-scale map. You can certainly get from Paris to Lyons by traveling along the little white roads, but it might take weeks and would certainly strain your budget for both maps and gasoline.

A MICHELIN GUIDE TO THE CHROMOSOMES

Like the maps that Michelin provides of France, the maps of our genes that have been made during these early years of the exploration of the human genome cover various scales. They range from very crude maps of whole chromosomes that give only a general idea of major features of their structure or of the positions of particularly important genes, to highly detailed maps of small stretches of the genome that show the precise placement of genes relative to each other. Because the making of these maps at every scale has been beset by great technical difficulties, dozens of techniques have been brought to bear on the problem. As a result of this multifaceted approach, the advances in the last few years have been considerable, but there is a long way to go before the chromosomes of our genome are covered as thoroughly as Michelin covers France.

No matter how coiled, folded, and complicated it may appear, a chromosome ultimately consists of a single, very long strand of DNA. Two different kinds of one-dimensional maps can be made of this strand: a genetic map and a physical map. Both of these maps will show the same genes in the same order, but the relative distances between the genes may differ greatly. This is because a physical map measures the actual amount of DNA that lies between genes, while a genetic map measures genetic *recombination.*

There are two copies of each of our twenty-three chromosomes in our cells, one derived from each parent, and in turn we bequeath one chromosome of each pair to each of our offspring. The chromosomes do not make this passage unchanged, however. Each generation, just before the sex cells are produced, exchange pieces of themselves with matching segments from the other member of the pair. The process is called genetic recombination, or crossing over, and it ensures that a chromosome is not simply passed down through the generations as a unit. Instead, it scrambles maternal and paternal alleles in unpredictable combinations. Indeed, crossing over is the whole point of sex, which may come as a surprise to those who think the point is something else entirely.

The further apart two genes are on a chromosome, the more likely they will recombine with alleles at the corresponding loci on the other chromosome of the pair. The frequency of recombination, however, is not simply a function of the physical distance between the genes. Genes recombine more readily out near the tips of the chromosomes and less readily around the regions (usually near the middle) where the chromosomes attach to the spindle fibers that move them around the cell. The frequency of genetic recombination is especially reduced in regions of the chromosomes where there are stretches of repeated sequences of DNA. In such regions large pieces of DNA are likely to be passed on to the next generation as a unit, while in regions where recombination is frequent the pieces of DNA that are passed on to the next generation will be smaller. The result is that there are often big differences between the physical and genetic maps.

Figure 6–1 shows a physical and a genetic map of human chromosome 17. The distances between the genes are sometimes very different on the two maps, though as you can see the order of the genes is the same. The genetic map is especially important, for it measures properties of the chromosomes other than simple physical distance between genes. With the information provided by the genetic map, the likelihood that two adjacent genes will be passed on to the next generation together can be determined. Suppose that one of these genes is a marker gene that can be detected and the other is a gene for a serious genetic disease that cannot. The genetic map can be used to make predictions about whether a child who inherits the marker gene will inherit the disease gene as well.

As exploration of the genome proceeds, both the genetic and the physical maps will become more and more detailed. The most detailed map of all, of course, is the sequence. It is the equivalent of a Michelin map at such a fine scale that it shows every tree, every fencepost, and every dog sleeping in the road. Such a map is the ultimate goal of the Human Genome Project, and we have already glimpsed the enormous amount of information it would provide. The biggest problem facing the project is the huge gap currently separating the few fragments we possess of this highly detailed kind of map from the next scale up. It is as if you had to plan your tour of France using only a very few such maps, each covering a square mile or two, along with a very crude map of the whole country.

Closing this gap will be very hard, because some regions of the chromosome are appallingly difficult to chart, either genetically or physically. These regions are equivalent to areas of our planet about as far removed from the pleasant fields and woods of the French countryside as one can possibly imagine. The closest geographic parallel I can think of is the Simpson Desert in central Australia. Lying to the south and east of Alice Springs, it covers about 60,000 square miles. Much of it is made up of a series of long ridges of red sand, ranging from 30 to 120 feet high. Many of these ridges extend for 300 miles, and all are oriented almost precisely

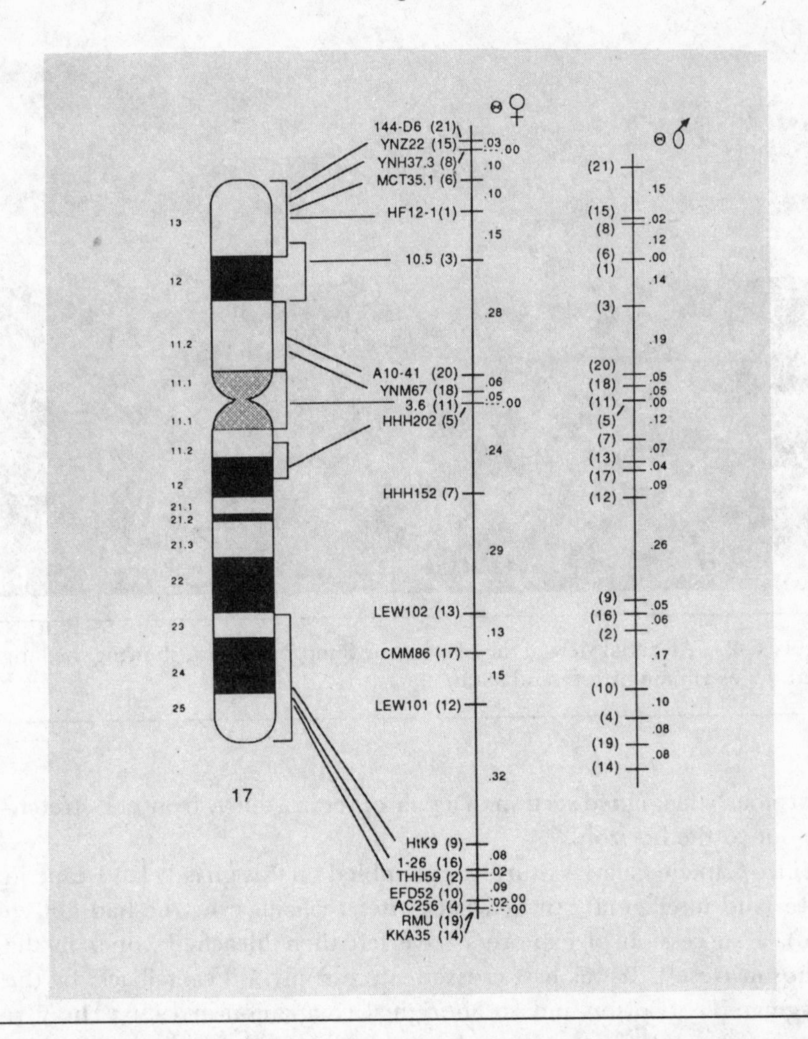

Figure 6–1. A physical map and female and male genetic maps of the human chromosome 17, with some of the many genes that have been mapped to this chromosome. You can see that while the order of the genes in the three maps is the same, the apparent distances between them can be quite different.

north and south. Their crests are spaced, like a series of frozen waves, at regular intervals of a quarter of a mile. The strips between the ridges are stabilized by clumps of spinifex bush, but the loose sand at the top of each ridge is constantly being blown to the next ridge, and then to the next. From the air, the Simpson Desert looks like nothing so much as an

Figure 6–2. An aerial view of the edge of the Simpson Desert, showing the long sand dunes running north and south.

enormously magnified section of a pair of red corduroy trousers, stretching out to the horizon.

Since Captain Charles Sturt first stumbled on this ghastly landscape in 1845 (and intelligently turned back after realizing what he had gotten into), a succession of explorers have left their bleached bones in the Simpson Desert. It was first crossed successfully, on camelback, by the swagman Ted Colson and an aboriginal companion in 1936. The first crossing through the center of the desert, where the ridges reach their greatest height, was made by Land Rover in 1966. The 300-mile trip involved climbing and descending 1,105 successive sand ridges.*

If you were dropped into the middle of this desert, you would have absolutely no idea where you were or which was the shortest way out. A long dry slog to the nearest crest would be no help at all, for you would find on the other side only another sand ridge. Had you been given a

*In keeping with Australians' urge to trivialize their terrifying landscape, a Queensland baker named Ron Grant has now *jogged* across the Simpson Desert—three times! The last time, in 1986, took almost four days. He beat his challenger by forty-five kilometers and swore he would never do it again.

map of the area, it would be of use only if it were sufficiently detailed to show every irregularity and every spinifex bush over the whole vast area.

To the frustrated builder of physical maps of the chromosome who is trying to map regions of repeated DNA, a stretch of sequence from such a region gives no more help than a detailed map of one part of a sand ridge would help an explorer of the Simpson Desert. Only a series of such detailed maps would be of any use, enabling one to detect small differences from one DNA repeat to the next: the genetic equivalent of the slightly different patterns of spinifex bushes that separate successive dunes. In the DNA, these differences are caused by mutations.

An intrepid explorer of this genetic equivalent of the Simpson Desert could use these small differences for orientation. But until ways are found to sequence long stretches of DNA rapidly and in one continuous piece, it will not be feasible to penetrate this chromosomal outback. No team of scientists has yet been brave or foolhardy enough to venture more than a little way into this most forbidding part of the human genome. Indeed, if they tried to do so with the techniques currently available, they might well succeed only in leaving the bleached bones of their scientific reputations for the next generation of explorers to find.

In spite of these enormous problems, laboratories around the world are trying to build maps of human chromosomes. Only a few have the resources to attack the task on many different fronts. One of these is at Livermore.

HOW TO MAP A CHROMOSOME

Lawrence Livermore National Laboratory is set in the rolling hills of California's coastal range, an hour southeast of Berkeley. The town of Livermore is small by California standards—a mere 50,000 people—but it is spread over a huge area; a sign on the freeway announces proudly: "Livermore, next 6 exits"!

The lab itself is some way out of town, and carefully guarded. Livermore is far more famous for weapons research, including work on SDI, than it is for biology, but a large and enthusiastic group of biologists works there on a wide range of problems. The young woman who issued my pass at the gate gave me a map carefully charting my route through the dusty complex, which was dotted with depressing one-story government-issue buildings. Some areas had signs advising me in the strongest terms not to enter, advice I heeded.

Tony Carrano runs the human genome research at Livermore. Only recently, and belatedly, was his group raised to the status of an official Human Genome Project center, although it has been working at full bore for a number of years.

Carrano is a calm, articulate man who keeps a careful eye on every part of his complex project. A chemistry major at Renssalaer, he joined the Marines after graduation and served four years, including a year in Vietnam. Faced with the prospect of another tour of duty there, he left to get his Ph.D. in health physics at Berkeley. During a postdoc at Argonne National Labs he was recruited by Mortimer Mendelsohn, who was assembling a biomedical group at Livermore. Carrano was given the chance to start a new genetics program, which was up and running by 1973.

The genetics program soon accelerated as a result of an important breakthrough by Livermore's biomedical group, one that has played a central role in shaping the exploration of the human genome. Another of Mendelsohn's earliest recruits to the program was Joe Gray, a hearty, bearded ex-physicist with a strong engineering background. Gray and his team found that it was possible to use a device originally developed for sorting living cells to sort far smaller objects—namely, individual chromosomes taken from these cells.

The procedure was a very high-tech one indeed. Human or animal cells were grown in liquid medium and stopped in their growth just before they would normally have divided. At this point their chromosomes were very short and thick. The cells were then broken open very gently to release the chromosomes, which could be labeled with fluorescing compounds. The broken cell material and chromosomes were fed into a narrow, rapidly flowing stream of liquid that was forced past a vibrating crystal. This broke the stream up into a series of tiny drops. The drops were so small that most of them contained no chromosomes, and most of the remainder contained only one. When they were hurled past a light source, the drops that happened to contain a chromosome fluoresced: The brighter the signal, the larger the chromosome. Upper and lower limits could be set on the signal, so that each time a drop containing a chromosome of a given brightness came hurtling by it could be briefly charged, breaking the stream and flicking the drop to one side into a waiting receptacle.

With the aid of this technique, the huge mass of DNA in the human cell could be subdivided into more manageable pieces for the first time. Gray and his group soon found, however, that if the DNA was to be useful they would need to increase the purity of the chromosome samples and the amount of DNA they contained.

Gray took me to a darkened room and showed me the product of fifteen years of further refinement of the chromosome-sorting process. It is a long black Maserati of a machine, covered with lights, toggles, and computer screens, that snarls softly to itself as it works. These days, the chromosomes are stained with two different dyes rather than one. When hit by two different laser beams, they fluoresce in both colors. The two dyes stain each particular kind of chromosome in a distinct pattern, creating a unique color signature.

In the latest machine the tiny drops torn from this suspension of chromosomes are hurled past the laser light sources at 50 meters a second, or over 100 miles an hour. Most of them contain nothing more than bits of cellular debris, but a few will carry a single chromosome. The fluorescence emitted by that chromosome is rapidly recorded by the computer, which must also decide whether it falls into the narrow box of intensities characteristic of the chromosome that is wanted at the time. If the computer decides the drop contains the right chromosome, the drop is diverted to the side by a sudden electrostatic pulse that flings it into a waiting tube. Otherwise, the drop is allowed to continue unimpeded into a waste receptacle.

The sorter can examine 20,000 chromosomes a second. On the day I visited, it was working on a sample containing human chromosome 11. During the course of the day it separated out 5.4 million of these chromosomes, yielding a total of 1.6 millionths of a gram of DNA.

The sorting is not perfect. Marvin Van Dilla, in charge of the sorting project, showed me a computer printout of a chart of the purified chromosome sample, with the colors plotted in two dimensions and the number of chromosomes in the third (see figure 6–3). The chromosomes stacked up like a mountain in the picture, but there were foothills around the mountain, probably consisting of fragments of other chromosomes and assorted bits of junk.

At the end of the run, the computer itself had used these data to make the sanguine estimate that the sample was about 75 percent pure. Van Dilla thought that a more realistic guess was about 50 percent. If the cells Van Dilla had started with had been human ones, then the remaining 50 percent would have been made up of bits of all the other human chromosomes and the sample would have been quite useless for further detailed

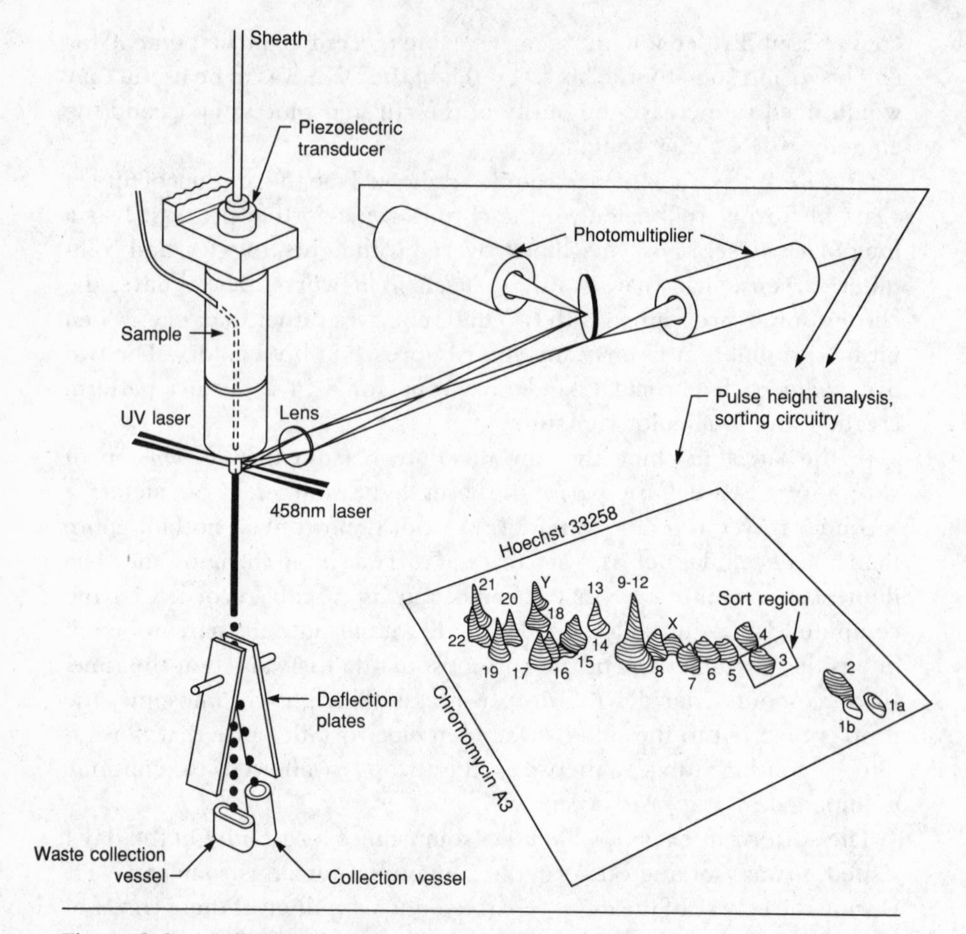

Figure 6–3. A diagram of how a chromosome sorter works. The series of little peaks shows how the human chromosomes can be sorted when they are labeled with the dyes Hoechst and Chromomycin. Most of the chromosomes, including the X and Y, can be separated, although some of the peaks tend to overlap.

work with the DNA. But the cells he had started with were not human. They were from an organism that never existed, derived from a kind of extreme miscegenation.

Some types of cells taken from humans and other mammals can be cultured outside the body, and grown in flasks and on petri dishes. Indeed, some cancerous or nearly cancerous cells can be cultured indefinitely, conferring a kind of immortality on the person from whom they were originally derived. It is beyond our capability to grow a complete person or animal back from such cells (though complete plants can sometimes be grown back from individual plant cells). Yet these cultures can nevertheless yield huge numbers of cells, which can then be manipulated in many ways, almost as if they were bacteria.

In one type of manipulation, cells from wildly different kinds of animals are mixed together, along with a harmless virus that causes the cells to fuse. Once the cytoplasms of two cells fuse, the two different nuclei contained within this single hybrid cell also fuse. Hybrids between human and hamster or human and mouse cells can be readily formed in this way.

These cells, with their double cargo of genetic information, are still able to grow and multiply normally. As time goes on, however, a kind of chromosomal warfare develops within their hybrid nuclei. During subsequent cell divisions, some of the chromosomes become lost. It is a bit of a blow to our *amour propre* when we realize that the chromosomes lost by these hybrid cells are mostly the human ones. It seems that our chromosomes cannot cut the mustard when they are thrown together in a nucleus with chromosomes from mice or hamsters.

After a while the chromosomal loss begins to slow, and finally stops. Frank Ruddle of Yale University was the first to take advantage of this phenomenon. He noted that occasionally a line could be established that was quite stable and that had all mouse or hamster chromosomes except for one human chromosome. It was possible in this way to produce cell lines that carried a single copy of most of the different human chromosomes. Lines were even isolated that carried only specific fragments of particular human chromosomes, along with a roughly normal complement of mouse or hamster chromosomes. These human chromosomes or pieces of chromosome seemed to be undamaged, despite being isolated in a cell primarily controlled by chromosomes separated from them by seventy million years of evolution.

The cells Van Dilla was working with had been made in this way, and carried one human chromosome, the number 11 that he wanted, in a sea of hamster chromosomes. Thus, the 50 percent contamination of the final sorted preparation was actually made up of hamster DNA, which is sufficiently different from human DNA to be separable by other methods. Van Dilla could be confident that any human DNA in his sample was from chromosome 11.

What could be done with this millionth of a gram of DNA? The first job was to disentangle it, as gently as possible, from the chromosomal proteins to which it was bound. This is hard to do because, as we saw in the last chapter, DNA is normally wound tightly around these proteins, and the resulting structure is folded back on itself many times. To make matters worse, when the cells are disrupted, enzymes are released that can damage or destroy the DNA. To free the DNA from its proteins without letting it get chewed up too badly by these enzymes, it is necessary to treat the chromosomes roughly with detergents and other strong chemicals. During all these processes the DNA is sure to be broken in many places. If Van Dilla were somehow able to prevent his tiny sample of chromosome 11 from being damaged by DNA-destroying enzymes, and managed to unwind the DNA from the proteins with the utmost care, he would end up with a collection of five million impossibly thin and delicate DNA molecules, each about an inch and a half long. These molecules would be so fragile that simply moving them from place to place would be enough to break them.

In reality, though, the molecules would already have been damaged at many points by even the gentlest treatment. If the breaks were random, this might not be a problem. The irritating thing is that the breaks are not likely to be random. Some parts of the DNA are stripped of their proteins before others, and these exposed parts are more likely to break or be digested by enzymes. The result is that even during this early treatment some parts of the DNA will already be lost or irretrievably damaged. So Van Dilla could not use his sample to construct a complete physical map of chromosome 11. For him to attempt it would be like approaching the end of a thousand-piece jigsaw puzzle only to make the dreadful discovery that a few pieces are missing.

Nonetheless, the sample could still be used to construct *part* of a map. Just as the defective jigsaw puzzle contains practically all the information in the picture, Van Dilla's DNA contained almost the entire sequence of chromosome 11. When these five million copies of the DNA of chromosome 11 were extracted as gently as possible, then centrifuged down to the bottom of a clear plastic tube and held up to the eye, they were just barely visible as a tiny white speck. Such a tiny amount of DNA was far too little to work with. It had somehow to be magnified, or *cloned.*

The first step in the cloning process is to break the DNA up into even smaller pieces. This can be done mechanically—slurping a solution of DNA back and forth through a narrow tube, or *pipette,* is a popular method—or by using restriction enzymes. These enzymes, you will recall, attack DNA at specific points and are one of the molecular biologist's most valuable tools.

Restriction enzymes are made by bacteria, which use them to digest foreign DNA. At last count, over two hundred of them had been purified and made commercially available. Like all enzymes, they have remarkably specific effects. One, AluI, made by a freshwater bacterium, breaks DNA whenever it finds the following sequence, or *recognition site:*

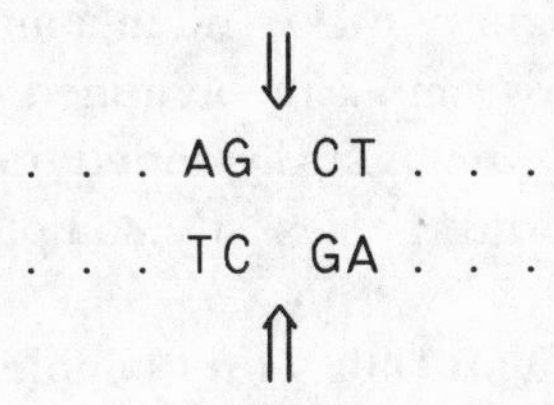

The arrows show where the DNA is broken. Note two things. First, AluI makes a "blunt end"—both strands of the DNA are broken at the same point. This means that it can be joined onto any other piece of DNA that has a blunt end, regardless of which organism the other DNA comes from. (You will remember that Paul Berg, who first realized this, shared the Nobel Prize with Gilbert and Sanger.) Second, the sequence AluI recognizes is only four bases long. Because it is so short, the sequence will occur fairly often along a stretch of DNA. There is about one chance in four that there will be an A at any position along the molecule, one chance in four that a G will follow it, and so on. So the chance that you will find the sequence AGCT somewhere along the DNA is about $\frac{1}{4}$ times $\frac{1}{4}$ times $\frac{1}{4}$ times $\frac{1}{4}$, which can be written $(\frac{1}{4})^4$. This equals $\frac{1}{256}$, so the sequence will be found on average once in every 256 bases.

Chromosome 11 contains about 70 million bases. If you were to attack it extensively with AluI, you would end up with the genetic equivalent of a heap of rubble: a quarter of a million different kinds of fragments, averaging 256 bases long. This is too many pieces to deal with, and they are all too short to be very useful.

Many other enzymes can be used to build a random library, however. One, HindIII, is made by a bacterium that causes bacterial pneumonia. HindIII looks for and breaks the following sequence:

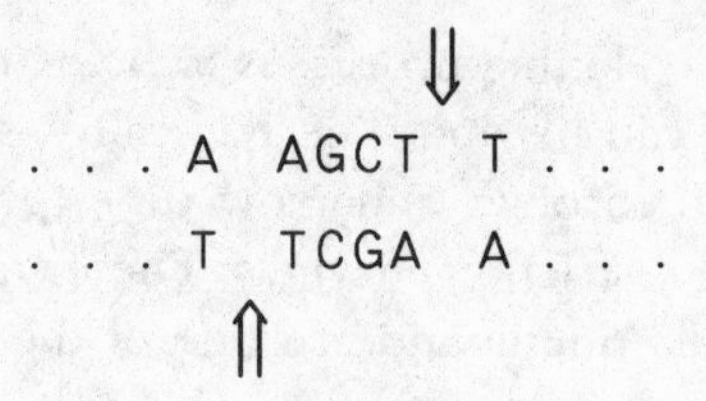

Because this recognition site is six bases long, it will occur much less often in the DNA than the shorter sequence of four recognized by AluI. The chance of finding such a sequence is only $(1/4)^6$, or about 1 in 4,000. Further, as you can see from the position of the arrows, HindIII breaks the DNA with a staggered cut, leaving "sticky ends." A sticky end will quickly find a complementary sticky end in a mixture of DNA molecules, and when it does the two can easily be joined or *ligated* together, using an enzyme called *DNA ligase.* A sticky-ended fragment will ignore all the sticky-ended pieces that are not exactly complementary, as well as any pieces with blunt ends.

If you were to digest Van Dilla's tiny sample of chromosome 11 with HindIII, you would end up with about 17,500 different kinds of fragments, averaging 4,000 bases long. Those numbers are a bit more manageable, but you would still have a problem. Even if you sequenced all those different fragments, you would have no idea of how they originally fitted together. They would be like the scattered leaves of an artist's notebook. You could gather them up, but you would have no idea of the order in which the artist had drawn the pictures.

To solve this problem, Van Dilla and his group took a leaf from Sanger's book. They made nested fragments of the DNA by using a light treatment with a restriction enzyme such as HindIII and stopping its activity before the digestion was complete. Suppose a stretch of DNA looks like this, with each arrow representing a recognition site for the restriction enzyme:

A complete digestion would consist entirely of fragments like these, each represented many times:

While a collection of fragments from a partial digestion might look like this:

You can see that a partial digestion is just what is wanted, because mapping or sequencing the overlapping fragments will give the order of the fragments on the original DNA.

The next step is to package the fragments into a so-called *random library* of DNA, made up of clones of each fragment. Although everyone uses this term, Van Dilla agrees with me that it is not very accurate, since after all a properly arranged library is anything but random. Perhaps a more descriptive term would be "clone collection." At the moment, there are three general ways to build a random library (or clone collection). The first is to insert quite short pieces of human chromosome into plasmids.

As you may recall from chapter 2, a plasmid is a very simple intracellular parasite, a small piece of free-floating, nonchromosomal DNA that lives inside a bacterial host—if we can really use the term "lives" to describe the plasmid molecule, which is even simpler than a virus. A common bacterial host for plasmids is the harmless *Escherichia coli* (*E. coli* for short) which lives in our intestinal tract. When an *E. coli* host is infected with a plasmid, the cellular machinery of the bacterium makes copies of the plasmid's DNA at the same time that it duplicates its own. As the *E. coli* divides, these plasmid copies are passed on to the daughter cells.

Unlike the DNA in a chromosome, plasmid DNA is not linear, but takes the form of a ring. It turns out that DNA can easily take this shape. Figure 6–4 shows the DNA snakes that we first saw on page 37, but bent around so that each can seize its tail in its mouth. The DNA in the figure, like the DNA of a plasmid, now forms a smooth ring with no breaks.

The great value of plasmids is that they can be used as carriers for foreign DNA. Like a wedding ring that is being resized, the ring of plasmid DNA can be broken with a restriction enzyme so that a piece of

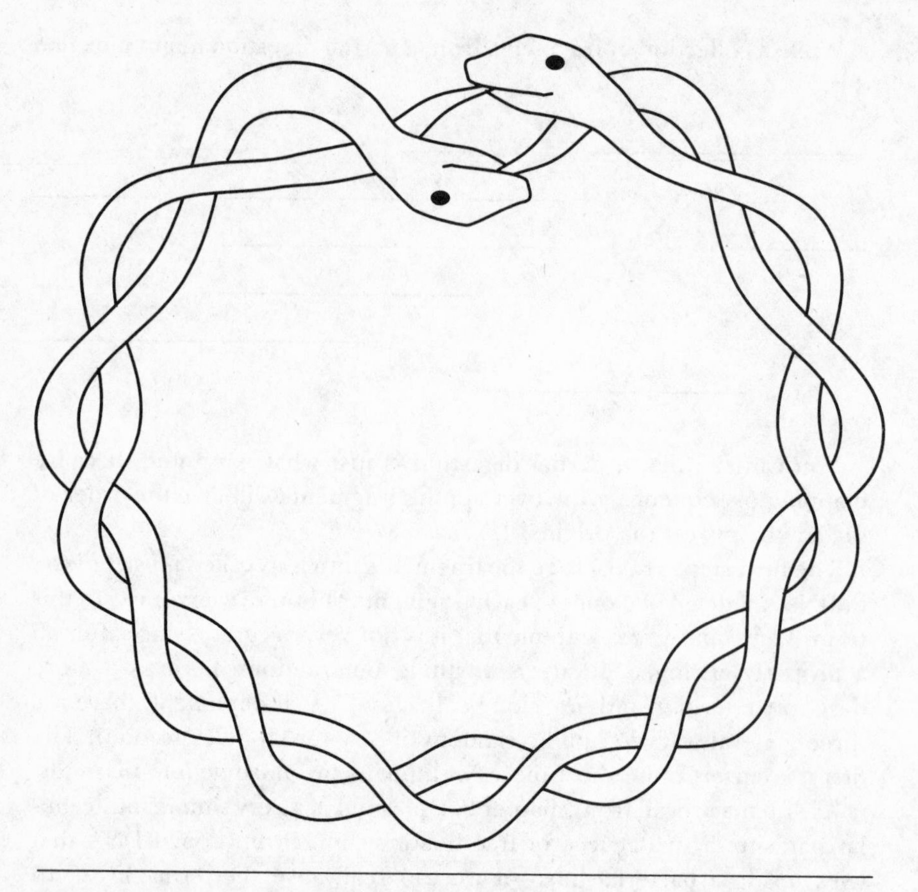

Figure 6–4. This shows what happens when the two DNA snakes from figure 2–4 are bent around and persuaded to take their tails in their mouths. You can see that this results in a seamless circular molecule of DNA.

foreign DNA can be inserted into it. One can easily purify billions of plasmids and then break their rings using the same enzyme that was used to make those overlapping fragments of chromosomal DNA. Then the chromosomal DNA fragments can be inserted into them. If the enzyme that is used happens to make sticky ends, this process is very easy. The result will be a huge and heterogeneous collection of plasmids carrying many different pieces of human DNA. Each plasmid can carry an insert of human DNA up to 10,000 bases long. (Much more than that causes the plasmids to become unstable, and they may lose or rearrange some of the inserted DNA.)

The next step is to grow up in a separate test tube tens of billions of cells of an *E. coli* strain that lacks a plasmid. So tiny is *E. coli* that this huge number of cells can easily be grown up in a few milliliters of culture medium. These cells are then mixed with the plasmid collection, carrying its vast assortment of inserted human DNA fragments, under conditions that allow the plasmids to infect the cells.

Even though most of the *E. coli* do not acquire a plasmid as a result of this treatment, the few that are infected will still number in the millions. Such cells are said to have become *transformed* by the plasmid DNA. Further tricks can be used to select only those cells that have been transformed by plasmids that carry foreign DNA, eliminating the great majority that are either untransformed or have been transformed with a plasmid that does not carry an inserted piece. It is the cells carrying the foreign DNA that will make up the random chromosome 11 library.

Checking out a volume from this "library" is relatively simple. A single transformed cell taken from the library can be isolated and then grown to yield any number of genetically identical cells, resulting in any number of copies, or clones, of a plasmid carrying a particular piece of human DNA. This cloned plasmid can then easily be purified, yielding as much of the human DNA fragment as is needed. Of course, because the single transformed cell has been chosen from the library at random, there is no way of knowing whether the cloned piece of DNA will turn out to be a new fragment or just another copy of one that has already been looked at. Occasionally, such a randomly picked clone will turn out to be a valuable bridge between two previously mapped fragments.

By grabbing books at random from the shelves of the local lending library, one can discover all sorts of new and interesting authors—but one need not then fit these books into any sort of a pattern. Determining the pattern of the DNA on the chromosome is by far the largest task facing the Livermore group. They must draw great numbers of volumes from their chromosome libraries, "read" them, and then fit them together into a vast mosaic that will eventually give the most complete physical map possible.

It would take a minimum of ten thousand plasmids, each carrying ten thousand bases of human DNA, to map chromosome 11 from one end to the other, and many times that number would be needed to be sure of finding all the overlapping pieces. Many plasmid libraries have been made of all the human chromosomes, but there have been problems with contamination and other difficulties that we will discover in a moment.

Van Dilla's tiny speck of DNA from chromosome 11 would meet a

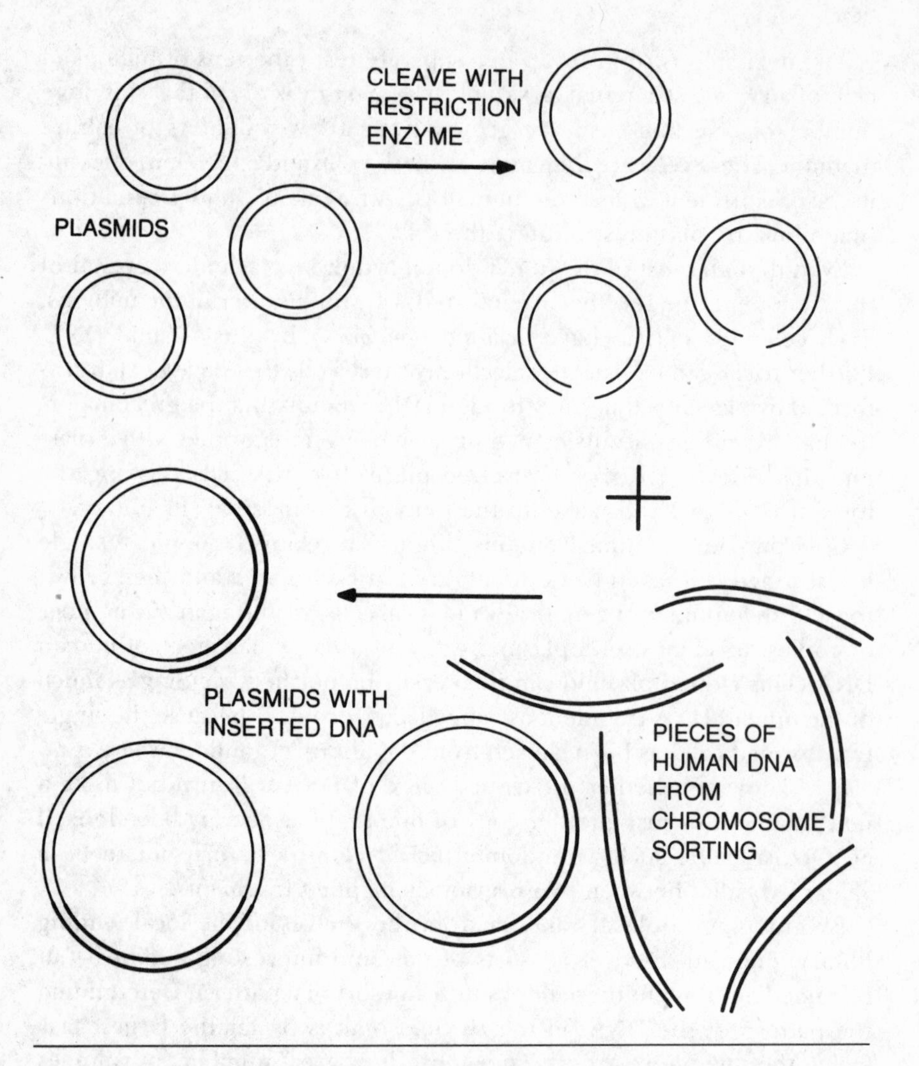

Figure 6–5. How a plasmid library is constructed. Circular plasmids are cleaved with a restriction enzyme that breaks the circle at only one point, producing sticky ends. The broken circles are then mixed with pieces of human DNA from a source such as a chromosome sort. These pieces have been cleaved with the same enzyme, giving them the same sticky ends. While many of the circles simply snap back together again, some of them join with the human DNA, forming larger circles that carry human DNA inserts.

slightly different fate. Following its isolation, his DNA was carefully purified and broken up over a period of a few days. The fragments were separated out according to size, and the ones about 40,000 bases long were isolated. These were then inserted into a special, more commodious infectious agent that also lives in *E. coli.* Made up of fragments of a virus, it is called a *cosmid,* because at the time cosmids were developed 40,000 bases seemed to be a cosmic amount of DNA. (Now that ways have to be found to clone *really* cosmic amounts of DNA—stretches that are hundreds of thousands or even millions of bases long—there is a whiff of irony about the name.)

Only about a third as many cosmids as plasmids are needed to cover a chromosome from end to end. As a result, cosmid libraries are replacing plasmid libraries in mapping efforts. Random cosmid libraries have now been constructed for about half the human chromosomes, chiefly at Livermore and Los Alamos. The two national labs have agreed to divide the task up, each taking twelve of the twenty-four human chromosomes (the twenty-fourth being the male-determining Y). While mapping with cosmids is a good deal more manageable than mapping with plasmids, sequencing the DNA of a cosmid unfortunately requires that it must first be broken up and inserted into plasmids. The mapping is easier, but the sequencing is harder.

At the moment, everyone is excited about the newest approach to making libraries, first developed by Maynard Olson of Washington University. It involves breaking up the human DNA into relatively huge pieces of up to a million bases, and turning each of these pieces into a brand new mini-chromosome. These new chromosomes can then be grown in yeast, which is much more manageable than human cells. In this way, the map of an entire human chromosome might be constructed by piecing together only a few dozen of these large *yeast artificial chromosomes,* or YACs for short.* A map resulting from this process will be many steps removed from the genetic equivalent of the map showing that charming little restaurant by the side of the country road; it might more closely resemble a crude map of the autoroutes of France put together for the same sort of tourist who does the Louvre in ten minutes flat. Nonetheless, YACs have made it possible to map chromosomes by using a "top-down"

*Olson has recently succeeded in putting the entire gene for cystic fibrosis, a stretch of DNA a quarter of a million bases long, into a single YAC. Remarkably, he did this by starting with two YACs that each carried part of the gene, putting them in a single yeast cell, and allowing them to recombine genetically to produce a YAC carrying the whole gene. The yeast cell carried out this process of recombination on the alien human DNA as readily as if it were its own.

approach, in which each YAC is mapped using cosmids, and each cosmid is then mapped in turn using plasmids.

BUILDING A CONTIG MAP

In the prewar England of my extreme youth, cigarette companies included in each pack a cigarette card, part of a series showing, say, antique automobiles, or famous Shakespearean actors. Children collected the cards, hoping to complete the set, but one or two members of each series were always rare. The most successful collectors were the most unscrupulous traders, and the luckiest kids were the ones with the largest number of chain-smoking relatives, who could be cajoled into smoking even more and buying the brands containing the most coveted cards. Fortunately for the state of England's public health, cigarette cards were discontinued during the war because of the paper shortage.

In the process of trying to construct what are called *contig maps* out of Van Dilla's collections of clones, Carrano's group is doing the equivalent of trying to complete a collection of cigarette cards. "Contig" is short for contiguous, and their goal is to fit adjacent clones together into long contiguous stretches, finally tying these stretches into a complete set of clones that span the chromosome from one end to the other. They hope eventually to end up with a collection of little tubes in a freezer, each containing a neatly labeled clone of DNA and all arranged in the order in which the DNA appears on the chromosome.

Mapping is far short of the ultimate goal of sequencing, but even so the logistics of the mapping process are daunting. Carrano's most productive idea involves an adaptation of a DNA-sequencing machine developed by Lee Hood's group at Cal Tech. He uses Hood's machine to map cosmids as well as to sequence them, and is currently focusing on cosmids from human chromosome 19, a fairly small chromosome of some sixty million base pairs. These cosmids were derived from a human-hamster cell line in which 19 was the only human chromosome.

The mapping process begins in a large, spotlessly clean room illuminated rather startlingly by lemon-yellow fluorescent tubes. Technicians tend robotized machines in which trays of plastic centrifuge tubes are subjected to computerized manipulations. The yellow light—people quickly get used to it, though they require a period of readjustment when they emerge into ordinary daylight—is necessary to prevent the fluorescent dyes that are used to label the DNA from glowing spontaneously.

DNA from each cosmid is divided into three samples, each of which is digested by a different restriction enzyme. The resulting fragments are then mixed with a fluorescent dye: a different dye for each tube. The dye binds tightly to the DNA. Then all three sets of fragments are pooled together in a single tube, along with a set of DNA fragments of known size labeled with yet another dye to serve as a reference. This mixture of fragments is then separated in a thin gel just like one of Sanger's sequencing gels. The clever feature of Hood's machine is that no radioactivity is needed to detect the fragments. Instead, laser beams are focused on the gel, so that the relative positions of all four sets of colored fragments can be measured as they move past the beams. The color of each band announces which enzyme the DNA was digested with. Forty or fifty such mixtures can be scanned at once by the machine.

These data can sometimes be used to build a map of the cosmid, in which the positions of the three sets of restriction enzyme recognition sites can be figured out relative to each other. Usually, however, the sheer number of fragments makes mapping them impossible. When this is the case, the numbers and sizes of the fragments from the cosmids are stored carefully in the computer. Thousands of cosmids have been treated in this way, and each time information from a new one is collected this information is then matched with that from all the cosmids in the growing data bank. A few other cosmids may appear to share some of their fragments with the new one. Does this mean they overlap the new one, forming a contig? Perhaps, but because only the sizes of the fragments are being measured, the apparent overlap may only be a coincidence.

The process of deciding whether it is a coincidence or not takes place in an adjacent room full of computers. There a cheerful, gray-bearded computer specialist named Elbert Branscomb presides over an enthusiastic collection of programmers who are trying to fit the cosmids together. At the same time, they are searching ceaselessly for ways to extract more information from the raw data, trying to separate bands that are almost the same size and might be missed, and trying to disentangle bands of different colors that tend to interfere with each other. Despite these difficulties, about a third of chromosome 19 has been grouped into contigs, though some overlaps have been verified with much less certainty than others. By the time of my visit in January 1990 the map already encompassed 3,000 cosmids, and new and improved protocols currently allow the Livermore scientists to process a remarkable 800 cosmids a month.

Unfortunately, some wrong guesses about which cosmids are joined to which are unavoidable. Just a few such mistakes could result in a hopelessly scrambled contig map with no relationship to reality. There are many reasons for this, and many ways to compensate for them.

First, while all cosmids are created equal, some are more equal than others. For a variety of reasons, cosmids carrying some kinds of DNA slow down the growth of their host *E. coli* cells. As a result, cosmids carrying these "poison sequences" tend to get lost. To make things more difficult, many parts of human chromosomes contain regions of DNA that, when cloned into *E. coli,* can goad their bacterial host into trying to recombine them, resulting in deletions and rearrangements. An *E. coli* host carrying a scrambled cosmid is useless for mapping and sequencing.

Fortunately, there are independent ways of checking the validity of the growing contig map. Carrano is in constant communication with scientists from all over the world who are working on genes that are known or suspected to be on chromosome 19. Some of these genes fall into the growing contig regions. If the location of such a gene can be pinned down on the chromosome, then the location of its cluster of contiguous cosmids can be pinned down as well. These collaborations are a two-way street. If a clinical researcher has cloned a disease gene that may be on chromosome 19, and Carrano's group can pin the gene down to a contig cluster, the possibility opens up of finding nearby genetic markers that the clinician can follow in families.

In a darkened room off the main lab, still another row of technicians sits at microscopes, checking the chromosomal locations of these genes and of puzzling cosmids. They are using a technique first introduced in 1969 during the paleolithic days of molecular biology, but greatly refined since. Developed at Yale University by Mary Lou Pardue and Joseph Gall, the technique relies on the astonishing ability of a single strand of DNA to find and pair with its complementary strand, even though it may be surrounded by enormous numbers of other strands of DNA with unrelated sequences. It also depends on the discovery that once chromosomes have been "fixed" chemically to precipitate their proteins, they can then be treated quite brutally by the experimenter and still retain their structure.

When I began working with chromosomes in the late fifties, it was axiomatic that preparations of chromosomes had to be treated with the greatest of care, in order to preserve as much detail of their complex structures as possible. But around that time it was discovered that harsh chemical treatment, and even enzymatic treatment, would often reveal

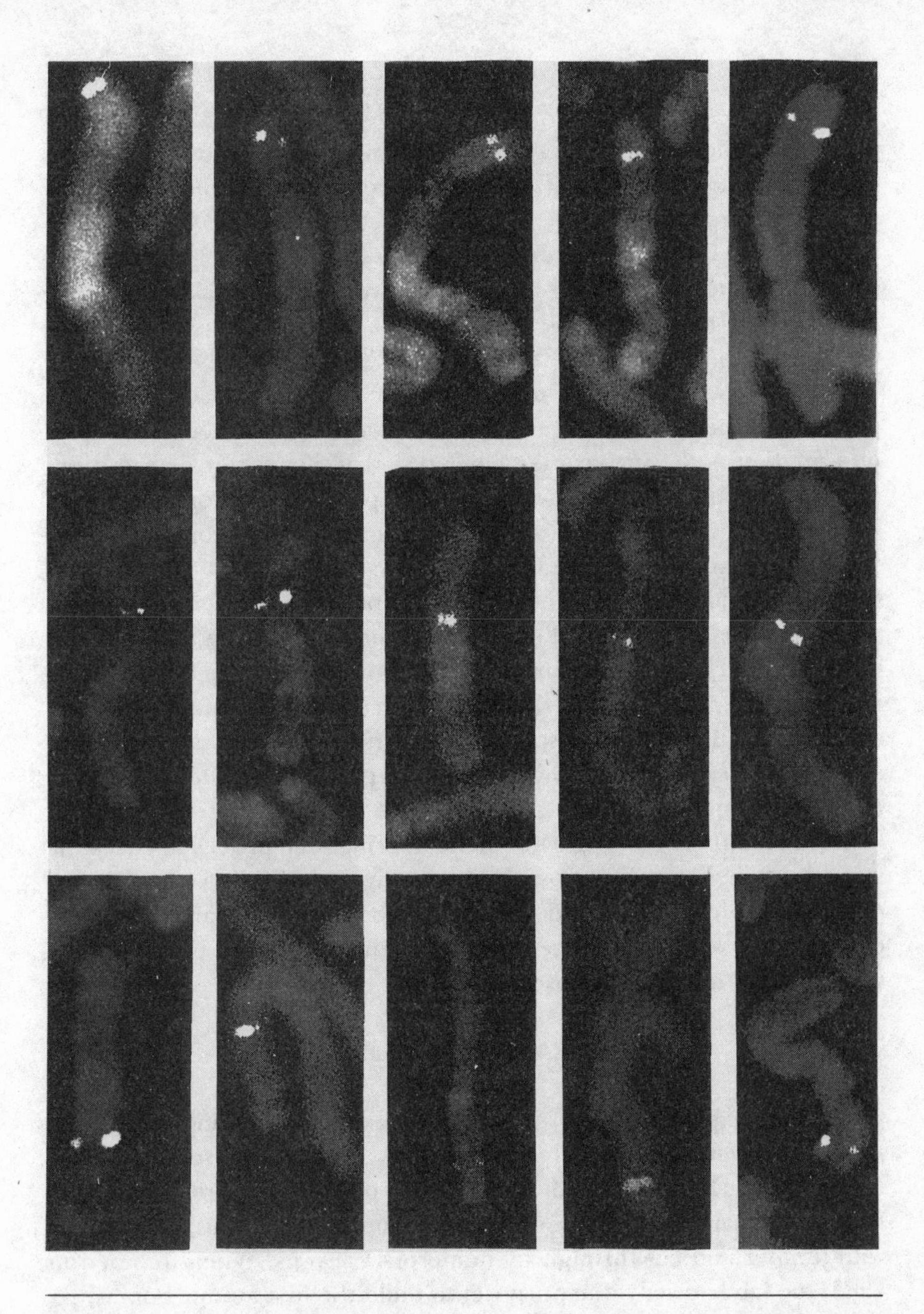

Figure 6–6. A series of pictures of the human chromosome 11 to which different probes have been hybridized. The probes fluoresce brightly against a duller background. Different probes hybridize to specific parts of the chromosome, giving at least a general indication of where the gene complementary to the probe is located.

previously hidden structures. Such treatments were used by Jorge Yunis to produce the beautiful, detailed pictures of chromosomes that we saw in the last chapter.

Pardue and Gall found that chromosomes spread on a microscope slide could be treated with strong alkali, which would break apart the double helix of their DNA into single strands. Drastic though this treatment was, the chromosomes remained quite recognizable under the microscope. The slide was then bathed in a solution containing cloned copies of a particular gene, which had also been treated to break their DNA apart into single strands. Such copies, labeled so that they can be detected, are called *probes,* because they can be used to probe the genome.

When one of the labeled probes in the solution finds its complementary sequence somewhere on one of the chromosomes, it anneals to it, re-forming the double helix. These days, the probes are labeled with fluorescing compounds. Even though only a single molecule of the probe itself can attach to the chromosome, a bright enough label can make it clearly visible. If all of the different fluorescent compounds currently available are used at once, spectacular pictures result in which four or more probes can be seen glowing in different colors on different parts of the chromosome.

Carrano and his group are not alone in using such probes to check the validity of their growing cosmid maps. Figure 6–6 shows a series of probes used by David Ward at Yale to label different parts of human chromosome 11. You can see how clearly the probes can be located on various parts of the chromosome.

HEALTHY RIVALRY

On a remote plateau in northern New Mexico, a mapping effort very similar to Livermore's is being carried out using chromosome 16.

Los Alamos National Laboratory sprawls over fifty square miles of the Pajarito Plateau, an extensive upland of piñon and ponderosa pine mixed with juniper and cut through by numerous canyons. When founded in 1943, the lab had only one purpose: to build the first atomic bombs. By now, it has become involved in a wide range of both destructive and constructive technologies, the latter including a substantial human genome project. It is also the home of GenBank, the world's major computer repository for DNA sequence information.

It is ironic that the lab itself sits on rock scooped out during an explosion far vaster than those produced by the bombs it built. The scenic route up to the labs from Albuquerque takes one through the edge of the Bandelier Basin. This enormous bowl-shaped depression, 20 miles across and as deep as 3,000 feet, looks deceptively peaceful, covered with grass and grazing cattle. In fact, it is a *caldera,* the remains of a volcanic explosion that took place a million years ago and deposited a hundred cubic miles of ash and pumice over the surrounding landscape.

The town of Los Alamos has metamorphosed since the war from a collection of raw wood huts into a typical suburban community. Most of the genome sequencing project is concentrated in a nondescript building not far from the corner of Trinity and Oppenheimer Drives—an address redolent with history but looking like any other small-town intersection. Ed Hildebrand, the deputy group leader, showed me around the basement rooms in which the sequencing is being carried out. Like Livermore, Los Alamos is concentrating on one technique, in this case an approach developed by group leader Bob Moyzis.

Moyzis's idea was to take what we have termed the "Simpson Desert" sequences—the highly repeated pieces of DNA that make mapping so difficult—and turn their annoying properties to advantage. It happens that many of these sequences are scattered throughout the chromosomes, which makes them useful as genetic signposts in what would otherwise be long featureless stretches of DNA. His group began by deliberately isolating from human DNA a set of clones that carry short pieces of the repeated sequences, and then using these clones to probe a cosmid chromosome library. So common are these repeated pieces throughout the genome that each of these probes can find a sequence to hybridize with somewhere on the DNA of about half the cosmids. Because each of his repeated-DNA clones is different, it hybridizes to a different, though overlapping, set of cosmids.

The experiments are designed to gather a great deal of information from each cosmid. After the cosmids are digested with restriction enzymes, the resulting fragments are separated on gels. Since gels are too fragile to withstand the next steps, a permanent and much more robust copy of the gel is made. The fragments of DNA are transferred to a tough nylon membrane in a way that preserves their relative positions. This process, invented at Edinburgh University by E. M. Southern, is called a Southern blot, because the DNA is transferred from the gel to the membrane in exactly the same way that wet ink from a document can be absorbed onto a piece of blotting paper pressed down on it. After the transfer, the DNA fragments can then be attached permanently to the nylon membrane.

One of Moyzis's repeated-DNA probes, made highly radioactive, is then hybridized to the plasmid DNA fragments on the membrane. When a photograph is taken of the membrane, any fragments containing sequences that can hybridize to the probe will show up as radioactive. The other fragments, though they retain their positions on the membrane, are invisible.

The membrane can then be washed free of the probe, leaving the cosmid DNA fragments still bound to it. The process is then repeated with another probe, which will bind to a different, though perhaps overlapping, set of fragments. As time goes on, a highly specific and detailed pattern emerges, a kind of molecular fingerprint that is unique for each cosmid. As with Carrano's method, it is eventually possible to find pairs of cosmids that have parts of these patterns in common, and to fit the cosmid pairs into a contig map. Moyzis gets more information from each cosmid, however, for not only can he find the pattern of fragments from a restriction digest, but he can tell which fragments contain a particular highly repeated sequence. He should thus, in theory, be surer than Carrano is about whether or not cosmids really overlap.

Because this approach is far more labor-intensive than Carrano's, however, fewer cosmids can be processed, and huge amounts of radioactivity have to be used. It is hard to tell at the moment which group is further along. There is a disconcerting vagueness about the progress of mapping projects, a vagueness that increases the further one gets from the foot soldiers who are actually doing the mapping. Several people in the higher echelons at Los Alamos stated that 40 percent of chromosome 16 had been mapped, with some contigs stretching along a million bases. Both Hildebrand and David Torney, a shy young man in charge of computer processing of the data, said that this was wildly optimistic. Torney told me that the current goal was to get even one contig a million bases long, and that they were very far from that goal.

The competition between Livermore and Los Alamos is intense, with each lab striving to make better flow-sorted chromosome libraries than the other and each racing toward the first complete contig map of a human chromosome. Many technicians and much equipment are needed to make such massive programs work. Meanwhile, some claim that there may be better and cheaper ways to map the genome.

OTHER WAYS TO MAKE A PHYSICAL MAP

Six thousand miles away, in London, an enthusiastic young scientist named Hans Lehrach is exploring simpler ways to make a physical map. Without the vast resources of Carrano or Moyzis, he has succeeded in putting together most of the genetic map of a yeast, representing almost as much DNA as a small human chromosome. He works in a small suite of labs in the Imperial Cancer Research Institute, an imposing building that faces Lincolns Inn Fields, a small island of greenery in the center of the city. His labs are in the back, well away from any glimpse of vegetation.

Lehrach has been involved in the Human Genome Project since its beginnings, attending both Sinsheimer's workshop at Santa Cruz and the first DOE human genome meeting at Santa Fe. He was invited because he was the inventor of a way to jump from place to place along a chromosome without the tedious business of sequencing or even mapping all the intervening DNA. This jumping method has been central to the process of tracking down a number of important genes, most prominently the one for cystic fibrosis.

Yet chromosome jumping is incidental to Lehrach's current concern: a search for better ways to map long stretches of DNA. During the course of a long and intense afternoon, he detailed for me his ingenious approach.

He began by pointing out that Carrano and Moyzis are limited by the number of cosmids they can digest and spread out on gels. Lehrach has developed a way to look at tens or hundreds of times as many cosmids, though each one is examined in less detail. He begins by manufacturing small pieces of DNA, eleven or twelve bases long, with a DNA synthesizing machine. He can specify the exact sequence of each of these pieces, and can make as much of each kind of fragment as he needs. These artificial fragments have never been exposed to natural DNA, but if they are mixed with human DNA they will find many complementary stretches and hybridize to them.

If Lehrach were to make a completely random eleven-base sequence of single-stranded DNA, it would have one chance in $(4)^{11}$ of finding a complementary sequence in human DNA. This is two chances in ten million, which seems very small. But remember that there are six billion bases in the human genome, if one counts both strands of the DNA. With six billion opportunities for a match, Lehrach's little piece of DNA should be able to find twelve hundred different perfect matches somewhere on the human chromosomes.

The distribution of bases in human DNA is not perfectly random. Knowing this, Lehrach was able to increase the likelihood of a perfect match. By constructing his little probe sequences to resemble human DNA much more closely than a random sequence would, he could increase the probability that his artificial probes would find complementary sequences in the genome. Lehrach now has a set of probes that can each hybridize to about one out of every twenty cosmids taken from a random human library. And of course, because each probe is different, each one will hybridize to a different though overlapping set of cosmids.

Using a special machine, he dots DNA from thousands of different cosmids onto a sheet of nylon membrane, to which it bonds tightly. By overlapping them slightly, he can fit as many as ten thousand dots onto a single small membrane. Then he hybridizes one of his little probe sequences, made radioactive, to the membrane, washes off all the extra probe that does not hybridize, and presses a piece of x-ray film to the membrane. About 500 of the dots will be radioactive, and thus visible when the film is developed. Although this approach resembles Moyzis's, Lehrach is looking at far more cosmids at once—though he is getting much less information about each one.

The probe DNA can easily be washed off, leaving the cosmid DNA behind. Then he hybridizes another probe to the membrane, and another. Gradually, using this method, Lehrach can build up a unique "signature" for each cosmid. The signature is not as legible as Moyzis's, but Lehrach can easily look at tens of thousands of cosmids at once. Given enough information, he can see which cosmids are identical, and which ones share parts of their signatures and therefore may overlap. No fancy chemistry or gels are required, and there is no need for rooms full of technicians.

Lehrach gets more excited as he ticks off the advantages of his method. Each signature is unique to a cosmid. If somebody else has a cosmid that contains a particularly interesting gene, it can be sent to Lehrach and he can easily find where it fits in his growing contig map. A known sequence of DNA can also be fit in, by using a computer to scan the sequence to find which of his set of little artificial probes would be able to hybridize to it. And if many different labs can be persuaded to use the same set of artificial probes, all this information can be freely exchanged.

There are problems, many of them similar to the ones Carrano and

Moyzis face. "Poison" sequences will be missing from his cosmid libraries, just as they are from everyone else's. The Simpson Desert regions of the genome are as much of a challenge to Lehrach as they are to Carrano and Moyzis; if a cosmid happens to come from such a region his little probes will either hybridize hundreds of times or not at all. (Lehrach's yeast mapping project has succeeded precisely because yeast does not have such stretches of repeated DNA.) Yet in spite of these difficulties he thinks, with good reason, that by using his technique even small labs with limited resources should be able to put together big chunks of the contig map of the human genome.

In other labs around the world, many other approaches are being tried. Several groups are pursuing variants of Lehrach's method. Maynard Olson, the inventor of YAC clones, and Cassandra Smith, who joined the new Human Genome Center at Berkeley in 1990, are working on ways to fit very large pieces of DNA together into contig maps. They hope to map these big pieces directly, so as to bypass both the poison sequences and the highly repeated regions that they might contain.

Who will be the first to build a complete contig map of a human chromosome? Will it be Tony Carrano, with his army of technicians, working smoothly and efficiently with the vast resources of the DOE behind him? Or Hans Lehrach, laboring to read his thousands of tiny dots, and still searching for the money and electronics experts to help him design machines that will speed up the reading process? Or someone else?

At the moment, as we will soon see, even the most intense efforts to make a complete contig map of small regions of a human chromosome usually do not succeed completely. Regions are always missing because critical overlaps cannot be found. Whoever succeeds in putting together a really large contig map will have to find a way to deal with problems like poison sequences and stretches of DNA that rearrange themselves maddeningly when they are removed from the chromosome. Ways will also have to be found to bridge those vast Simpson Deserts of repeated sequences that make the DNA of higher organisms such as ourselves so peculiarly difficult to map.

WHY SEQUENCE IT ALL?

The magnitude of the sequencing task, and the fact that over 90 percent of the genome consists of repetitive regions, introns, and regions between the genes that have no obvious function, is now leading to a reevaluation of the Human Genome Project. The DOE, for example, has announced that their part of the project will give the highest mapping and sequencing priority to the genes, rather than to the long stretches of noncoding DNA that lie within and between them. Craig Venter of NIH is already moving quickly in this direction: he is carrying out a kind of preemptive strike against the genes themselves, ignoring the regions that surround them. To do this, he relies on a kind of DNA library that is very different from the whole-chromosome libraries we have talked about up to now.

Venter's library is made with the aid of one of the most remarkable enzymes in the arsenal of molecular biologists. Called *reverse transcriptase,* it was discovered independently in 1970 by Howard Temin of the University of Wisconsin and David Baltimore, who was then at MIT and is now the president of Rockefeller University. Reverse transcriptase can partially reverse the flow of genetic information, which normally goes from DNA to RNA to protein. It does this by transcribing the information from an RNA molecule back into DNA, which is the opposite of the process by which messenger RNA is made. Evidence of this enzyme's activity was greeted with widespread disbelief at first, for by 1970 the idea that genetic information always flowed from DNA to RNA to protein had reached the status of holy writ among molecular biologists.

The chief source of reverse transcriptase is an important class of RNA viruses called *retroviruses,* which form a subclass of the retroposons mentioned briefly in the last chapter. Retroviruses use reverse transcriptase to insert DNA copies of their RNA genomes into the chromosomes of their hosts, an ability that has made them the favored tools for gene therapy.

Soon after the discovery of reverse transcriptase, scientists realized that the enzyme could convert not just the viral RNA but any collection of RNA molecules back into DNA. It could thus be used to overcome a major problem facing molecular biology.

There are many reasons why molecular biologists would love to clone pieces of RNA, especially messenger RNA, directly into plasmids. Unfortunately, RNA's single-stranded structure, along with other properties, make it chemically incompatible with the double-stranded plasmid DNA. DNA copies of the RNA, however, *can* be cloned into plasmids. Using reverse transcriptase, one can make such copies from purified samples of messenger RNA from a cell culture or a tissue, break up the DNA with restriction enzymes, and then clone the resulting pieces. The result is a *cDNA library*—cDNA standing for "copy-DNA."

A cDNA library differs from a whole-genome DNA library in two important respects. First, it consists of only those genes that are being transcribed in the cell line or tissue from which it was derived. Second, all the introns have already been removed from these genes, so that they consist primarily of exons, the parts of the genes that carry the genetic information. We have seen that many genes are broken up by dozens of introns and as a result they can extend for hundreds of thousands or for millions of bases along the chromosome. Such genes will be reduced in a cDNA library to the two or three thousand bases that actually carry the information for the protein. A cDNA library is thus like a *Reader's Digest* Condensed Book: the editors have both chosen the book they will present to you and been thoughtful enough to take out all the parts that they think you will find boring and irrelevant.

Venter began with a particular cDNA library carrying only genes expressed in the hippocampal area of the brain. The function of the hippocampus, a seahorse-shaped region of the archicortex of the brain, is unclear, although it has been implicated in emotional responses, memory, and the perception and mapping of the surrounding environment. Whatever its purpose, thousands of genes are expressed in the hippocampus, and its RNA is a rich source of genes that are presumably responsible for brain differentiation and function.

Using a technique called PCR (polymerase chain reaction), which allows specific short pieces of DNA to be multiplied millions of times, Venter has reached into his cDNA library for dozens of random bits. Most of these bits, when sequenced, yield genetic gold rather than dross, since they will give the code for short segments of protein. If the same procedure were carried out with a random library, more than 90 percent of the pieces would not carry any information about proteins, and most of the remainder would be pieces of gene that are not expressed in the brain.

Searches of GenBank for genes containing sections homologous to Venter's short pieces have yielded a treasure trove of genes, from creatures as far-flung as the fruit fly. While some of these homologous genes are expressed in the brains of this zoo of creatures, others seem to have no particular connection with the nervous system. Venter intends to sequence the whole length of some of the more interesting human genes that he discovers in this way. Some fraction of these, he feels confident, will turn out to be important in brain growth and differentiation, and perhaps lead to an understanding of mental illnesses.

A good argument can be made for Venter's approach, which skims the cream from the human genome, as opposed to blindly sequencing along the whole length of a chromosome. Can a counter-argument be made for blind rather than selective sequencing: an argument for sequencing the whole genome? I think it can. Take, for example, the problem of Down syndrome, with which I opened this book.

Down syndrome results from having three copies of chromosome 21 rather than the normal two. This is caused by a genetic accident—a kind of mutation. Not all mutations involve sudden changes in the DNA of the genes; some occur at higher levels, and can involve entire chromosomes. While a normal sperm or egg carries only one copy of chromosome 21, the cell it is derived from has two copies. The halving of the number of chromosomes during sperm or egg formation results from a special kind of cell division called *meiosis*—Greek for "diminishment." The full number of chromosomes is then restored when the sperm and egg fuse.

Very occasionally, however, a mistake occurs in meiosis, and two copies of one of the chromosomes end up in a single *gamete* (sperm or egg). When this gamete fuses with a normal one, the resulting zygote will have three copies of this chromosome. The consequences of having three copies of a chromosome are usually so disastrous that the fetus aborts spontaneously long before coming to term. Many fetuses with three copies of chromosome 21, however, survive to term and beyond.

Currently, Down syndrome accounts for about a third of the cases of mental retardation in western countries. Because of improved medical care, the numbers of children and adults with Down syndrome are increasing dramatically, putting further strains on many already overburdened health care systems. In the United Kingdom, the percentage of Down syndrome children alive after four years rose from 45 percent among those born between 1942 and 1952 to 75 percent among those born between 1966 and 1976.

It has been observed that both women in their teens and women at the end of their reproductive life are more likely to have babies with Down syndrome. Various explanations have been offered for this odd phenomenon, among them being the age of the sperm at the time fertilization took place. This can now be ruled out, for recent molecular studies have shown that the extra chromosomes 21 come from the mother in 95 percent of the cases. The age effect seems to be connected with the fact that the cells slated to become egg cells in a woman's ovary remain in stasis partway through moiosis, for as long as forty years until the egg matures.

Regardless of the mechanism, older women remain the most at-risk segment of the population. A woman aged forty or older has a one percent or greater chance of having a Down syndrome child, compared with a chance of less than one in a thousand for the general population. This is one reason why screening programs to detect affected pregnancies have concentrated on older mothers. Screening currently employs the surgical techniques of amniocentesis or chorionic villus sampling to obtain a sample of the cells of the developing fetus. The cells are then cultured and the chromosomes prepared and examined. These techniques, particularly villus sampling, are expensive and pose some risk to the fetus. Thus, it is simply not feasible to extend screening to the whole population.

Unfortunately, this leaves the majority of Down syndrome pregnancies undetected. The overwhelming majority of pregnant women are young and do not routinely undergo amniocentesis or chorionic villus sampling. The risks for women during their peak reproductive years may only be a tenth as high as those for older women, but the sheer numbers of such pregnancies means that most of the affected pregnancies will not be detected by current methods. Further, of course, not all those in whom the problem is detected elect to terminate their pregnancies by a therapeutic abortion. The proportion of all Down syndrome pregnancies that were both detected and then therapeutically aborted ranged from 20 percent of the total in a small Danish study to 5 percent in an extensive study carried out in Queensland between 1981 and 1983. In the Queensland study, two out of twelve couples elected to proceed with the pregnancy despite the positive outcome of the test. Physicians' attitudes as well as the attitudes of society at large play an important role. A recent study of physicians in the south of France found that 78 percent of them would favor termination of a Down syndrome pregnancy, while only 21 percent would favor termination of a pregnancy in which hemophilia had been diagnosed.

Yet are amniocentesis and allied types of screening the only possible means of detecting these pregnancies? A number of other prescreening tests now being investigated involve the monitoring of enzymes and other proteins that may register unusually high or low levels in Down syndrome pregnancies; one particularly effective test measures the levels of the hormone chorionic gonadotropin. Such tests could help to pinpoint pregnancies at risk among women in the middle of their reproductive years. A combination of several tests might greatly narrow the number of pregnancies requiring amniocentesis to be carried out.

There is some indication of underlying genetic reasons for these chromosomal problems, reasons perhaps connected with the structure of chromosome 21 itself. Studies have shown that if a young woman has a child with Down syndrome, the likelihood that her second child will be similarly affected rises from less than one in a thousand to two in a hundred. Sometimes there are complicated genetic reasons for this, but alternatively, the woman or her spouse may carry a type of chromosome that, while carrying the normal complement of genes, is particularly likely to give rise to a problem during meiosis.

It turns out that those regions of the chromosomes that determine how they will function in cell division are precisely the regions that seem to be empty of genetic information—what we have been calling the "Simpson Desert" sequences. Some of these are known to be important determinants of how chromosomes pair with each other during meiosis. It is known that genetically quite normal chromosomes can differ dramatically in size from person to person. Barbara Trask of Lawrence Livermore National Labs has found that different chromosomes 21 can vary by as much as 25 percent. The size differences are primarily due to the fact that they possess different amounts of these highly repeated regions.

Understanding these differences, and how they influence the behavior of chromosomes during cell division, should be a high priority of the Human Genome Project. It may not be necessary to sequence every last base, but we certainly will need to explore these regions in detail in order to determine which classes of chromosomes are most likely to misbehave. Eventually it may be possible to pinpoint the people most at risk of carrying a Down syndrome child or a child with another chromosome abnormality. A concentration on monitoring the pregnancies of these people will be far more cost-effective, and far less dangerous, than monitoring every pregnancy in the general population.

So there are good reasons for sequencing it all. But how can this possibly be done if sequencers have to plod through these endless reaches of DNA 300 bases at a time?

CHAPTER 7

The Search for a New Sequencing Technology

At the marriage of Charles the Bold and Margaret of York in 1468, an entire whale was hauled into the banquet hall. It was not meant to be eaten, of course, but rather was filled with various eye-catchers and circled the table to the sound of bugles and trumpets.
—Jean-François Revel, *Culture and Cuisine*

A cosmid, or worse yet a YAC, is the equivalent of a whale at a banquet. A little bit of it can be cut off and dealt with, but most of it has to be discarded. This is because current DNA sequencing technology, wondrous though it is, can only be used to sequence about 300 bases of any piece of DNA, no matter how long or short. As a result, most of the DNA being generated in laboratories where sequencing is being done is wasted. It is frustrating, when one wishes to sequence a piece of DNA only a few thousand bases long, to be faced with the tasks of having to chop it up into many different plasmids, each carrying a slightly different piece, each of which must be dealt with separately. If only there were a way to zip across the length of a cosmid or even a YAC, obtaining the sequence of thousands or millions of bases in a single operation!

TRYING TO BREAK THE 300-BASE BARRIER

The Sanger and Gilbert methods, despite their 300-base limitation, have already generated tens of millions of bases of DNA sequence. Any smallish piece of DNA can be made to yield up its secrets with these techniques. They are cheap to set up initially and easy to do, though expense lies in the labor—it is currently calculated that the real cost of obtaining each base lies between three and five dollars. But it works! Scientists hug themselves with joy as they contemplate beautiful gels that give them stretches of error-free DNA sequence. The problem is that the explorers of the human genome have now set themselves such immense tasks that these techniques are becoming inadequate.

Anything that is to replace them should have one of two properties. First, if some new technique is developed that is able to read a few thousand bases at a time rather than a few hundred, it will gain wide acceptance only if it is cheap and can easily be adopted by many labs. Otherwise, as the technicians and postdocs in several sequencing labs have remarked to me, why bother? If you can sequence a piece of DNA ten times as long as the present techniques allow, but it takes a quarter of a million dollars worth of finicky machinery to do it, forget it: very few people will pay that sort of price for such a slight gain.

Of course, the shoe will be on the other foot if a technique is developed that will allow stretches of DNA hundreds of thousands or millions of bases long to be sequenced at one go. Such a technique might be the only way to sequence those huge stretches of Simpson Desert that are sure to separate the oases of contigs in the genome. Even if it takes millions of dollars worth of machinery and only a few centers in the world can afford it, the gain will be great enough to justify the expense.

The best strategy might be to pursue both kinds of improved techniques: cheap ones that are marginally better and expensive ones that are hugely better. Incremental changes in technology may not gain wide acceptance, but can sometimes help in unexpected ways. You will remember that Tony Carrano's large mapping operation depends on a method of labeling pieces of DNA with fluorescent dye, a method originally developed for sequencing. It works wonderfully, but with its attendant lasers, computer, and elaborate software, the apparatus costs a hundred thousand dollars or more and needs somebody to nurse it full-time.

The machine was invented in a rabbit warren of Cal Tech labs inhabited by the swarming postdoctoral fellows, graduate students, and technicians of Leroy Hood's group. At any given time there are between twenty-five and thirty-five postdocs in his group, working in a huge variety of areas ranging from protein chemistry and immunology to cell culture and mass spectrometry. It is one of the few places outside the national laboratories in which such a large range of expertise is available for tackling biological problems. To the discomfort of some on the Cal

Tech campus who prefer their science to be a little smaller and more genteel, Hood's lab is soon scheduled to metastasize and spill over into a nearby new building of neo-Stalinist design.*

Over a hasty lunch, Hood remarked to me that while the available set of techniques for dissecting the genome are all very well, to get the job done we should spend the next ten years trying to improve the techniques a thousandfold in all directions. The Human Genome Project legitimizes this goal, and changes the perception of how people should be trained. Hood has tried to capitalize on this by helping to set up a graduate program in biotechnology at Cal Tech. By assigning each student a set of mentors from a variety of different areas, the program tries to encourage the kind of interdisciplinary thinking essential for the required breakthroughs.

Hood's lab began using a laser approach for sequencing in the early 1980s, just as the Sanger and Gilbert techniques were coming into wide use. Lloyd Smith, a postdoctoral fellow in the group, wanted to get around the dangerous use of radioactivity to label the nested fragments of DNA as they moved through a sequencing gel. He suggested focusing a laser beam on the gel to try to light up the bands of DNA fragments themselves.

The machine Smith built was a room-sized monster, using a huge paraboloid mirror salvaged from the laser-ranging device of a tank. Hood's associate Tim Hunkapiller, the designer of the special-purpose computer chip for matching sequences, whom we met in chapter 5, had joined the lab not long before. Regarding the machine, he told me, "I was the least skeptical of the group, but I gave it no chance of working!"

While the bands of DNA could indeed be made visible by focusing the great laser beam on them, glare and diffusion made them appear all smeared together. Faced with this impasse, the group began to experiment with ways of separating the bands further. They tried running the DNA fragments through a long and complexly folded thin capillary tube rather than a regular sequencing gel, and found they could then distinguish individual bands. This was encouraging, but they knew that if the technique was to be useful for sequencing, four different capillaries would have to be set up in parallel, one for each of the sets of nested fragments that ended in G, A, T, and C. When they tried clusters of four capillaries, however, they were unable to make them all behave the same

*As I write this, Hood has just decided not to move to Berkeley to head up the troubled Human Genome Center there. He would have replaced the able scientist but less-than-able administrator Charles Cantor.

way. The fragments traveled through the separate capillaries at very different rates, making it impossible to figure out the sequence.

During a lunchtime discussion at the Cal Tech Faculty Club in 1984, Hood, Lloyd Smith, and Tim and his brother Mike Hunkapiller talked about how only one capillary would be needed if the fragments ending in G, A, T, and C could somehow be distinguished. Perhaps, they realized, they could be made to fluoresce in different colors. Dyes were quickly found that could accomplish this, and they discovered, to their pleasure, that this approach eliminated the need for the clumsy capillaries: the differently colored fluorescent bands did not scatter laser light as badly as did unmodified DNA bands. The separate bands illuminated by the laser beam were very sharp, and could easily be picked out on a regular sequencing gel.

The building of the machine that resulted from all these experiments was licensed to Applied Biosystems, a company run by Mike Hunkapiller that has been very successful in adapting the ideas and instruments developed by the Hood group to the scientific market (Hood gets no direct financial benefit). Over two hundred of the machines have since been sold, a good indication of how successful they have been. Their strength is that because of the four different colors employed, a sequence can be read from a single lane rather than the four needed in a regular sequencing gel. Further, the machines require no dangerous radioactivity—something that makes Tony Carrano's mapping lab at Livermore a much safer place to be than the mapping lab at Los Alamos. Their weakness is that just as with any other gel-based sequencing method, only three or four hundred bases can be read before the bands get too weak and diffuse to be read accurately. The length of sequence that can be read is essentially the same as with the original Sanger technique.

How well are the machines working? In early 1990 I talked with Rick Wilson and Ben Kope, two postdocs in Hood's group who are engaged in using this technique to try to sequence a million-base region found in both humans and mice that is centrally important to the workings of the immune system. At that time they had sequenced about 40,000 bases in each organism, and were finding the genes of this part of the immune system to be more numerous and complex than had been thought. But the sequencing was not proceeding at the speed they had originally hoped. Indeed, many people feel that at the present time a well-organized team using the original Sanger method can produce as much usable sequence as one using the much more expensive fluorescence sequencing machines.

Still, the machines are ingenious, fairly user-friendly, and adaptable. As Carrano and others have found, they can be used for mapping as well as sequencing. Many of the machines have been bought by groups or departments, with the idea of setting up central sequencing facilities to which people can bring their DNA to be sequenced. Sometimes this has worked well, but more often it has not, as people tend to bring in DNA of wildly different quality.

What is certain is that the machines will soon be improved or superseded. Wilson took me to the crowded room where two of the machines share space with some earlier and less successful instruments. He put his coffee cup down on one of the earlier instruments and remarked with a grin that in a few years these latest machines would no doubt be demoted to the status of very expensive tables for coffee cups.

Other, very different approaches to sequencing are being taken at Livermore. There, Joe Gray talked excitedly about a new sequencing method that he is developing. Some years before, Tom Maniatis at Harvard had managed to convert solutions of pieces of DNA up to five hundred bases in length into something called *quasicrystals.* He found that if a concentrated solution of such DNA pieces is put in a strong magnetic field, the molecules line up. If the solution is then dried, the molecules stay aligned.

Gray has decided to couple this trick with some space-age technology available at Livermore. One convenient spinoff of Livermore's Star Wars research is the ready availability of coherent X-ray sources and even of powerful X-ray lasers. (Edward Teller had suggested long ago that such instruments could be used to explore the structure of whole cells.) Gray plans to shine a beam of coherent X rays through a quasicrystal of DNA molecules, and get sequence information from the diffraction pattern that results when the X-ray beam collides with the molecules and spreads out. If one of the four bases is labeled in some way—such as with iodine—the pattern will in theory give the pairwise distances between all the labeled bases. This is too much information to be useful, but the system can be simplified by using another labeled base, separated by some distance from the rest of the labeled bases on the molecule, as a reference point. All the pairwise distances between that point and the labeled bases could then be measured. If the experiment were then repeated with each of the three other bases, the entire sequence could be determined.

Given an intense X-ray laser and a supercomputer, Gray estimates that each measurement and calculation should take milliseconds. With more mundane equipment, it might take minutes. Compare this with hours for a sequencing gel and the advantage is obvious.

Unfortunately, the more bases that are measured simultaneously, the more crowded and confusing the diffraction pattern will become. It seems unlikely that Gray will be able to penetrate much beyond the infuriating limit of three or four hundred bases that all current sequencing approaches seem to run up against. The technique will have a huge advantage in speed, but the sample preparation may prove cumbersome and the machinery employed will be dauntingly expensive.

The same barrier is faced by workers in many labs who are using an expensive and sophisticated device called a *mass spectrometer* to sequence DNA. The mass spectrometer works on the principle that molecules of different masses, when accelerated by an electric field, will travel different distances through a vacuum before they touch down on a charged plate. The principle is rather as if you had a bucket of balls of similar appearance but different weights. You could separate the balls into their weight groups by throwing them all as hard as you could and seeing how far each one traveled.

The mass spectrometer is wonderful at separating smallish molecules, but poor at separating the very large ones that sequencers have to deal with. It is possible to get fairly large pieces of DNA to fly through a vacuum, and to label all the bases in them so that they can be distinguished. Using heroic means, workers at the biotechnology firm Genentech actually succeeded in sequencing about eighty bases before the fragments became too large to be distinguished. Then they ran into the same old problem: the larger the fragments, the more difficult it is to distinguish fragments that differ in size by only one base. It will be some time before this technique even approaches the 300-base barrier, much less breaks through it. And a good mass spectrometer costs $300,000, putting it well out of the reach of small science.

FAR FROM THE MADDING CROWD

All these methods butt up against a fundamental problem. They deal with large numbers of molecules. A piece of DNA, say one hundred bases long, that travels down a sequencing gel does not do so alone. It is traveling with a collection of identical molecules, each one hundred bases long, that are moving through the gel together. Crisp-looking though the band may be when their collective picture is taken, they are still a huge and madding crowd. Each band on the gel is made up of about twenty million molecules.

Even though all of these molecules are the same size and have the same properties, they cannot all be expected to behave exactly alike. Some will move a little more quickly and some a little more slowly through the gel. If the bands are crowded together this means that they also inevitably tend to blur together.

What is needed is a way to sequence a single molecule. If you could shrink yourself down to the size we achieved in chapter 5 and swim along a DNA molecule, reading the bases, it would be just as easy to read the sequence at the end of the molecule as at the beginning.

Of course, in the real world this is impossible. To see DNA at that magnification would require a beam of gamma rays, which would certainly destroy everything in their path.

But suppose we could *feel* our way along the DNA!

Just down the hall from Joe Gray, Rod Balhorn and his group are trying to do just that. His is one of a number of groups around the world using a remarkable new instrument, the scanning tunneling microscope, which can give images of single atoms on a smooth surface and has yielded remarkably detailed pictures of the DNA molecule. This microscope was invented by two IBM researchers, Gerd Binnig and Heinrich Rohrer, who received the Nobel Prize for their invention in 1986. They invented it to look at crystal surfaces in great detail, but like all seminal technologies it is now being used for a great variety of tasks.

The scanning tunneling microscope does not look at all like the microscopes you are familiar with. It has no lenses or eyepiece or light source. In spite of this it can perform what are surely the most delicate set of operations ever achieved by a machine.

First, a finely pointed probe is positioned a millionth of a millimeter from the surface that is to be investigated. It can be moved toward and away from the surface by means of a crystal that expands and contracts delicately according to the amount of current passing through it. The crystal is carefully insulated from the probe, and a separate, tiny charge difference is set up between the probe and the surface.

Given such a charge difference, and such a tiny gap, electrons will sometimes "tunnel" from one side of the gap to the other. This is not real tunneling. If the gap is small enough, then in the probabilistic world of electrons there is a real possibility that an electron may appear on the "wrong" side of the gap. It will seem to have tunneled across the intervening space, even though in quantum terms no real movement took place. The particular electron in question is simply more likely to be on the other side of the gap. A gentle "flow" of these tunneling electrons can be set up if the probe is moved close enough to the surface.

If the probe is then moved *across* the surface, the flow of tunneling electrons will remain constant if the surface is smooth. It will increase or decrease if the probe traverses a bump or a valley. So fine is the structure of the minuscule world revealed by the probe that the bumps and valleys turn out to be due to individual atoms and the spaces between them.

The delicacy with which the probe operates is astonishing. It can be made to traverse the tiny bumps of the surface in a scanning pattern. If at the same time it is moved toward or away from the surface so as to keep the flow of tunneling electrons constant, these tiny movements can be plotted by the computer to give an accurate topographical map of the atoms of the surface. Alternatively, it can be scanned across the surface at a fixed height, and as the surface undulates beneath it the charge difference between the probe and the surface can be increased or decreased to keep the flow of tunneling electrons constant. Remarkably, these two methods give different pictures of the surface. The first method shows the true conformation of the atoms, while the second also measures the relative contribution of the atoms themselves to the electron flow.

The first picture in figure 7–1 shows a surface of carbon atoms with two silicon atoms embedded in it. They have been mapped using the second method, in which the probe was kept at a constant distance. The silicon atoms seem to tower above the surface because the electron flow from the probe is dramatically increased when it is positioned above these atoms. Fascinatingly, those carbon atoms that are near the silicon atoms also seem to project above the surface; they are excited to a greater degree than their neighbors by the proximity of the silicon atoms. This picture is a vivid example of the kind of fine-structure analysis of surfaces that the scanning tunneling microscope makes possible.

The second picture in the figure shows an image of double-stranded DNA that Balhorn's group obtained recently. The molecule doubles back on itself like a hairpin and at several points parts of the double helix can be seen, though no finer details are visible.

Working at the University of New Mexico in Albuquerque, Carlos Bustamante has actually succeeded in producing scanning tunneling microscope images of individual adenine molecules. To accomplish this he used long single chains of DNA consisting only of As. Such mole-

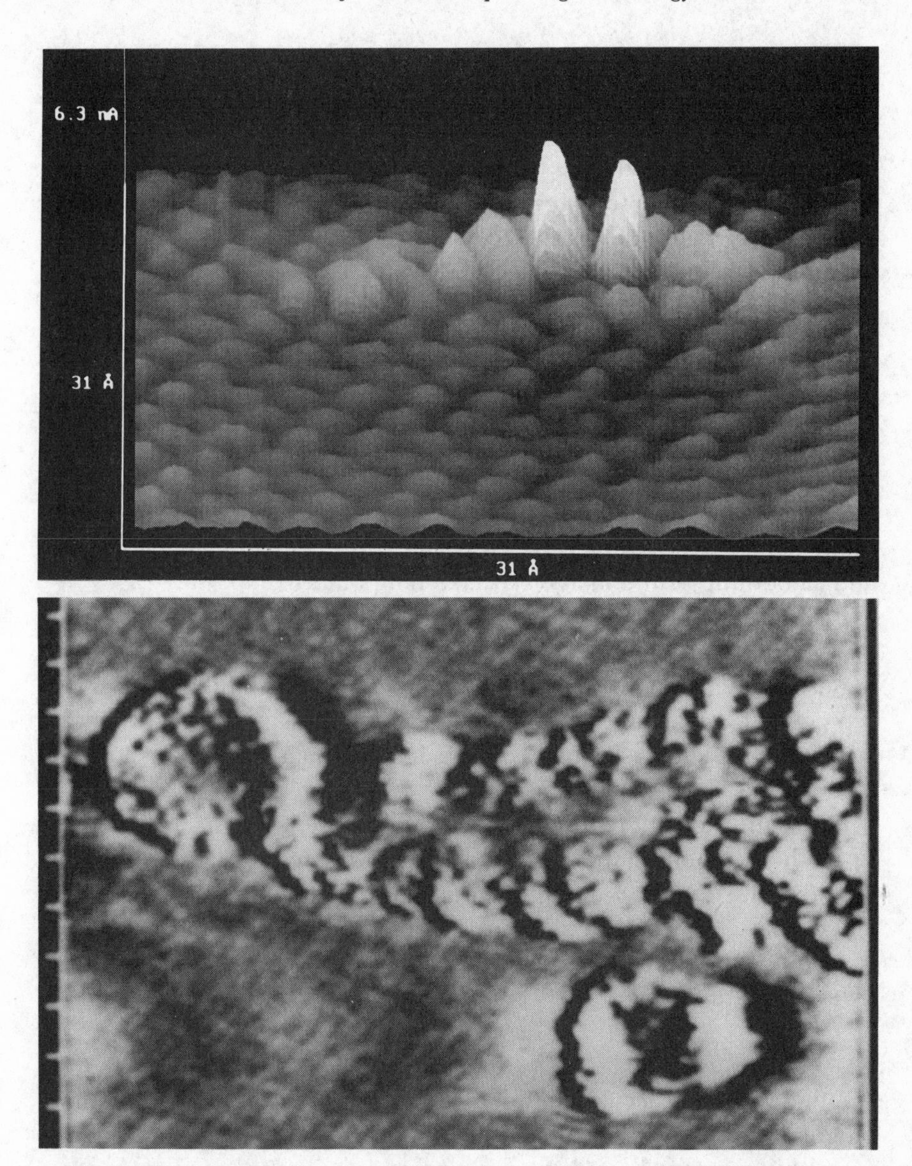

Figure 7–1. Two pictures made by the scanning tunneling microscope. The first shows two silicon atoms embedded in a smooth surface of carbon atoms. The greater conductivity of the silicon atoms causes them to appear to stand out above the surface, and the carbon atoms near them share this increased conductivity. The second shows a picture of a DNA double helix lying on a carbon surface. The coils of the double helix can be seen clearly, though the details of the bases in the inner part of the molecule cannot be distinguished.

cules are biologically uninteresting but physically tractable, in the sense that the adenines along their backbones can be pinned down flat on a surface like butterfly wings. This enabled Bustamante to see details of the adenines, including the two-ringed structure of the molecules, that would be impossible to detect if they were all crowded into the interior of a double-stranded DNA molecule. Recently, Balhorn has used a similar approach to obtain images of DNA made up of adenines and thymines.

Obtaining pictures of these giant molecules is extremely difficult. First, it is hard to attach the DNA to any perfectly smooth surface, so the microscopist must be content with whichever molecules happen to get caught in tiny cracks. Second, as the probe passes repeatedly back and forth, it hovers so close to the molecule that it pushes it about and can hopelessly blur the picture. And third, obtaining each image can require hours of patient scanning with the probe, while computerized enhancing and smoothing of the image can take hours more.

Balhorn is optimistic about the future of the technique, but he points to enormous problems. The interior of the DNA is so crowded with atoms, all interacting with each other and exciting each other in different ways, that the small differences among the four bases will probably always remain invisible unless some kind of signaling markers can be attached to the bases.

He is currently attempting to get around these problems by using the tip of the scanning tunneling microscope to excite the DNA at the same time that the tip moves along the molecule. This would cause differently labeled bases to emit different numbers of photons. If the method works, it should increase the rate of DNA sequencing by a factor of a hundred or even a thousand.

Another single-molecule methodology is being developed in a remote building in the Los Alamos complex, by a physicist named Jim Jett. Jett began his career at Los Alamos as a postdoc in nuclear physics, but was soon lured into biology.

His first biological ventures involved harnessing radiation to cure cancer. Then he became fascinated by chromosome sorting—in particular, how large a chromosome or piece of chromosome needed to be before it could be detected as it zipped through a chromosome sorter. It turned out that the ability to detect tiny chromosomes was directly tied in to the question of how many fluorescent molecules they could be labeled with.

The limit, he realized, was a single molecule: could a single fluorescent molecule be seen as it hurtled past the photodetector?

When Jett and his collaborator, Dick Keller, started their project, they could see a chromosome only if it was labeled with at least 1,800 molecules of the highly fluorescent dye rhodamine 6G. Now they are able to see one labeled with as few as 25. Indeed, their photodetector is now so sensitive that it is able to pick up a single molecule of the very fluorescent molecule phycoerythrin.

In 1988, Keller asked whether it might be possible to sequence a single strand of DNA using these methods. Together he and Jett came up with a daring approach. The idea was to label each of the four types of base with a different fluorescent molecule, then suspend a single strand thus labeled in a gentle flow of liquid. If an enzyme was added that would grab on to the end of the DNA molecule and begin to chew it away base by base, the labeled bases could be read in the proper order as they came free and flowed past the detector.

Easier said than done. Jett and his group are currently wrestling with a number of state-of-the-art problems, not the least of which is how to hold a single DNA strand motionless in a liquid flow. One possibility is to attach the DNA to a tiny plastic bead, then position the bead in the flow using focused lasers—a technique known as optical tweezers. The further problems—of labeling the DNA, of finding the right enzyme to cling to the end of the DNA molecule so that it gets chewed away at a regular rate, of being able to detect the resulting single bases as they float away—still loom ahead of them.

All this might have seemed like science fiction a few years ago, but researchers are moving with greater and greater boldness into the world of the single molecule and even the single atom. The scanning tunneling microscope has recently been used to grab single xenon atoms and slide them about. This has made it feasible to spell out the letters I.B.M. using individual atoms on a submicroscopic surface. Since the message can only be read by another scanning tunneling microscope it is of limited usefulness as advertising, but such forays into the ultra-small provide practice for tougher jobs like reading DNA.

Single atoms and molecules have also been trapped motionless in a supercooled state with lasers, to be studied at the investigator's leisure. New variants of the scanning tunneling microscope—such as the atomic force microscope, which can determine the shape of individual molecules by pushing against them—are being invented almost daily. One intriguing possibility that Balhorn and I had fun exploring involves mounting a single DNA polymerase molecule on the end of an atomic force microscope tip, pinning it down, and measuring with this molecular stethoscope the quiverings and palpitations of the polymerase as it makes a new strand of DNA. If one knew just which of these palpitations to monitor, perhaps it would be possible to tell the order in which bases were being added by the enzyme onto the growing DNA chain.

None of these technologies seem to hold the answer in themselves. The winning technology will probably marry these techniques with new approaches only just emerging. Japan has recently announced a $200 million program to explore the manufacture and uses of microminiaturized machines—tiny robots that could perform a variety of tasks such as surgery within the body. Such tiny machines, which can be made by some of the same techniques used to construct computer chips, might be adapted to DNA sequencing technology. Cleverly designed, these machines might be capable of handling and positioning individual DNA molecules. Scanning tunneling or atomic force microscopes, with their lack of complex optics and few moving parts, could easily be designed to be part of the machines, and could be precisely positioned to read the bases. It is out of just such crazy ideas that the new technologies capable of sequencing long stretches of DNA will surely be born.

A CAUTIONARY TALE

One of the reasons Congress has shown such enthusiasm for the Human Genome Project is fear that if the United States does not devote significant resources to the effort, the Japanese, with their well-known superhuman capabilities, will do the whole thing while we dither about. It seems on the face of it to be the sort of thing in which the Japanese might excel: a very large project, lending itself to automation, in which the preliminary methods have already been worked out, and with copious spinoffs for industry.

Unfortunately, simply scaling up existing methods without a clear determination to improve them markedly can lead to disaster. The Japanese were unlucky enough to be the ones to demonstrate this. A group of Japanese scientists and companies made a massive attempt to improve the efficiency of DNA sequencing by automating it. The group was established by the government's Science and Technology Agency in 1981, just four years after the publication of the sequencing techniques of Maxam and Gilbert and of Sanger. It was headed up by a biophysicist with much experience in automation, Akiyoshi Wada of Tokyo University.

The idea was simple: to automate the Maxam-Gilbert technique on a massive scale, eventually achieving a million bases a day of relatively error-free sequence. Sequencing reactions would be carried out by robots designed by the Seiko company. Fuji Film would develop prepackaged sequencing gels which could be used to separate the resulting radioactive nested fragments. The films would be developed and read automatically.

Yet after eight years and several million dollars, the elaborate machines were achieving at best 10,000 bases a day, or 1 percent of the rate expected. Even this required intense servicing efforts on the part of a small army of technicians. The amount of interest expressed by the participating companies has thus cooled considerably, particularly since it has proved impossible to prefabricate and market the huge, thin, and correspondingly fragile sequencing gels with any success.*

The whole story is reminiscent of a monstrous, robotized typesetting machine, the brainchild of the inventor James W. Paige at the end of the last century. Samuel Clemens (Mark Twain) sank his entire fortune into it, and in the end it drove him to bankruptcy. At its best the machine performed brilliantly, but it was rarely at its best. It kept breaking down, flinging parts in all directions. And almost as soon as it was built it was superseded by new machines that cast type out of molten metal in whole lines rather than fitting together individual letters. These machines soon did away with mechanical typesetting altogether. The experience of Wada's group provides a similar cautionary tale.

Despite such setbacks, the advent of the Human Genome Project has focused intense interest on the problem of how to sequence massive amounts of DNA. As a result, the problem has received such wide publicity that scientists in fields other than biology are starting to hear about

*This sobering experience has made the entire Japanese government cautious about joining with the U.S. and with European countries on the Human Genome Project. Japanese caution and unwillingness to commit resources led to a recent outburst on Watson's part, in which he suggested that they either contribute more money or be denied the results from the U.S. project. This exercise in Japan-bashing made others associated with the project very uncomfortable. Victor McKusick in particular had to spend a good deal of time mending fences and smoothing feathers. Watson has not been as vituperative about the Germans, who have also been less than enthusiastic about the project—though for ethical rather than economic reasons.

it. Whenever that happens a breakthrough is sure to occur. The NIH part of the project seems the least likely source of such a breakthrough, with the DOE part more likely, and a currently unknown source most likely of all. I am betting against the NIH because the biomedical community has different priorities; indeed, the whole subculture of biomedical science militates against an intensive push to develop such techniques. As Bruce Alberts put it in his testimony at the 1988 House subcommittee hearings: "I think you should realize—it's important to realize that biological scientists have a sort of inherent bias—I'm giving you my own opinion now—against people who develop technology. . . . NIH study sections want to know what biology you're going to accomplish, not what methods you're going to develop."

For most biomedical scientists, the fun lies in finding out the ways that organisms work, not in developing new techniques. The mechanics of sequencing DNA does not interest them in the least; they want the results. A single gene can tell a fascinating story. In many cases—as we will see in coming chapters—trying to understand one gene is quite enough to keep hundreds of scientists busy for decades.

Further, the NIH contribution to the exploration of the human genome has been less focused than the DOE contribution, a lack of focus that became apparent soon after the agency reluctantly joined the effort. This is because of the very nature of biomedical research—the multitude of opportunities that have opened up to use DNA sequencing techniques to explore the genes of other organisms, ranging from bacteria to fruit flies to mice. The rapid clustering of all kinds of biologists around the NIH part of the Human Genome Project has led to the bestowal of much of its money on worthy and fascinating projects that have little to do with the human genome.

The DOE has kept a much narrower focus, and has traded on its ability to foster a multidisciplinary approach. While it has a much narrower set of interests in biomedicine, its strengths have always been in the development of wonderful gadgets.

My guess, however, is that neither of these sources will be the one that funds the necessary breakthroughs. There is a much greater chance that these advances will come from somewhere else, somewhere quite unexpected. Two safe predictions can be made, however. The first is that the breakthrough technology will involve the sequencing of single molecules. The second is that by the magic year 2005 this new technology—whatever it is—will have made sequencing a human genome's worth of DNA a relatively simple task.

Even this will only be the beginning. The map is not the territory. Now that we can see how the genome will be mapped and sequenced, perhaps even sooner than the original planners of the project predicted, it is time to ask how this information is being used. Its primary use, of course, is to hunt for genes that cause disease, though we will see in a moment that this is not its only use.

CHAPTER 8

Sherlock Holmes Meets the Human Genome

A hair perhaps divides the false and true.

—Omar Khayyám

There is no single human genome. There are five billion of them, one for virtually every human being on the planet—ten billion if you count each set of chromosomes separately. Identical twins do of course share a genome, but even in twins the inexorable process of mutation causes the DNA of the different cells in their bodies to diverge gradually with time.

As noted earlier, we differ from other members of our species by one base in a thousand, creating an astonishing average of six million differences between unrelated individuals. A few of these differences can have enormous effects, condemning their carriers to lives of suffering. Many others must influence the innumerable ways in which humans differ from each other, both mentally and physically. The overwhelming majority, however, seem not to matter at all. They are like the tiny defects in an old-fashioned analog sound recording. Each copy that was pressed of such a record had a different set of defects, producing a different pattern of hisses and pops. So long as they were not so large that they disrupted the music or caused the needle to jump out of the groove, the defects could be ignored by the listener.

Remarkable new techniques now exist that can, in principle, detect all the differences between any two individuals. The remainder of this book will explore the enormous subject of what these differences are, how they are being detected, and the many ways in which they are being put to use in order to explore and understand the genome—or, to be more precise, the five billion genomes possessed by our species. Let me begin with one class of variation, the study of which has led to enormous legal and ethical complications. So far these variations seem to have no detectable biological significance, though this may soon change, as we will see in chapter 12.

Jo Kiernan, a San Diego district attorney, is chief of the city's Major Violator Unit. She heads a group of six DAs who concentrate solely on major crimes of violence, building the strongest possible cases against the perpetrators so as to put them away for as many years as possible. The group works on only a hundred or so cases a year, supported by a special grant from the state. No plea bargaining is allowed. The conviction rate in their unit is very close to 100 percent.

Kiernan laid out her current case for me in ghastly detail. In two busy evenings, a young man robbed two elderly women living in nearby houses in a quiet neighborhood, raping one of them and attempting to rape the second. He broke into a third house and robbed an elderly man, beating him over the head with an air pistol until the pistol shattered, then fled into the night. A friend of the rape victim, who had been in the victim's house during the crime and had been kept at bay by the assailant, spent the next several hours driving the victim from hospital to hospital until he finally found one that would examine her. By the time the hospital obtained a vaginal swab, hours had passed since the rape.

A prime suspect was soon arrested: a young man who had twice before been convicted of robbery with rape but who had been released on parole soon after each crime. His fingerprints were found at the scene of the rape. When the broken pieces of pistol were discovered nearby it was established that the gun had been stolen from the friend of the rape victim. No direct identification was possible, however, since the crimes had taken place in the dark.

Jo Kiernan's problem was to try to link the suspect to all three crimes. Conviction on only one would result in a relatively short sentence, and he would soon be on the streets again. If, as she explained to me with relish, she could get him on all three, this would put him away for "life." In practice this would probably mean thirty years, still long enough that upon his release the man would be unlikely to take up his career as a rapist with quite such vigor as before.

Kiernan's job was to amass every possible bit of evidence. One of these involved the remarkable new technique of DNA fingerprinting. Oddly, her life before going into law had prepared her for this case. She had obtained a master's degree in biology from Cornell in the late 1960s, and had become a technician in the lab of the Nobel Prize–winning biochemist Robert Holley at the Salk Institute. Fascinated by the interaction of science and society, she soon joined a new discussion group at the Salk that was studying bioethics and law. The group worried about such avant-garde problems as *in vitro* fertilization years before the English physicians Patrick Steptoe and Robert Edwards finally succeeded in carrying it out in humans.

Soon Kiernan was taking law classes at night. Realizing she did not have the patience for science, she soon switched to law full time and found herself—to her surprise—enjoying the challenge of being a prosecutor. Being no stranger to DNA, she reveled in learning about the new field of forensic molecular biology. Her cluttered office was piled high with transcripts from some of the many courtroom battles in which DNA fingerprinting itself had already been put on trial.

Simply put, DNA fingerprinting is a way of obtaining information from the tiny amounts of biological material that are almost always left behind at the scene of a crime: a few hairs, a drop of blood, some semen, even skin cells from under the fingernails of a victim. The procedure cannot establish absolute guilt, for many reasons may explain the presence of particular DNA at the scene. Also, while real fingerprints are unique to each individual, there is a small possibility that two people can have—or appear to have—the same DNA fingerprint. This possibility increases if the material left at the scene is difficult to work with, or if there is not enough of it to extract much information. Thus, apparent matching of a suspect's DNA fingerprint with a DNA fingerprint found at the scene is only meaningful if the probability that the two could be identical by chance is very small.

The technique can, however, establish innocence. If the DNA fingerprint of a suspect does not match the criminal's DNA found at the scene, then the suspect is exonerated. In March 1990 a Washington, D.C., man who had been convicted of rape and sent to prison was exonerated on the basis of DNA evidence obtained with the greatest difficulty by the public defenders Neal Kravitz and Sharon Styles in the face of numerous obstructionist tactics by Assistant U.S. Attorney Ronald Dixon.

As I penetrated into the single-minded world of prosecuting and defense attorneys confronted by this new technology, I was reminded of a wonderful spoof of C. S. Forester's equally single-minded naval character Horatio Hornblower. In the original Hornblower stories, which take place during the Napoleonic wars, the hero's only aim is to foil the French. In the spoof, written by Harry Harrison, the ship of the hero Honario Harpplayer is visited by a highly advanced creature from outer space, whom he immediately co-opts into aiding his latest campaign against the dastardly Frogs while showing a complete lack of curiosity about the alien himself.

Jo Kiernan is an exception because of her background; the majority of DAs and defense attorneys who have been handed this new tool show a Harpplayeresque lack of inquisitiveness about what it means. As we explore the implications of this and similar techniques over the next few chapters, we will see what the DAs do not: that these are not only high-tech ways to catch crooks, but also an entrée into the detailed genetic mapping of the human genome. They also tell us a great deal about the genetic variation that exists in our species.

MINDLESS VARIATION

DNA fingerprinting takes advantage of the DNA variation that is found everywhere on the chromosomes. In Chapter 5 I remarked that unrelated individuals differ by about one DNA base in every thousand. Now it is time to see how this estimate was actually obtained.

The technique used is something called a whole-genome Southern blot, a variation of the Southern blot technique used by Bob Moyzis and his group to hybridize DNA fragments derived from cosmids to their repetitive-DNA probes. In the whole-genome technique there is one important difference: Rather than probing fragments obtained from a single cosmid, one uses probes to explore the mass of fragments obtained when the whole human genome is digested by a restriction enzyme.

The process begins with the purification of a millionth of a gram or so of DNA from a single individual. Tiny as this amount may seem, it contains 100,000 complete copies of the genome. The DNA is then completely digested by a restriction enzyme, usually one that recognizes a six-base sequence. Such an enzymatic treatment will result in about 1.5 million different kinds of fragments of various sizes, each of which is present in 100,000 copies.

The fragments are then separated on a gel similar to a sequencing gel. There are so many different kinds of fragment that the result is a smear of DNA rather than the discrete bands seen on Carrano's or Moyzis's gels.

The smear is not a completely random jumble, of course, since all the fragments it contains will at least have been separated out according to size. Still, if you were to look at the DNA at this stage it would be quite uninformative, since the DNA smear from one person looks much like the smear from another.

Southern's blotting technique is then used to transfer the entire smear of DNA from the fragile gel to a tough nylon membrane. Further treatment binds it immovably onto the membrane, and breaks the double helices apart into single strands that can be hybridized to a complementary probe. The membrane is then bathed in a solution containing such a probe, which has been made radioactive. The probe used is normally quite short, much shorter than the genomic DNA fragments. It will quickly find and hybridize with the longer piece or pieces of "target" genomic DNA that are complementary to it. There are about a hundred thousand of these targets waiting on the membrane, which hybridize to an equal number of probe molecules. The surplus probe molecules are then washed off. X-ray film is laid on the membrane and it is put in a dark place for two weeks.

Bob Moyzis is able to develop his pictures after only a few hours because each cosmid consists of only a few fragments and he can load his gel with many more than 100,000 copies of each fragment without fear of spoiling the gel by running too much DNA through it. The reason the picture of a whole-genome blot takes two weeks to develop is that there are so many more types of fragment in a whole genome than in a cosmid, and consequently fewer copies of each type that can hybridize to the probe. One hundred thousand is a small number in the world of molecules, and it takes a long time for enough of these probe molecules to decay radioactively to make a visible band on the film.

When the picture is developed, a series of bands appear. If a probe has incautiously been used that happens to be a piece of highly repeated DNA, then the bands will still be a hopeless smear, because there will be so many different pieces of genomic DNA on the membrane to which the probe can hybridize. But if the probe is a stretch of DNA that occurs only once or a few times in the genome, it will hybridize to just one or a few types of fragment in the genomic DNA smear, producing recognizable bands. The result is that the original uninterpretable smear of millions of genomic DNA fragments has been replaced by a much simpler picture—just those few fragments that hybridize to the probe are visible. You can appreciate the enormous power of this technique—with the right probe, one small piece of DNA can be singled out from the entire genome. The piece need not even be a gene. Any probe can find its complementary piece of DNA in the genome and hybridize to it, whether the probe is part of an exon, an intron, a repetitive region, or any other part of the genome.

There are some very large caveats. This technique tells you nothing more than the size of the genomic fragment to which your probe hybridizes. All the DNA fragments start at the top of the gel and are then spread out by electrophoresis in decreasing order of size from the top of the gel to the bottom. If the fragment to which the probe can hybridize is large, it will migrate slowly to a place high on the gel. This is because a large piece of DNA can move through the gel only with difficulty. If the fragment is small, it will migrate much further and appear lower on the gel. Indeed, if it is very small it might run off the gel completely and be lost, so the probe might not appear to hybridize at all. Size is the *only* information that the whole-genome Southern blot provides; if you know nothing about your probe, you will have learned nothing about the piece of DNA to which it hybridizes except how big it is. Further, because of the rather low resolving power of this system, you will only know the approximate sizes of the fragment. At the low end of the gel two pieces of DNA might differ by as much as ten or twenty bases and yet be indistinguishable. At the high end, where the larger molecules move very slowly, fragments of DNA can differ in size by hundreds or even thousands of bases and still appear to migrate to the same place.

Figure 8–1 shows the results of a remarkable experiment. The figure shows whole-genome Southern blots for two parents and two children of an African-American family. The family had already had one child who suffered from the serious recessive condition sickle-cell anemia. When the mother became pregnant again, Judy Chang and Yuet Wai Kan at the University of California Medical Center at San Francisco addressed the question: Could a test be constructed that would detect whether the baby in the womb was also homozygous for the disease? There was one chance in four that it was.

Chang isolated a small amount of DNA from fetal cells taken from the mother's amniotic fluid, cleaved it with a restriction enzyme called DdeI, and ran the resulting collection of fragments out on a gel. Along with it she ran DdeI-digested DNA from the two parents and from an unaffected sibling of the baby in the womb. Then she blotted the DNA smears onto

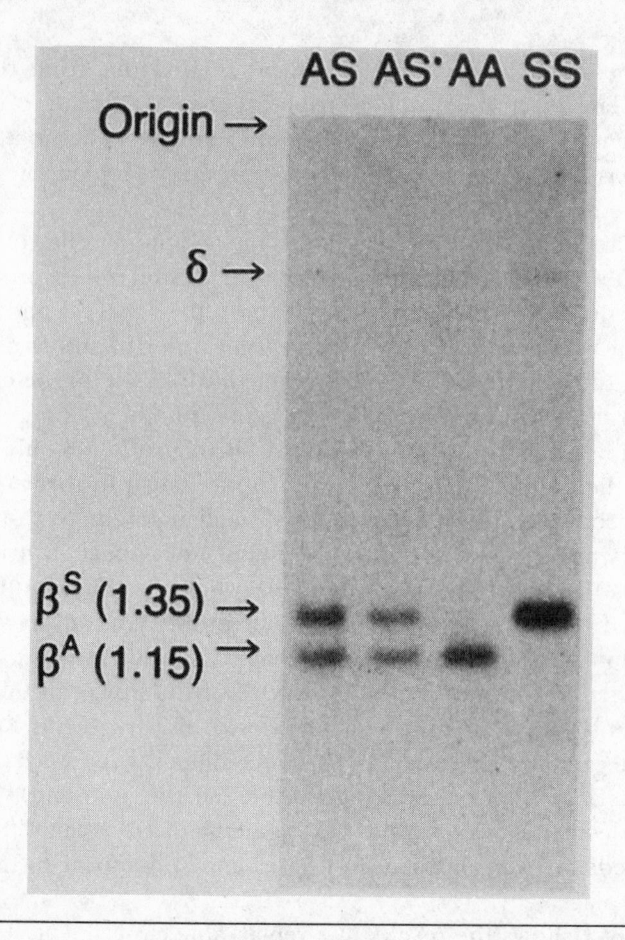

Figure 8–1. Whole-genome Southern blots of the DNA of two heterozygous parents (AS), a homozygous normal child (AA), and their unborn child, who is homozygous for the sickle-cell mutation (SS).

a nylon membrane and hybridized a specific radioactive probe to the membrane. The probe was a part of the hemoglobin gene that is affected by the sickle-cell mutation.

In the normal gene the region of DNA near the mutant site is this:

. . . GACTCCTGAG . . .
. . . CTGAGGACTC . . .

The corresponding sequence in the mutant gene is this:

```
. . . GACTCCTGTG . . .
. . . CTGAGGACAC . . .
             ⇑
```

The arrow shows where the mutant differs from the normal gene. DdeI recognizes the following sequence:

```
      ⇓
. . . CTNAG . . .
. . . GANTC . . .
          ⇑
```

The N in the middle of the sequence stands for any base, and the arrows here show where the DNA is cleaved. You can see that this sequence is the same as the five-base sequence that I have underlined in the normal gene, and that the mutant gene differs by one base. As a result the enzyme will not cleave the mutant DNA at this point (though of course it cleaves DNA from people with and without the mutation at many other points).

Given this information, the pattern of bands seen in the Southern blot in the figure becomes clear. As expected, the probe hybridized only to the hemoglobin gene and not to the other DNA fragments in the gel. The gel showed that both parents were heterozygous for the sickle-cell mutation. In each of them the piece of DNA carrying the mutant allele was missing a DdeI site, so it moved more slowly through the gel. As a result, the probe hybridized to two bands of their DNA, once at the lower point where the smaller normal allele had migrated and once at the higher point reached by the larger and slower mutant allele. The gel also showed that the unaffected child was homozygous for the normal allele.

The critical part of the gel, however, was the lane in which the DNA from the developing fetus had been run. Here, only a single slow-moving band was seen. The fetus turned out to be homozygous for the sickle-cell mutation; both its hemoglobin alleles were larger, slow-moving pieces of DNA. The one genetic chance in four had, unfortunately, happened to this unborn child.

Chang and Kan had found a quick and easy new way to screen for sickle-cell anemia in the unborn. In the process they had also revealed a tiny bit of the genetic variation found everywhere in the human species.

This takes the form of *restriction fragment length polymorphisms,* or RFLPs. The term "polymorphism," Greek for "many forms," refers to the fact that there can be many different allelic variants in the DNA that are detectable by restriction enzymes. This particular RFLP is highly unusual, for it happens to occupy a site that is also the site of a damaging and common mutant allele.

It is very unusual for a restriction fragment length polymorphism to be related to a disease in this way. In the case of most RFLPs that are detected on Southern blots, there is no indication of what effect, if any, they might have. In addition to aiding in the detection of diseases and the mapping of genes, genomic Southerns are a way of sampling the genetic variation in the human species at the most fundamental level, that of the DNA itself. In the course of the evolution of the human species, mutations have caused potential restriction sites to appear and disappear all over our genomes. When I remarked earlier that any two human beings differ by about one base in every thousand, this number was not just a guess. It can be estimated in several ways, and one is by looking at several RFLP sites in each of a number of people and determining how many of them are different.

THE REMARKABLE VNTRS

RFLPs are, at least potentially, of tremendous importance for identifying people. If you look at enough RFLPs, the chance that two human beings will be the same at all of them is vanishingly small (unless, of course, they are identical twins). But most of the RFLPs you look at will not be very informative, because there are only two alternatives—people either have the restriction site or they don't. If, for a given RFLP, most people have the restriction site and only a few do not, then most of the time when you look at two individuals both of them will have the site and their patterns on that particular genomic Southern will be identical.

RFLPs have other limitations as identifiers. In her rape and assault case, Jo Kiernan might have tried to use RFLPs to see whether the DNA left at the scene of the crime matched that of the suspect. Given the time and money, she could have asked the crime lab to digest the suspect's genomic DNA with many different enzymes, then separate the digests on gels and hybridize them to an assortment of probes. If the technicians used probes known to hybridize near predetermined human RFLPs, they would discover that the suspect carried a certain allele at one site, a certain allele at another, and so on. Eventually, they would develop a unique DNA "fingerprint" for the suspect.

A problem would arise, however, when they tried to determine whether the DNA from the assailant matched that of the suspect. The only available DNA from the assailant was a bit of dried semen on a swab—a sample that was hours old before any attempt had been made to preserve it. It was mixed both with cells from the victim and with bacteria, and if there is anything bacteria love to eat, it is DNA. This tiny quantity of partially degraded DNA could never have been carried through so extensive and elaborate a procedure. DNA fingerprinting using RFLPs can more easily help to determine paternity, since the mother, the putative father, and the child are all available to provide large fresh samples of blood for unlimited testing.

Until 1985, however, it was not possible to do forensic DNA typing of the kind Jo Kiernan needed. In that year a way to determine a nearly unique DNA fingerprint was found, a way that could be useful even when only one tiny, maltreated sample of DNA was available. Alec Jeffreys of the University of Leicester reported that he had discovered a different class of RFLP, one that met the exacting requirements of forensic analysis. These new RFLPs were soon given yet another bewildering acronym: VNTR, which stands for "variable number of tandem repeats." VNTRs are miniature representatives of those long stretches of repeated DNA that we referred to earlier as genomic Simpson Deserts. Instead of being repeated thousands or hundreds of thousands of times, they may only be repeated a dozen or two times, forming a kind of genetic stammer here and there on the chromosomes. Jeffreys tracked down these regions by finding probes, made up of one or more of these repeated segments, that would hybridize to specific VNTRs.

VNTRs are found in many parts of the genome. Currently about a hundred of them have been mapped to the chromosomes, compared with four thousand of the more ordinary kind of RFLP. They do not code for proteins, and their role in the genome is unclear. They would remain only mildly interesting were it not for the fact that they are genetically highly unstable and mutate at an enormous rate. At the most highly mutable VNTR loci, one person in every ten will pass a new VNTR allele on to the next generation. This astounding mutation rate enables us to infer that the exact state of our various VNTRs is not important to our survival, for were VNTRs really important genes very few people could endure this mutational onslaught, and those who did would suffer from an appalling variety of genetic diseases.

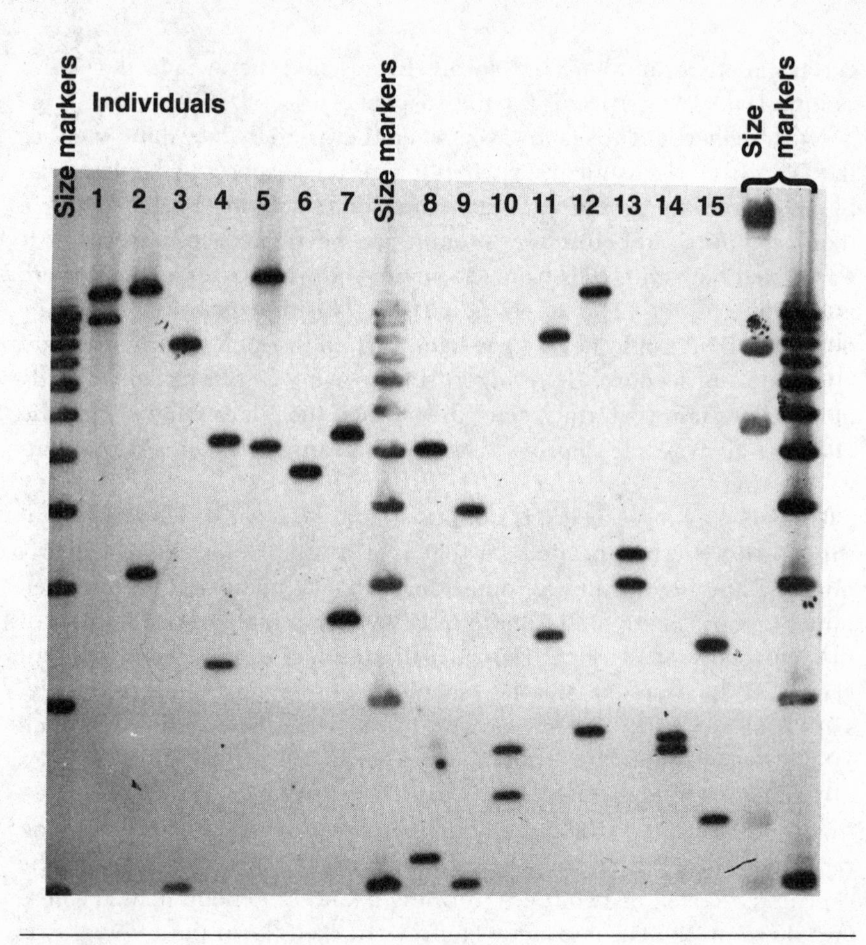

Figure 8–2. DNA from a random series of Caucasians, hybridized to a VNTR probe. At both ends of the gel and in the middle are ladders of DNA, which are simply size markers that tell the experimenter the sizes of the pieces of DNA from each individual.

These mutations are not only very common; they are of a specific kind. They cause the number of repeated segments in a VNTR to change, which results in an increase or a decrease in its overall length. It seems not to matter exactly how many repeats there are in each of our VNTRs—as long as we have some, since VNTRs seem neither to disappear entirely nor to grow without limit in size.

Some VNTRs happen to be flanked by convenient pairs of restriction sites that are both recognized by the same enzyme. These are the ones that turn out to be most useful for forensic analysis. If a person carries a long VNTR made up of many copies, the restriction sites will be far apart, and a probe specific for that VNTR will hybridize to a large piece of DNA. If the VNTR is shorter, the sites will be closer together, and the probe will hybridize to a shorter piece.

Most ordinary RFLPs have only two possible states—the site is there or it is not there—and so different individuals often have the same alleles. The highly polymorphic VNTRs are quite different. Because of the high mutation rate, a VNTR at a given genetic locus can come in dozens or hundreds of size variants. Just as with any other genetic locus, no individual can have more than two alleles at any VNTR locus, but mutation has produced so many VNTR alleles in the population that most pairs of individuals chosen at random will be found to carry quite different alleles.

Determining the precise number of alleles is very difficult, since there are so many at some VNTR loci that they can form an almost continuous spectrum of sizes. The only sure way to find out would be to sequence them all, looking for tiny variants of the kind that have been found to distinguish repeated segments of DNA in the more extensive Simpson Desert regions of the genome.

Figure 8–2 shows a collection of randomly picked individuals probed by one of Jeffreys' VNTR probes. Though each individual shows only two bands, representing the two alleles, the enormous amount of variation from one individual to another is clearly apparent.

YOU LEAVE IT, WE CLEAVE IT!

VNTRs are just what the forensic analyst needs. Several different VNTRs can be surveyed at once by probing a single sample of genomic DNA with a mixture of probes. The "fingerprint" that is obtained is not absolutely unique like a true fingerprint (even identical twins have different true fingerprints), but it is very unlikely that another person will have exactly the same VNTR pattern. From the forensic standpoint, the question is: How unlikely?

Two private companies are the current leaders in using this technology. Lifecodes, based in Valhalla, New York, was the first on the scene, testing DNA samples taken from crime scenes as early as 1986. It was followed by Cellmark, based in Germantown, Maryland, and a subsidiary of Britain's Imperial Chemical Industries (ICI). Cellmark, which has the great advantage of direct access to Jeffreys's best probes, received in early 1991 a contract from the state of Arizona to carry out its forensic analyses.

These companies have been joined in VNTR analysis by the FBI and by state laboratories in Florida, California, and a number of other states. The FBI labs are already swamped, processing an estimated 1,500 cases in 1990 with a growing backlog.

Initially, the introduction of DNA evidence simply bowled over the defense. Peter Neufeld, a New York defense lawyer who has been involved in DNA cases, said in an interview, "In the first two dozen cases where DNA evidence was introduced, the opposing attorney[s] did not even challenge the evidence. They felt scientifically illiterate and unable to even perceive of questions. No adverse experts were even retained by the counsel. Everyone just sort of lay down and died."

This soon changed, as a result of a notorious murder case. In early 1987, Joseph Castro, a Bronx handyman, was accused of murdering a pregnant neighbor and her two-year-old daughter. Circumstantial evidence was suggestive but not conclusive. A small blood stain was found on Castro's watch, from which Lifecodes managed to extract about half a millionth of a gram of DNA. After analysis, the company announced that the VNTR bands from this DNA were identical to those of the dead mother, and that the probability that two unrelated people would have the same pattern was 100,000,000 to one.

This DNA evidence was presented in an extensive pretrial hearing, to determine whether it should be admitted on the usual legal grounds of being "generally accepted by the scientific community." As the hearing progressed, so many gaping holes became apparent in the evidence that expert witnesses who had been brought in on both sides became uncomfortable. Richard Roberts, a leading expert on restriction enzymes from Cold Spring Harbor Laboratories and a witness for the prosecution, took the unprecedented step of meeting outside the court with three other witnesses from both sides. Freed from the adversarial atmosphere of the courtroom, they considered the evidence and found it wanting.

A defense witness, Eric Lander (a mathematical geneticist from Harvard who will appear again in a later chapter), was so horrified by the low scientific quality of Lifecodes' data that he wrote a devastating article in the June 1989 issue of *Nature.* He pointed out that Lifecodes had ignored or misrepresented some of its data, and had kept changing its story about the source of the controls that had been run on the gels and the criteria it used to determine whether there was a match between bands. He also found many reasons to quarrel with their calculation giving a likelihood of 100,000,000 to 1 that such a match would occur by chance. In August 1989 the judge in the case ruled some of the Lifecodes evidence impermissible.

He agreed that the company had shown that the blood on the watch was not the suspect's, but not that it was likely to be the murder victim's.

Though Castro later confessed to the crime, even admitting that the blood on his watch had come from the victim, the myth of infallibility about DNA evidence had been dispelled. Around the country defense lawyers took crash courses in DNA technology, and recruited expert witnesses. These ranged from some with honest reservations about the technique to others who felt for a variety of reasons that DNA evidence should not be used under any circumstances.

Both Cellmark and Lifecodes had built up private data bases of VNTR fingerprints from a number of ethnic groups, to which they could compare the fingerprints of suspects. These have been challenged by defense lawyers on various grounds—the data bases are too small, they are not representative of the diversity of groups found in the United States, and they have odd statistical properties. All these problems are real. Unfortunately, determining their seriousness and how to correct them has been made very difficult by the fact that until very recently geneticists and statisticians have not been able to examine the raw data on which Lifecodes and Cellmark base their statistics. Lisa Foreman of Cellmark explained to me that she and her company mean to publish thorough analyses of the data, but have been so swamped by endless demands for court appearances that they have been unable to do so. I was assured by Ivan Balasz of Lifecodes that their data base has always been available to the scientific community, but other sources suggested that this had only been true since about the end of 1989. In any case, information has gradually begun to trickle out to the scientific community, and a picture has started to emerge.

The worst problems have been with the data bases themselves. Consider the problem of ethnicity: Are there appreciable differences between different racial groups in the frequencies of VNTR alleles, differences that might have some bearing on the identification of forensic samples? The answer seems to be yes and no. While some isolated populations have unusual VNTR allele frequencies, differences between major ethnic groups appear at first sight to be rather small.

As it turns out, however, this apparent similarity is probably more a function of the tremendous amount of mutational turmoil at these loci than anything else. To understand the difficulty of making comparisons between such data bases, it might be helpful to compare them to the New York and Tokyo stock exchanges, which are very similar at one level and yet very different at another. Both are enormously active, with companies appearing and disappearing all the time. Most of the companies listed on the two exchanges have different names and histories, but the general breakdown into *type* of company is roughly the same. In both exchanges there are banks, automobile companies, shipping companies, energy companies, and so on. If listings on the two exchanges were classified only according to these general types, both institutions would appear very similar. If they were classified according to the company names, they would appear to be very different.

In the same way, both within and between ethnic groups, different representatives of an "allele" which all migrate to a certain position on the gel might really be a variety of different alleles with different mutational histories. As a result, the true amount of differentiation among ethnic groups can only be determined by extensive DNA sequencing of the VNTRs, just as the true amount of differentiation between the New York and Tokyo stock exchanges can only be determined by a detailed listing of the different companies that make them up. What this means in practical terms is that there may be far greater differences among racial groups and even within different groups within the same racial category than a casual survey of the data would indicate. This poses a problem, since both Cellmark and Lifecodes routinely claim that the chances of finding a match to a particular VNTR pattern are millions to one. If the suspect and the real criminal were to come from the same ethnic group, the chances might be far greater than that.

A favorite defense witness is Laurence Mueller, a geneticist at the University of California at Irvine, who claims to have found a pattern that shows there are indeed detectable differences between ethnic groups, and that these might change the odds of a match considerably if both the suspect and the real culprit have the same ethnic background. These differences, Mueller maintains, are not so much in the allele frequencies as in the fact that certain alleles tend to be inherited in groups. His argument depends on a phenomenon called linkage disequilibrium, or, more properly, *gametic-phase disequilibrium.* The existence of this disequilibrium, he contends, indicates that there could be a great deal of substructuring into small groups. Unfortunately for his argument, neither he nor anybody else has been able to demonstrate convincingly that such disequilibrium exists and can be measured, particularly since nobody knows just how many VNTR alleles might be concealed within a single size class of DNA that migrates to the same place on a gel.

In the pretrial hearings of Jo Kiernan's rape case, Mueller was brought in as a star witness for the defense. I spent a long and frustrating afternoon feeding questions to Kiernan in the courtroom as she cross-examined him, the two of us trying to get him to admit that it was impossible to measure disequilibrium using VNTR data with their multiplicity of constantly changing alleles. At times the argument became very arcane indeed. Yet despite Kiernan's efforts, the testimony of Mueller and other witnesses raised some doubts in the mind of the judge. At the end of the pretrial hearing she ruled that the DNA evidence could be admitted, but testimony about the likelihood of a match could not.

Another difficulty with the data is the random mating problem. Since people do not pick their mates on the basis of their VNTR alleles, it would be expected that mating is random with respect to this set of characteristics. If so, then one can apply a simple algebraic principle called the Hardy-Weinberg Law. The law says that, given all the allele frequencies and random mating, it is easy to calculate the frequencies you would expect for all the various homozygous and heterozygous combinations of these alleles in the population. For instance, if an allele is found in the population at a frequency of 1 percent, the Hardy-Weinberg Law predicts that homozygotes for that allele should be found at a frequency of 1 percent of 1 percent, or one in ten thousand.

Unfortunately, examinations of VNTR data bases reveal an excess of homozygotes—by as much as thirtyfold. When this was discovered, Richard Lewontin at Harvard and others pointed out that this might mean that the population is not mating randomly with respect to VNTRs. The only rational way this might happen would be if the human population were divided into tiny groups with little exchange between them, so that people with similar genotypes would tend to marry more often than expected. If a suspect and the real culprit were drawn from one of these groups, the probability of identity would be greatly increased.

It now turns out that this excess of homozygotes is an artifact, apparently due entirely to the fact that it is difficult to distinguish homozygotes from heterozygotes when the two pieces of DNA involved have similar lengths. This was shown in September 1990 by a group at the Yale School of Public Health, who finally obtained the raw data from Lifecodes and did a very simple analysis to show the problem. So far as this group could tell, people really were mating at random with respect to their VNTR genotypes.

As soon as this paper appeared, the defense lawyer in Jo Kiernan's rape case changed tactics. He began to concentrate on what he claimed were problems with the quality of the gels obtained from the assailant's DNA, and dropped his objections to the introduction of evidence about the likelihood of a match.

As a scientist observing these cases, I could not help but feel annoyance that so much argument, money, and paper had been wasted during literally hundreds of pretrial hearings about the quality of DNA data bases and the meaning of the apparent homozygote excess. Much of this could have been avoided if population geneticists had been allowed to analyze the full sets of data years earlier. On the other hand, I could not help but be impressed by the resourcefulness and tenacity of defense lawyers who seized on this and every other perceived weakness of the technique and attacked them with every tool at their command. Judges and attorneys at pretrial hearings have been forced to grapple earnestly with such arcana as the Hardy-Weinberg Law and linkage disequilibrium. In the courtroom, however, it really boils down to the way the evidence is presented to the jury.

I watched Lisa Foreman from Cellmark present DNA fingerprint evidence in a different rape case. This was about the fortieth time she had testified in such a case, and she and the prosecuting attorney moved through questions and answers in a superbly choreographed *pas de deux.* The defendant sat hunched at one side of the courtroom, ignored by the jury, who stared with rapt attention at the X-ray films displayed on a portable light box.

Foreman began with a picture showing the VNTR polymorphisms from a variety of people. While there were a few shared bands, most of the patterns were very different. (You saw a similar picture in figure 8–2.) She then showed a gel with the patterns of the victim and the defendant, and a series of extracts of DNA from a vaginal swab. In one of the vaginal extracts the DNA of both victim and assailant showed up clearly, and the match between the assailant's and the defendant's bands was glaringly obvious.

To bring the point home, she followed up with a series of pictures made with individual probes, in which the matches were equally clear. She assured the jury that the matches had been made not by eye but by a computerized scanner. Cross-examination concentrated on the fact that neither Lifecodes nor Cellmark are currently being overseen by any government agency, or had to pass any kind of competency test. Foreman replied serenely that when such oversight was mandated, Lifecodes would pass any requirements with flying colors.

The outcome of this particular rape case is instructive. Confronted with the DNA evidence, the defense switched from claiming that the accused had not done it—since he almost certainly had—to claiming that he already knew the victim. This raised enough doubt in the mind of one juror to hang the jury. The case may or may not be retried. The result vividly demonstrates that DNA evidence is only one of a complex of factors in every trial.

Jo Kiernan had a more difficult problem with the DNA evidence in her rape and assault case. In the pictures she showed me, the match between the defendant's DNA and the DNA from the vaginal swab was very clear, but in the swab only six out of the eight possible bands were visible.* The two largest pieces of DNA had either been so badly degraded that they were no longer visible, or—as she was afraid the defense would claim—they might never have been there in the first place. Luckily for the prosecution, in the material from the swab the victim's DNA was also degraded, and the larger pieces were more badly degraded than the smaller ones. (On such gels, if there has been any degradation, the larger pieces of DNA are usually the ones to disappear first.) On the basis of the six visible bands, Lifecodes had calculated the probability of a match at 1 in 800,000, using information from the data base that matched the defendant's ethnic group. Would such odds be enough to impress the jury, or would the defense argue that there was too much uncertainty in the calculation?

During the trial, one of the defense witnesses called was Paul Hagerman, an expert on the physical properties of DNA from the University of Colorado. He contended during direct examination that there could be explanations other than degradation for the missing band. Throughout his testimony he said nothing about the fact that the larger pieces of the victim's DNA appeared to be degraded as well. He considered that contamination with bacterial DNA, to which probes do not hybridize, was a likely explanation for the fact that the bands of both victim and rapist were so faint.

On cross-examination, Kiernan bored in on this. She asked him to examine each gel and tell the jury what he saw there that would support the possibility that degradation had occurred. It turned out to be quite a long list. Then, in the closing seconds of her cross-examination, she asked him what he would need to do to support his theory of bacterial contamination. It turned out that this would involve taking half the sample and running a very different type of gel.

"Does this mean, then, *Doctor,*" she demanded icily, "that you would

*Four different VNTR probes were used. This should have given a total of eight bands, since the suspect was heterozygous at all four loci.

have to use half of the only available DNA sample, in which the bands are already faint, to look for bacterial contamination?"

Yes, he admitted.

"Then you would run the risk of destroying the only available sample in order to test your pet theory, and perhaps cause a rapist to go free?"

"Objection!" yelped the defense attorney. "Sustained!" cried the judge. But Kiernan had made her point. She looked as pleased as Punch as the court recessed.

Shortly after this exchange the case went to the jury, who within a few hours found the defendant guilty on all twelve counts. When Kiernan interviewed the jurors afterwards, she found that they had not been impressed by any of the defense witnesses. They were primarily impressed by the match between the bands of the defendant and those of the assailant, something that they felt was obvious even without the aid of fancy machines to digitize the data. One juror said that he was "sick of all that stuff about degradation." So much for the fine points of molecular biology and population genetics!

VNTRs and other RFLPs are moving into the legal world with remarkable swiftness. So far as explorers of the human genome are concerned, however, this is only a sideshow. The most important current use for these and the millions of other polymorphisms that are scattered everywhere in the genome is to track down the genes that kill, disable, or damage their victims. And the most important future use, as we will see, will be to map and explore our five billion genomes in unprecedented detail.

CHAPTER 9

Racing After the Killer Genes

> *I have ever held that any accession whatever to the art of healing, even if it went no further than the cutting of corns, or the curing of toothaches, was of far higher value than all the knowledge of fine points, and all the pomp of subtle speculations; matters which are as useful to physicians in driving away diseases, as music is to masons in laying bricks.*
>
> —Thomas Sydenham, *Medical Observations*

About one male child in every three thousand worldwide is born with a terrible genetic disease called Duchenne muscular dystrophy. These children appear normal at first, but quickly they develop weakness in their muscles. The muscles swell abnormally, then begin to waste away. A child with this disease has great difficulty getting up off the floor, and shows a characteristic pattern of movements in which he must use his legs as supports, climbing painfully up them.

The disease is named after Guillaume Duchenne, the nineteenth-century French physician who first described it and who tried to cure it using electric shocks. From the time of its discovery to the present, the prognosis has remained uniformly grim. Eventually, the cardiac muscles become affected—more slowly than the skeletal muscles, but inevitably. Most of these children are dead by age twenty of heart failure or of a progressive inability to breathe as the muscles that control the movements of their rib cages waste away. About a third of them suffer from mental retardation, although nobody knows why. The rest keep their mental faculties intact to the end, keenly aware of each progressive stage of the disease.

Children with a milder and less common form of the disease, called Becker muscular dystrophy, can sometimes survive to middle age. While symptoms are the same as in the Duchenne form, they develop much more slowly. When the heart muscle weakens, heart transplants can be used to prolong the patients' lives. In addition, the muscles of these more mildly affected children waste away in a very odd fashion, some of the muscle fibers remaining intact while others break down. Some sufferers can actually slow the onset of the symptoms by exercise; one man, troubled by muscle weakness, became a body builder and developed huge, though not very effective muscles.

Almost all the children affected by both kinds of muscular dystrophy are male, for the same reason that Lesch-Nyhan syndrome primarily affects males. The gene, which is the same for both Duchenne and Becker, is recessive and located on the X chromosome. Mutant alleles are thus more commonly expressed in males, because males, having only one X chromosome, lack a normal X to cover up the effects of the mutant gene.

Why, when other severe X-linked diseases like Lesch-Nyhan are so rare, is muscular dystrophy comparatively common? There is no obvious reason. After all, children with the Duchenne form never live long enough to reproduce, so the mutant gene they carry dies with them. In order to keep the numbers of these alleles in the population so high, new mutant alleles must be arising all the time at a very high rate. This rate turns out to be between ten and a hundred times higher than the rate for most genes.

The numbers are grimly impressive. Each generation, one in every ten thousand X chromosomes acquires a new Duchenne muscular dystrophy mutation. There are 7.5 billion human X chromosomes on the planet at the moment, the majority of which will be passed on to the next generation. Seven hundred and fifty thousand of them carry these new Duchenne muscular dystrophy mutations, and another 2,000,000 or so carry mutations that arose a generation or two earlier and have yet to be revealed in males. Most of these will eventually result in one or more slow death sentences. This is an incredible load of suffering and heartbreak for the victims, and for their families in the generations yet to come.

Why is the X-linked muscular dystrophy gene so strangely fragile? The discovery of the answer to this question is one of the most remarkable success stories in the exploration of the human genome.

AN ENORMOUS GENE

The patients who have provided clues to researchers about the location of a gene or its effects are the unsung heroes and heroines in the war against genetic disease. In the late 1960s, when a child known as B. B. was born in Seattle and given up by his mother for adoption, little was understood about the nature and location of Duchenne muscular dystrophy, except that it was located on the X chromosome. By the time B. B. died at age sixteen in an auto accident, the general position of the gene on the X was known. His cells, preserved in the laboratory, have lived on after him, and this genetic legacy, more than any other, has enabled the muscular dystrophy gene to be pinned down and cloned.

Were it not for this contribution, the story of B. B. would be unrelievedly sad. He seemed at first to be a normal child, but as time went on not just one but many things began to go wrong. He began to develop chronic granulomatous disease, which affected his immune system. Lacking certain molecules on the surface of his red blood cells, he developed anemia. By the age of two, the retinas of his eyes had begun to show the progressive disorder and damage that signal retinitis pigmentosa, a disease that leads to night blindness. And even at that early age he was displaying the first signs of Duchenne muscular dystrophy. When he finally managed to begin walking, he could only walk for short distances.

Grim as this collection of problems was, nothing else was wrong with B. B. Through the remainder of his short life he showed normal intellectual development, and actually became quite a good musician until progressive muscle weakness prevented him from playing the organ. The murderous circumstances fate had handed him seemed random, but they were not. All the diseases he suffered from were genetic, and all of them were known to be due to mutant alleles of genes that reside on the short arm of the X chromosome. There was obviously something wrong with that part of the chromosome, something that had affected not one but several genes. To a geneticist, this could mean only one thing—a deletion. A piece of B. B.'s X chromosome must have been removed, carrying with it a number of genes that would have enabled him to grow and develop normally.

Ute Francke, a brilliant cytogeneticist at Yale University, found this deletion just where it would have been expected, on the short arm of B. B.'s X chromosome. It was very small, just barely detectable under her microscope, though as you know already even such a small deletion can remove millions of bases of DNA. Francke and her collaborators realized immediately that this discovery could lead them to the locations of all the diseases B. B. had suffered from. The greatest prize by far would be the muscular dystrophy gene.

Their first move was to isolate the deleted X chromosome and confer a kind of immortality on it. This involved hybridizing B. B.'s cells to hamster cells and then isolating a hybrid cell that carried only one human chromosome—that precious X with the tiny deletion. Meanwhile, and quite independently, small pieces of DNA known to be located on the short arm of the X chromosome had already been cloned in many laboratories. Francke and her collaborators collected these, turned them into radioactive probes, and looked to see which of them could be hybridized to the DNA of B. B.'s X. Any probe that did not hybridize would presumably come from the part of the DNA that had been deleted.

Nineteen probes out of twenty hybridized, but one did not. This was a probe that had been isolated by Louis Kunkel, a quiet, determined pediatrician who has worked for years on Duchenne muscular dystrophy at Children's Hospital of Harvard Medical School in Boston. He leaped early into the search for the gene and managed to move permanently ahead of the competition by a simple expedient: he found more and better DNA probes than anybody else, and found them earlier. So numerous did the probes become, and so certain was Kunkel of their location, that he was able to give them out to other labs around the world, recruiting these labs to help in the localization of the gene. These donations have given him an enviable reputation for altruism, though Kunkel cheerfully admits that had his own lab not been quite so far ahead, "this would have brought out my own competitive instincts more strongly!"

The probe Kunkel had given Francke had been derived from one of the early X-chromosome libraries produced at Livermore by chromosome sorting. He soon found from genetic studies that the probe was not in the muscular dystrophy gene itself, but it was obviously very close. He needed more probes, the more the better. If he saturated the deleted region with probes, some of them were sure to be in the muscular dystrophy gene. He and his coworker Anthony Monaco had a bright idea about how to do it. They asked Francke for a different cell line, one that had been isolated directly from B. B. and contained all the child's chromosomes, including the deleted X. They then proceeded to purify large amounts of DNA from this line.

They also isolated DNA from another human cell line, one carrying several copies of a normal X chromosome. They knew that this "X-enriched" cell line contained a small amount of DNA that would not match up with B. B.'s DNA, because its X chromosomes did not have the deletion. These were exactly the stretches of DNA that they wanted, because some of them would surely be parts of the muscular dystrophy gene. The trick was somehow to get rid of the huge number of pieces of DNA from the X-enriched line that *would* match up with B. B.'s DNA.

They broke up B. B.'s DNA into fragments by slurping it back and forth in a pipette. The small pieces of DNA produced were of random lengths and had blunt or ragged ends. Then they digested the X-enriched cell line's DNA with a restriction enzyme, producing small restriction fragments of very specific lengths, all having the same sticky ends. These ends would allow some of these fragments to be inserted later into a plasmid library.

Kunkel and Monaco heated both of their samples, breaking the double helices of the DNA apart to make all of it single-stranded. Keeping the DNA hot, they were ready to go after the muscular dystrophy gene. They mixed together a large amount of B. B.'s DNA, many genomes' worth, with a small amount of the X-enriched cell line's DNA, then allowed the sample to cool slowly. As the frantic thermal activity in the tube slowed, the fragments of DNA began to find their complementary strands and pair up, recreating the double helices that the heating had destroyed.

Because all these various pieces of DNA were of different lengths, all kinds of odd pairings happened—short pieces with short, long with short, long with long. There were almost always ragged bits of single-stranded DNA left over that stuck out at their ends. The odds were very great that no two of these fragments would pair up to give ends exactly like the sticky ends produced by the restriction enzyme. Because there was so much of B. B.'s DNA in the tube, both complementary strands were usually DNA from B. B., but sometimes one of them was a piece of DNA from the X-enriched cell line.

There was one exception to all this promiscuous pairing. Among the crowd were a few pieces of the X-enriched cell line's DNA from the part of the X that had been deleted from B. B.'s X chromosome. Thus, they had no complementary strands to pair with among B. B.'s DNA. Like wallflowers at a dance, they floated about until long after most of the other DNA fragments had finished pairing. Very occasionally, they would find a complementary strand among the other wallflowers. When these strands paired up they did not do so higgledy-piggledy, for their complementary strands were from the same restriction fragment. As a result, when they paired, they did so perfectly, even recreating the sticky ends. (As we all know, when wallflowers find love, the love is somehow deeper and truer than the casual pairings that take place elsewhere on the dance floor.) Of all the pairings that occurred in the tube, these were the only ones that produced the same size and kind of fragment that Monaco and Kunkel had originally produced with their restriction digestion of the X-enriched line.

When Monaco and Kunkel took this mass of DNA and tried to insert it into plasmid chromosomes, the only pieces that could be inserted successfully were these perfectly reconstituted fragments, for they were the only ones with the right sticky ends. And of course these were exactly the pieces they wanted, for they were pieces that came from that segment of chromosome that was missing in the B. B. DNA. Figure 9–1 shows a simplified diagram of how this extremely clever experiment worked.

Monaco and Kunkel ended up with nine new probes. Armed with these, and with some other probes isolated by different methods, they were ready to begin looking for the gene. They had a mapping problem, however, for while they knew that all their probes were in the deleted region, they did not know where they were relative to each other or which ones (if any) were in the muscular dystrophy gene.

Here luck played a role. They knew that, because of the high mutation rate for Duchenne muscular dystrophy, most patients carry different mutations. Perhaps—just perhaps—some of these mutations were small deletions, large enough to damage the gene but still so small that they removed just a part of the gene, as opposed to making the kind of huge deletion that caused B. B.'s overwhelming genetic troubles. If so, Monaco and Kunkel reasoned, some of their probes might fall within these tiny deletions. They began to collect DNA from Duchenne patients from all over the world, and matched these samples with all their probes. Out of fifty-seven such samples, most matched with every probe. There were five, however, that only hybridized with eight out of the nine probes. The probe that these five refused to hybridize with was always the same one, a little piece of DNA 200 bases long.

At last, Monaco and Kunkel had a toehold on the gene itself. These five patients each had a deletion of a part of the muscular dystrophy gene that included the region of the probe—a deletion far too small to be detected under the microscope. They quickly sequenced the probe. As it turned out, the toehold was enough. Beginning with this little fragment they could "walk" along the chromosome in both directions, looking for bigger pieces of DNA that hybridized to their precious fragment, then for pieces that hybridized to those pieces, and so on. Gradually they were able to move away from the vicinity of the probe, collecting the

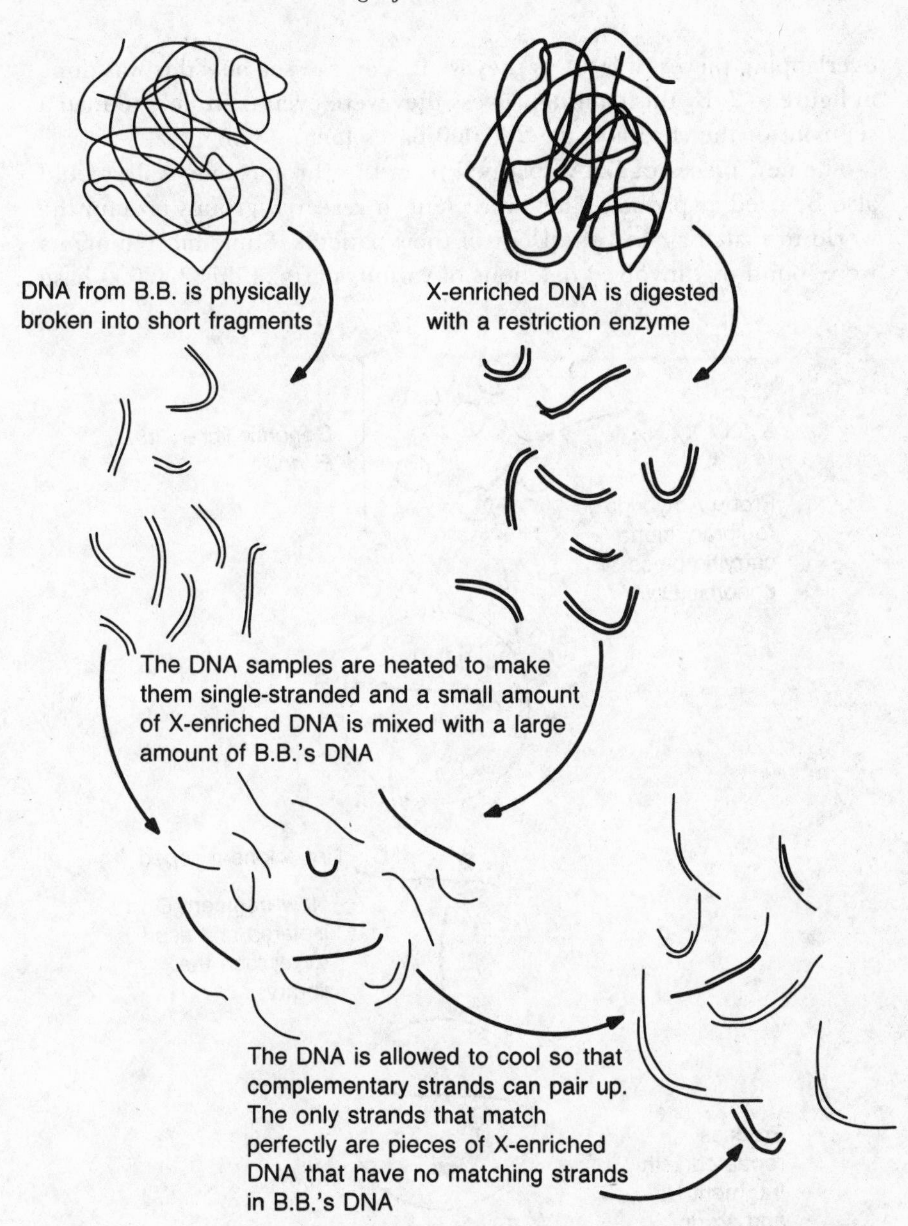

Figure 9–1. A sketch of the Monaco and Kunkel experiment. When the DNA from a cell line carrying several X chromosomes is mixed with B. B.'s DNA, the only pieces that anneal properly to give the right sticky ends are pieces from the part of the X chromosome that is missing in B. B.'s DNA. Because of their sticky ends, these are the only pieces from the mixture that can be cloned into plasmids.

overlapping pieces of DNA as they went. You can see how this was done in figure 9–2. By this painful process they were eventually able to map a segment of the chromosome 200,000 bases long.

The new pieces of DNA obtained from the chromosome walks could also be used as probes. They were sent to research groups around the world to match against the DNA of their patients. Some more mutants were found that involved deletions of various parts of this 200,000-base

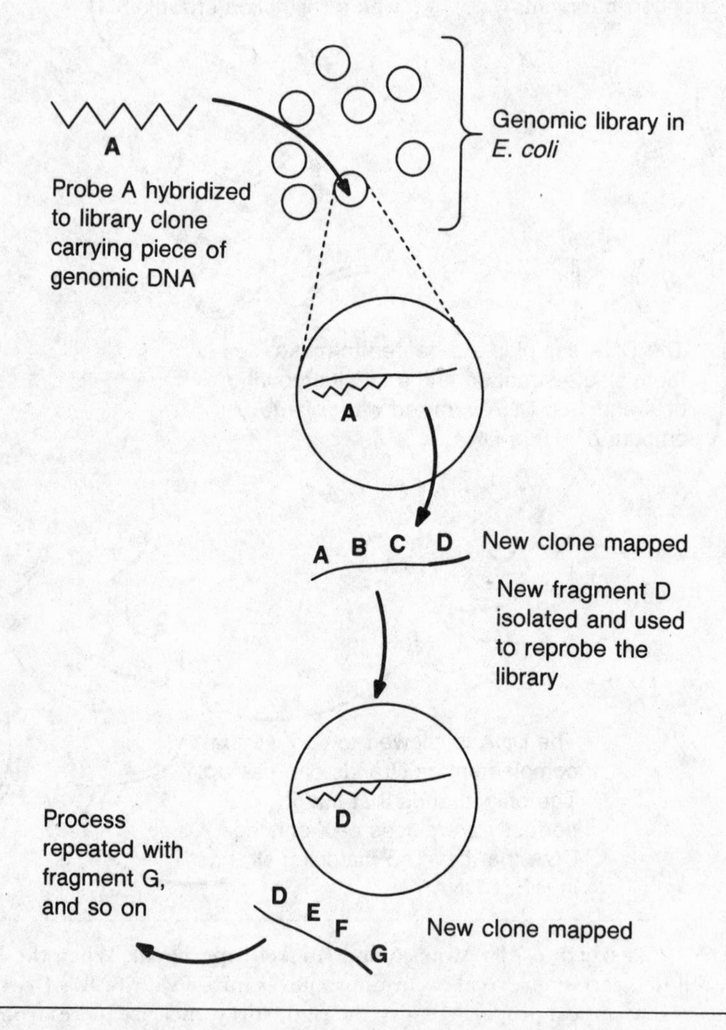

Figure 9–2. One can "walk" along a chromosome by isolating a series of overlapping cosmids.

region. In the meantime, however, genetic evidence was growing that the great majority of muscular dystrophy mutations seemed to lie outside this long mapped region—in both directions! The muscular dystrophy gene had to be huge, far bigger than any other gene yet discovered.

We have already seen that a gene can be much longer than it needs to be, because it is broken up by those long and apparently meaningless stretches of DNA called introns. It looked to Kunkel as if most of the muscular dystrophy gene must be made up of immense introns, which did not interest him. He wanted the *transcript* of the gene—the messenger RNA made up of the exons that carry the gene's message after all the introns have been snipped out. (You will remember that we watched such a snipping process being carried out by the little molecules called snurps during our exploration of the nucleus in chapter 5.) The transcript would be much shorter and easier to work with, and would allow Kunkel to determine the sequence of amino acids in the protein itself. He had every expectation that this would give a clue to the function of the protein, and perhaps even to the nature of the disease.

Separating introns from transcript was hard, however; without more information, it was like sitting in a dark room and trying to figure out which pages of a magazine carry stories and which are simply advertisements. There turned out to be a quicker way to find out more about the gene, by relying on how genes evolve.

Kunkel knew that humans are not alone in suffering from the ravages of muscular dystrophy. About one in three thousand males in other mammalian species also have the disease, although it affects different animals in markedly different ways. Dogs waste away and are dead within a few months. Cats exhibit hypertrophy of the muscles and stiffness, but live a normal life span.

Kunkel also knew that the exons, the transcribed fragments of genes, evolve much more slowly than introns. While the introns of even closely related organisms are often quite different from each other, the exons are very similar. To pursue our magazine analogy, comparing the same gene in two different organisms is like comparing two editions of a magazine in different languages. The stories might be on different pages, and all the advertisements would certainly be different, but even without knowing the languages it would be relatively easy to separate story from advertisement and even to determine which part of a story was on which page.

Thus, Kunkel began to hybridize the probes from his 200,000-base region of the gene to the DNA of other animals. As he expected, most of the probes did not match up with the DNA of any other animal he tried—which meant they had to be pieces of intron. But two did match, and they matched the DNA of many different mammals. He knew these must be pieces of exon.

The final step was to match these probes to a cDNA library constructed by reverse transcriptase from the mixture of messenger RNAs found in muscle cells. This was, of course, the logical place to look for the message from the muscular dystrophy gene. This search was successful; the rest of the gene was quickly pulled from the cDNA library and sequenced.

The success of this story is a tribute to the young patient B. B. Parts of him live on, in the form of cells and fragments of DNA that enabled Louis Kunkel and his hundreds of collaborators from around the world to track down the gene causing his devastating disease. It is a remarkable memorial.

THE PAST IS PROLOGUE

In the few brief years since Kunkel's breakthrough in 1985, the muscular dystrophy gene has yielded up many of its secrets. It is now known to extend an astonishing 2.5 *million* bases along the chromosome. The protein coded by the gene is huge, but even so it accounts for only about eleven thousand bases, half a percent of the total, scattered in tiny exons along the gene's length. The rest of the gene is made up of more than sixty-five introns, many of them big enough to conceal dozens of other genes, although nobody yet knows whether they do. A similar gene is found in many other organisms going at least as far back in evolutionary time as fish and probably much further. It seems to have retained its enormous size all that time.

The protein coded by the gene is now known as *dystrophin.* It is rod-shaped, and the complete amino acid sequence shows that it resembles many other proteins that make up a kind of framework of the living cell, known as the *cytoskeleton.* We glimpsed the cytoskeleton in the form of those fibers we saw everywhere during our tour of the cell. Almost all the mutations leading to muscular dystrophy are deletions. Duchenne deletions destroy the gene completely because they throw off its "reading frame" of triplet codewords. A fragment of protein terminated by a string of incorrect amino acids results. The milder Becker deletions snip out a chunk of the gene in such a way that the protein is shortened, but the reading frame is preserved.

Much is now known about the distribution of dystrophin in the muscles. Its function seems to resemble that of the giant molecule titin; it may form a kind of anchor for muscle filaments, so that when it is missing the muscle fibers lose their integrity and become susceptible to various kinds of damage. If one thinks of dystrophin as being like a bungee cord with hooks at both ends, one can imagine the Duchenne mutations removing one end of the cord along with its hook, rendering it useless. The Becker mutations snip out a piece from the middle, shortening the cord but allowing it to function after a fashion.

Why should we have such an enormous, fragile gene? If it were not so immense, the mutation rate would be far lower, simply because the gene would present a smaller target for the mutational mistakes that lead to deletions. There must be some very good evolutionary reason why this gene has stayed so huge and vulnerable over hundreds of millions of years, but we have no idea what it might be.

Kunkel's group is not resting on its laurels. Alan Beggs, a postdoc in the lab, has examined hundreds of Duchenne and Becker mutants, and is developing rapid screening methods for them. These methods are already proving useful in intrauterine diagnosis when the mother or a close relative has already had a child with the disease so that the nature of the mutation she carries is known.

It is possible that children with muscular dystrophy will be helped in the near future by a primitive but promising therapy: the injection of healthy immature donor muscle cells into weakened muscles. There they will grow, mature, and colonize the wasted tissue. Experiments with mice, some of them carried out by Kunkel's group, have shown this treatment to be effective, and teams of doctors in the United States and Canada are now cautiously trying the therapy on human patients. Results have been promising, but how well the new muscle tissue will survive the body's effort to reject it and how many such injections will be needed to improve the patient's ability to breathe and move are questions that have yet to be answered.

Kunkel, dissatisfied with these crude therapies, is moving on to the next set of problems. What proteins do dystrophin attach to? Other kinds of muscular dystrophy are known, including one type common in Japan that is caused by a gene on another chromosome. Victims of this disease make normal dystrophin, so they must be missing another important link in muscle structure. Kunkel's group is trying to pin down this link.

Eric Hoffman, a postdoc in the group involved in the cloning of the dystrophin transcript, is fascinated by the fact that cats and mice are so much less affected by mutations in the dystrophin gene than are humans and dogs. If the muscles of a normal cat are injured repeatedly, he points out, they continue to regenerate—perhaps one source of the folk belief in a cat's nine lives! The muscles of a normal dog, by contrast, will soon stop regenerating if they are treated in the same way, resulting in a permanent injury. Hoffman would love to find and understand the genes that permit the regeneration in cats, for this might open up a path to genetic therapy.

Kunkel himself is excited about these new directions, but admits that the toughest work is yet to come and the competition will be intense: the gene's discovery permitted hundreds of labs to join in the chase to understand the disease. He is also unhappy about the Human Genome Project. Like Martin Rechsteiner, he characterizes it as a mindless attempt to sequence enormous stretches of DNA. He resents the fact that good research proposals to attack specific human diseases are being turned down. Had he waited for the Project to hand him the sequence of the dystrophin gene, he would still be waiting. Instead, through a combination of skill and luck, he gained a fifteen-year head start on conquering the disease.

A GENETIC ICEBERG

When I met him, Kevin Riley was a slight young man with fair hair and a face with the transparency of a Botticelli angel. Born in 1960 in Jacksonville, North Carolina, he faced a situation in the three decades that followed that for most of us would be beyond imagining.

Kevin's father, a Marine, was a dominant, hard-drinking man given to bouts of depression. This frail and listless boy hardly fit his idea of a real son. Kevin's appetite seemed unimpaired, but he gained weight only slowly. At age one, his health problems came to a head; he contracted severe pneumonia. The doctors at the base hospital became increasingly worried by his lack of response to antibiotics. His mother, frantic, read through the chapters on childhood diseases in Dr. Spock, and she came across one that seemed to match some of Kevin's symptoms. There was a simple test for it, which she persuaded the reluctant doctors to perform. They wrapped Kevin's arm tightly in plastic and collected some of the sweat that accumulated underneath it. The sweat was salty—about twice as salty as normal. Dr. Spock's long-distance diagnosis was accurate: Kevin had a genetic disease called cystic fibrosis. With this discovery, the pattern of his symptoms fell into place.

Cystic fibrosis, known as CF for short, is a disease of secretion. The major problem seems to be that cells affected by the disease are unable to transport chloride ions properly from one side of their outer membranes to the other. First described in the 1930s, CF is the commonest serious genetic ailment affecting Caucasians, though it is rare in other racial groups. About one in every 2,500 Caucasian babies is born with CF, making it slightly more common than the X-linked Duchenne muscular dystrophy, though its commonness is due to a very different genetic reason. You will remember that because the muscular dystrophy mutations are X-linked and recessive, they can be masked by a normal allele on the other X in females, but are soon revealed whenever they are passed on to males, because males have only one X chromosome. The males die before they can reproduce, quickly removing their mutations from the population. The disease remains common, however, because new mutations arise at such a high rate.

Cystic fibrosis mutations, in contrast, lurk in the population like a kind of genetic iceberg, and can persist for thousands of generations. Also recessive, they can be concealed in both sexes rather than just one, because the CF gene does not lie on the X chromosome. Instead, it is located on one of the other twenty-two chromosomes that make up the greater part of the human genome. Unlike the X, these other chromosomes are found in two copies in both males and females. As a result, both males and females can be symptomless heterozygous carriers of the mutant CF allele. One in every twenty-five Caucasians is such a carrier.* For the disease to appear, the victim, either male or female, must have two copies of the mutant allele. This is a hundred times less likely than being a carrier. Yet this is what happened to Kevin.

After his diagnosis, Kevin's disease developed in the classic CF pattern. Although only his CF gene was affected, several of his organ systems fell victim to the disease. Unable to secrete digestive enzymes properly, his pancreas developed cysts full of broken-down enzyme products, which in turn led to fibrous degeneration of the entire organ. (Indeed, the disease gets its name from these effects on the pancreas.) As a result, he required special enzyme pills to digest his food properly. And he was sterile, for a similar degeneration affected his ability to produce sperm. Given modern therapy, neither of these conditions is

*The arithmetic of the situation is straightforward, and once more illustrates the Hardy-Weinberg Law. Unsuspecting carriers may marry each other. The likelihood of such a marriage is $1/25$ times $1/25$, or one in every 625 marriages. One quarter of their children will be affected, giving the 1 in 2,500 rate for the disease itself.

life-threatening; the killer component of CF is its effect on another organ, the lung.

Kevin's lungs, like those of all CF patients, were his weakest point. Again, the problem can be traced to secretion. The lungs normally secrete sticky mucus in which foreign particles brought in with the air become embedded. The mucus is carried out of the lungs by the beating action of billions of tiny cilia, and when it reaches the back of the throat it is swallowed. But Kevin's mucus had the wrong composition. It was too heavy and sticky, and accumulated in the network of narrow passages called bronchioles. Some of these passages became blocked off, and behind them bacterial infections developed. Indeed, because of these ever-present islands of infection, the bacterial populations of the lungs of CF patients are quite different from those of healthy people.

The primary treatment for CF is time-consuming and exhausting. A vibrator is used to break up the mucus in the lungs, which must then be brought up by painful coughing. This has to be done as often as the patient can stand it. Kevin's youngest brother also has CF, and was introduced to this disagreeable ritual at a younger age. As a consequence, he is healthier than Kevin was at the same age.

Early neglect may have accelerated the course of Kevin's disease, but the affected lungs of CF patients, no matter how much treatment they receive, gradually become weakened as a result of continual infections. Kevin's next severe crisis came when he was a student in graphic design at San Diego State University. He began coughing up large quantities of blood; it turned out that a bronchial artery had ruptured. The prognosis was grim: he was told that if he didn't stop bleeding he would die. Luckily, he was seen at that point by Ivan Harwood, an expert in pulmonary medicine at San Diego's University Hospital. Harwood recommended a drastic treatment—the closing off of the bronchial arteries (which turn out to be dispensable) with a plastic material.

The treatment worked, but Kevin's abused lungs continued to fail. Although he was able to work briefly after college, he was soon on total disability. He used oxygen almost continuously, particularly when driving, a therapy that probably accelerated the degeneration of the lung tissue. Finally he took a dramatic step and applied for a heart-lung transplant. He carried a beeper at all times, to alert him the moment a potential donor became available.

A heart-lung transplant is the final therapy for CF patients. The operation can be performed at relatively few places around the country. Not enough of these operations have been done on CF patients to determine success rates, but Kevin talked to one survivor of the procedure who told him, "Just for the first breath, it's worth it!"

When I talked to him, Kevin was relaxed and quite unafraid of his coming ordeal. For him, life had been reduced to its starkest terms. He could afford to be dispassionate about his condition, to the point where he could even talk with wry amusement about the blessings that CF had provided. He ticked them off on his fingers: His father was an alcoholic, as were both of his father's parents and two of his own brothers. He recognized the need for dependency in himself, and felt sure that were it not for CF he would have been an alcoholic too. The disease had also made his emotional life richer and more fulfilling than he thought might have been possible otherwise. And—here he grinned—he was the only member of his family with a college degree.

Like Kevin, sandy-haired Steve Shepherd is slightly built, but he has a vigorous air about him. Although he too is a victim of CF, he has been far less marked by it, although a trained eye would note immediately the enlarged barrel chest and splayed fingertips characteristic of the disease. At the time we talked, Steve was thirty-five years old, and could look forward with some caution to a nearly normal life span. "Though I don't expect to spend long sunset years fishing on a lake!" he remarked.

Steve was diagnosed at age one in England, where his father, a university professor, was spending a year with his family. As an infant he ate voraciously but gained no weight—the sign of pancreatic failure, though his lungs were not involved at that time. His failure to thrive was corrected by digestive enzyme tablets, and the disease did not affect his lungs noticeably for many years. Remarkably, he was even able to run cross-country in high school. His first real bout with the disease came in college, when he suffered an attack of acute pancreatitis. His doctors attributed this to alcoholism, even though Steve assured them he did not drink. He tells of his frustration at not being believed when he told them he had CF; after all, the disease had been diagnosed decades before in a country afflicted by socialized medicine, and no records were available. To make the circumstantial case for his alcoholism worse, he had been putting himself through school by working nights as a bartender.

When, a year after this episode, he began coughing up large amounts of blood, Steve was finally seen by Ivan Harwood. Harwood was able to diagnose him immediately by his appearance alone. Steve's form of the disease was an unusual one. In most affected people, CF takes one of two courses. In both types the lungs are affected, but in one of them the pancreas is not. Steve's form did not resemble either of these; the pancreas was affected early, but damage to the lungs was slow to appear.

Now a science writer for local media, Steve is knowledgeable about the disease and has written a good deal about it. It happened that we had independently stumbled on a paper by an Australian doctor, J. G. O'Shea, who had suggested that Frédéric Chopin had died not of tuberculosis but rather of CF, and we talked at some length about the suggestive but fragmentary evidence O'Shea had turned up.

Though the disease has threatened his life, Steve is less involved in activities centered around CF patients than he was at first. At one time he volunteered as a counselor for the CF summer camp, and helped to edit Harwood's videotapes of young patients, which he found a moving and intense experience. The fact that he can lead an approximately normal life has tended to separate him from most CF patients. He feels both guilt and relief that the finger of fate has touched him so lightly.

To find out why Kevin and Steve were affected so differently, and why some children with the disease are even more severely affected, it is necessary to examine the gene itself.

THE BIG PAYOFF

Were the human genome already sequenced, tracking down the cystic fibrosis gene would be quite straightforward. As it is not, the result has been a frantic race that has wasted resources, opened the door to endless litigation, and left a legacy of bitterness that may actually endanger some aspects of the Human Genome Project itself. But behind all the acrimony and quarrels reported in the newspapers is a remarkable story of hard work and ingenuity that shows it is quite possible, even at the present time, to track down any gene in the human genome if the desire is great enough.

We know that the gene for muscular dystrophy is probably located on the X chromosome because almost all the disease's victims are males. The X carries only about 5 percent of the genes in the human genome, however; the majority of genes are located on the other chromosomes, known as *autosomes.* There are twenty-two of these, and the cystic fibrosis gene is on one of them. How can the CF gene be tracked down on these vast stretches of DNA without any obvious clue, like X-linkage, indicating its whereabouts? At the outset of the search for CF, family studies provided the only signposts.

While thousands of family histories of children with CF have been collected, most of them present problems for the geneticist. Humans marry whom they choose, and most have few children. As a result, in most cases the affected child appears to be alone in an otherwise healthy family. Usually this is not because the child carries a new mutant, as in muscular dystrophy, but rather because the mutant CF alleles are recessive and have been passed down in hidden form through heterozygotes on both sides of the family.

A few families are far more informative. These are large families in which several children are affected. In such families, the genotypes of the victim's unaffected parents, grandparents, uncles and aunts, and sometimes even cousins can often be inferred. It becomes possible to know with some assurance whether these people are heterozygous for a normal and a mutant allele, or homozygous for two normal alleles.

At the Hospital for Sick Children in Toronto, Canada, there is a collection of records of over 600 CF families, many of whose children have been treated at the hospital. A large proportion of the family members, both with and without the disease, have donated blood samples to the hospital collection. Since the hospital serves all of Eastern Canada and parts of the United States, this represents the largest single data base of family histories of this disease, although many similar data bases exist in other parts of the world.

How can such a data base be used to search for the location of the CF gene? Ideally, it can be used to find out whether CF is *linked* to some other gene on a chromosome—that is, whether another gene is so near to CF that alleles at these two loci are seldom broken up by genetic recombination and so tend to be passed together to the next generation.

It would be very nice if there were some obvious functional arrangement of the genes on the chromosomes. If all the genes affecting the lungs were on one chromosome and all those affecting the eye were on another, the geneticist's task would be greatly simplified. This is not the case, however; most of the time, genes that are near each other on the chromosomes have totally unrelated functions. Until recently, this meant that the human geneticist was forced to look for the simultaneous occurrence of several very unlikely events in order to detect genetic linkage. First, families had to be found in which two different genetic diseases or characteristics could be followed from one generation to the next. Second, and more improbably, these two different characteristics had to be genetically linked to each other—no more than twenty million base pairs apart in a genome of three billion pairs. Third, several such families had to be found in order to have enough data to perform a convincing statistical test. You can see how unlikely it is that all these conditions could be met simultaneously.

In view of all this, it is not surprising that the first linkage between two autosomal genes was not detected until 1955.* And it is also not surprising that when the hunt for the CF gene began in earnest in the early 1980s, nobody had any idea which chromosome it was on, in spite of all the family data.

The first big breakthrough in locating genes on the chromosomes came with Frank Ruddle's development of the human-hamster and human-mouse hybrid cells that have been so useful in chromosome sorting, and that played an important role in tracking down the muscular dystrophy gene. In a mouse or hamster cell line with only one human chromosome, or better yet just a fragment of a human chromosome, it is sometimes possible to detect whether a human gene is on that fragment. At the outset, this approach worked only if the protein coded by that human gene happened to be made by these tissue culture cells, and if the human protein was distinctly different from the same protein coded by the mouse or hamster. Advances in molecular biology have now made it possible to detect many more human genes in these cell lines.

Gradually, using these lines, more and more genes have been assigned to various autosomes. By 1974, the number of genes whose autosomal location was at least roughly known surpassed the number of known sex-linked genes for the first time. The location of the cystic fibrosis gene, though, remained a mystery. The protein product of this gene could not be detected in cells grown in tissue culture, since nobody had any idea what to look for. As a result, the hybrid cell trick could not be used. Linkage to CF could only be searched for in those few families in which another gene with a known location could also be followed. Such slender evidence was used to locate the CF gene on chromosome 4 and later on chromosome 13. Both these guesses turned out to be wrong.

What was needed—what has always been needed in human genetics—was a greater number of *markers,* genetic signposts in the genome. With enough markers, any gene can be tracked down. The race to a gene has

*Even so, the dice were heavily loaded in favor of detecting it. One of the two genes, the one that controls the ABO blood groups, has four common alleles, several of which can be followed in most families. The mutant allele at the other gene, which causes simultaneous malformations of the fingernails and kneecaps, is dominant. This means that every occurrence of it can be followed in the few families that carry it. Thus, almost every family exhibiting this nail-patella syndrome could be used in the search for linkage to the ABO gene.

always been won by the lab with the most markers. This was where RFLPs came into their own, for they provided needed markers. The first RFLPs were detected in the late 1970s, as the genomic Southerns that we met in the last chapter began to be performed widely on humans and many other organisms. Population geneticists, knowing how genetically variable the human species is, had expected RFLPs to be everywhere, but their prevalence came as a surprise to molecular biologists, who were nevertheless very quick to exploit them.

In a paper published in 1978, Yuet Wei Kan of the University of California at San Francisco had suggested as something of an afterthought that RFLPs could be used as markers to track down genes. Later, in 1980, David Botstein, Ray White, and several collaborators made the same suggestion. Theirs was the first paper to catch the attention of the human genetics community. They pointed out that RFLPs have enormous advantages for such a hunt, advantages which they ticked off.

First, they were numerous and seemed to be everywhere on the chromosomes. Second, they were easily detectable using DNA from blood. Third, most of them did not make their carriers sick, like so many of the mutant genes that human geneticists had been forced to use up to that point. This meant that perfectly healthy families could be used to gather data on linkage, for RFLPs could be linked to other RFLPs just as easily as they could be linked to disease genes. Any disease gene could of course be fit into the RFLP linkage map, and as the map grew more and more dense with RFLP markers, this task would become progressively easier.

Botstein and White took this terrific idea and ran with it in different directions. Botstein, a professor of molecular biology at MIT, became one of the scientific advisers to a company specifically designed to hunt for RFLPs. This company took the name of Collaborative Research.

White, a human geneticist at the University of Utah, had already spent years working with the genetics of Mormon families. These families were as ideal as any human material could be for linkage studies. They tended to be very large, to stay in one area, and to keep in touch with each other. Many of their members were well educated, and they all tended to be cooperative in nitty-gritty matters like donating blood and filling out questionnaires. As we will see later, the geneticist Eldon Gardner had pioneered the study of these families soon after World War II, setting in motion one of the early triumphs of human genetics by tracking down dozens of members of one extended family that carried a gene predisposing them to colon cancer. Early removal of the colon, while drastic, stopped the developing cancer in its tracks and saved their lives. As a consequence, the whole Mormon community was very receptive to White's endeavor.

The work was mind-numbing. It involved hunting through thousands of genomic Southerns for useful RFLPs. Any clue to where the RFLPs might be was seized. Many promising probes were isolated from the early chromosome-sorted libraries obtained from Los Alamos and Livermore. Because these libraries, crude as they were, were greatly enriched for the DNA of a particular chromosome or part of a chromosome, RFLPs recognized by probes from these libraries could be tentatively assigned to that same chromosome or chromosome fragment.

Hybrid cell lines, too, could be used to locate probes on individual chromosomes. Whole sets or *panels* of human-mouse and human-hamster cell lines were developed that carried human chromosomes and parts of human chromosomes in various combinations. An unknown probe could be hybridized to a series of DNA samples from these panels, and the pattern of successful hybridizations would give its chromosomal location through a process of elimination. Once the assignments were made, RFLPs were followed through dozens or even hundreds of family members to look for linkage. (One of White's families consists of hundreds of members extending over six generations.) Promising probes were sent all over the world to groups investigating families that carried disease genes, and information that came back was used to fill in the map.

While White concentrated on building maps of particular chromosomes, Collaborative Research under Botstein's guidance took a more global approach. A large group headed by Helen Donis-Keller began to screen a random human library of probes in an attempt to build a linkage map of every human chromosome. This immense project, started in 1981, was eventually successful, though it consumed some ten million dollars of Collaborative's resources. One of their most important goals, from the very beginning, was the cystic fibrosis gene.

Why would a commercial company spend so much money on a project that at first sight looks simply like an exercise in pure science? Because the payoff is potentially huge. This is due to the genetics of cystic fibrosis—so very different from the genetics of muscular dystrophy.

There is at the moment no easy way of predicting the appearance of muscular dystrophy. Because of the high mutation rate at this gene, many cases appear without warning in families with no previous history of the disease. With such a heterogeneous collection of different muscular dystrophy mutations in the population, it is impossible to devise a single test to determine whether a mother carries a mutant gene or not. Consequently, such tests are confined to families in which one child has already been born with the disease, alerting parents or other relatives to the possibility that the next child might be affected. Alan Beggs, the postdoc in Kunkel's lab who is trying to develop tests, must use a great deal of ingenuity in each such case to figure out which of thousands of possible mutations the mother might be carrying. Such tests could not feasibly be performed on the population at large in order to pick up the 1 woman in 2,000 or so who carries a mutant allele.

Cystic fibrosis, on the other hand, is autosomal. One Caucasian child in 2,500 develops the disease. Because both sexes can be heterozygous for the gene, however, 98 percent of CF mutations are concealed. A cystic fibrosis mutation will be hidden in the population for an average of fifty generations before it is revealed, as compared with two or three for a muscular dystrophy gene.

This raises the possibility that cystic fibrosis, like sickle cell anemia, is caused by a single kind of mutation, a mutation that might have occurred once in the distant past, and from which all the current alleles are descended. Which in turn means that a simple test might reveal all the heterozygotes. Every Caucasian on the planet who is thinking of having children would be a potential customer for such a test. A market like that is well worth the expenditure of a paltry ten million dollars. This potential payoff explains why Ray White freely sent out his probes to anyone who asked for them, while Collaborative Research only sent theirs to people with whom they had signed elaborate agreements. One of these was Lap-Chee Tsui.

WHICH CHROMOSOME?

At the time that Botstein and White published their paper on RFLP mapping, Lap-Chee Tsui (pronounced Choy) was a young postdoc at Toronto's Hospital for Sick Children. He spent the subsequent ten years there, rising to the position of senior scientist while closing in on the CF gene and racing several other groups for the prize.

Tsui's tiny office opens directly off his crowded lab on the top floor of the hospital, a massive building in downtown Toronto. The whole lab rocks alarmingly every few minutes as helicopters carrying emergency patients land and take off. Tsui, however, is relaxed and open as he fills me in on the dozens of projects the lab is in the middle of now that the gene has been found.

Things were not always this relaxed. Tsui's first five years on the project were spent tracking down which chromosome the gene was on. Out of the 600 families in the hospital data base, he picked 50 that would be likely to provide the most genetic information. Then, working entirely blind, he tested all the probes Ray White could send him, developed other probes himself, and begged probes from every other source he could track down. None of the RFLPs detected by these probes showed any linkage with CF.

By 1984, Collaborative Research had developed 200 probes scattered all over the genome—four times as many as Tsui had managed to put together up to that point. Tsui had the families and they had the probes, so getting together was natural. The subsequent course of their collaboration was traced in detail some four years later by Leslie Roberts in a series of investigative articles in *Science.* By an incredible stroke of luck, one of the first probes Collaborative sent to Tsui showed linkage to CF in his families. Collaborative had not yet determined, however, which chromosome the probe itself was on. Tsui was told that he should sit tight and they would locate it shortly.

Localization involved hybridizing the probe to a panel of DNA from hybrid cell lines that contained all the human chromosomes in various combinations—a very simple experiment. Weeks went by, and Tsui, who had a similar panel in his possession, seethed with impatience. Helen Donis-Keller eventually told him she thought the probe *might* be on chromosome 7. Tsui, maddened by uncertainty, could stand it no longer and did the experiment himself. When he told Donis-Keller that he had confirmed the probe's location on 7, she was furious.

Then followed a rather ghastly period, in which everybody ended up mad at everybody else. Collaborative insisted that Tsui, scheduled to give a talk a month later at a meeting in Utah, should say only that he had found linkage to CF, without mentioning which chromosome the probe was on. By the time Tsui got to the meeting, however, rumors were already flying. Both Ray White and Tsui's other chief competitor, Robert Williamson of St. Mary's Hospital in London, had heard the rumors. White cheerfully admitted that as a result he had immediately reoriented the priorities in his lab, concentrating on probes that he knew to be on chromosome 7. Williamson later insisted that he was led to chromosome 7 simply by a process of exclusion; probes on most of the other chromosomes had shown no linkage, making 7 one of the few likely chromosomes left. Tsui still disputes this claim.

Regardless of the role played by rumor, White and Williamson soon had papers ready to submit to *Nature.* In a showdown with Collaborative, Tsui insisted that something be done. Donis-Keller then phoned the offices of *Nature* and threatened a major blow-up if they did not give Tsui and Collaborative priority over the White and Williamson papers. The editors at *Nature* refused to delay the papers, but pointed out that there was still time for a hasty submission from Tsui and Collaborative. Donis-Keller frantically wrote a paper, and all three were published simultaneously. Although the papers themselves, like most scientific writing, showed no trace of the intense rivalries involved, they were accompanied by an editorial in which assistant editor Peter Newmark recounted the role played by rumor and clearly awarded the victory to Tsui and Collaborative.

Several positive things came out of all this. Collaborative, for one, had learned that they did not have an exclusive license to stretches of the human genome—the technology was too straightforward and too many probes were generally available for that. As a result, when Donis-Keller and her coworkers at Collaborative published their magnificent mapping paper two years later in 1987, a paper in which 400 markers were mapped to all 23 chromosomes, they announced at the end of the paper that all the probes and data would be made freely available.

The frantic mapping activity of White and Williamson had also paid off. Each group had found probes that mapped much closer to the CF gene than Tsui and Collaborative's original probe, which was a remote fifteen million base pairs away. Both groups were now poised to swoop down on the gene itself.

LEAPFROGGING TO THE GENE

The probe that White had found to be so closely linked to CF was not an RFLP at all, but a well-known oncogene called *met,* which was widely available. Williamson knew that his own RFLP probe was on the other side of the CF gene from *met,* and so he took a shortcut. He started with a human cell line in which the *met* gene had been activated. When an oncogene like *met* is turned on in a line of cells, they take on some of the attributes of cancer cells, piling up into great heaps in their petri dish instead of growing in a single layer like noncancerous cells. This process is known as *transformation,* a term used by students of tissue culture cells that should not be confused with the process of bacterial transformation by which bacteria take up plasmids or cosmids. In this cellular kind of transformation normal cells are transformed into cancerlike cells.

Williamson broke up the chromosomes from these *met*-activated cells and persuaded untransformed mouse cells to take them up. Remarkably, the versatile *met* gene will transform mouse cells as well as human ones. Some of the mouse cells did grow in heaps, so he knew that they must have picked up the human *met.* He isolated DNA from them in order to determine whether this DNA carried a sequence corresponding to his RFLP probe, which he knew was on the other side of CF. One of the lines did. This line, Williamson hoped, would carry all the DNA lying between the two probes, including the CF gene, and very little else.

This scheme was ingenious but doubly dangerous. The *met* gene, he knew, could scramble DNA, and so could the method he used to get human DNA into mouse cells. After all this, would the CF gene still be there? His critical line carried about a million bases of human DNA, far too much to sequence. Most of it, he knew, was introns and the long stretches of DNA between genes. He had no idea how many genes in addition to CF were contained in this chunk of the chromosome.

He began an intensive search for some "islands" of unusual DNA that are often characteristic of genes. One of these islands had the characteristics he was looking for. Followed in families, RFLP markers near the island always behaved as if firmly linked to the CF gene. The island itself, he found, was expressed in the right tissues: the pancreas and the lungs. And it was so tightly linked to the CF gene that one of his RFLP markers was nearly always associated with mutant CF genes in his collection of families.

Delighted, Williamson rushed into print in *Nature* in April 1987 with what he termed "a candidate for the cystic fibrosis locus." This caused consternation in the scientific community. Had the gene been found? The preliminary data looked convincing, and Ray White, for one, gave up his search. But Tsui did not. While Williamson's paper showed that his candidate gene was closely linked to CF, there was a glaring omission. None of the sequence of the gene had been published. Was there any difference between the mutant and the normal gene? Only the DNA sequences themselves would show that. Tsui also knew that while Williamson had charged into the midst of that part of the chromosome using facile tricks and in the process *might* have scooped up the prize, his own slow and methodical approach would certainly find the gene sooner or later. When he did, he would know it beyond a shadow of a doubt.

Like Williamson, Tsui began with the two flanking markers: *met* and the marker Williamson had isolated that lay on the other side of CF. He wanted to fill in the gap, however, and so he looked for more probes. He started with four thousand different clones from a chromosome 7 library from Los Alamos. To winnow them down, he made use of a human cell culture that had lost a small piece of one of its two copies of chromosome 7, a piece that included *met* and Williamson's marker and therefore also had to include the CF gene. When he hybridized his probes to DNA from this cell line, some of them gave bands that were only half as strong as those of others. These, he knew, had to be in the deleted region. Of the twenty-five clones in this category, two were associated with informative RFLPs, and when he followed these RFLPs in his families they mapped between *met* and Williamson's marker. In fact, the two probes were only about ten thousand bases apart. They lay half a million bases from *met* in one direction and a million bases from Williamson's marker in the other. The CF gene had to be somewhere out there in one of those huge stretches of unknown DNA.

Starting with these new probes, Tsui began the painful process of "walking" along the chromosome in both directions. Genetic evidence soon indicated that CF was in the direction of Williamson's original probe, so he was thankfully able to abandon one of the two walks and concentrate on the other. The process of "walking," you will recall, requires the systematic isolation of cosmids that overlap the original probe, then the isolation of cosmids that overlap those in turn, and so on. It is possible in this way to inch down the chromosome, carefully saving the pieces of DNA as you go.

Unfortunately for the would-be hiker along the chromosome, there are pitfalls. They come in the form of those "poison" or "unclonable" regions which are not found in cosmid libraries, and which make the lives of Tony Carrano and the others who are trying to construct contig maps so difficult. When such a region is reached the walk stops, for the next cosmid in the walk cannot be found. The only thing to do is to jump over this region and continue. Tsui needed the help of a chromosomal athlete, somebody with the expertise to jump but not to jump too far. Hans Lehrach, while at Heidelberg, had pioneered methods for jumping, but he was not alone. Francis Collins of the University of Michigan had developed a similar technique, and it was to him that Tsui turned for help.

The trick is to take a big piece of DNA that spans the poison region and insert it into a plasmid. The plasmid that results will be far too big to survive in *E. coli,* but it can be reduced to a more manageable size by immediately removing most but not all of the inserted DNA. This smaller plasmid will now include only the two ends of the big piece of DNA. The walker can then use this plasmid as a probe, effectively jumping from one end of the big piece of DNA to the other while ignoring all the difficult and unclonable regions in between.

The technique does have its dangers. It is easy to make the mistake of joining two unrelated pieces of DNA together. There were times when Collins and Tsui jumped and found themselves on some unknown chromosome—sometimes even on a hamster chromosome! In order to keep track of where they were going, Tsui used the pieces of DNA from his walk and mapped them in a different way. He digested human DNA with enzymes that cut it only rarely, producing huge pieces. Then he separated these pieces on a gel using the technique of pulse-field electrophoresis, which is very slow but can separate pieces of DNA up to a million bases long. Finally, he hybridized the pieces of DNA from his walking and jumping clones to these gels and determined the size of the pieces to which they hybridized. If they did not hybridize, or if they hybridized in the wrong place, then he knew that they had jumped too far. Walking and jumping, Tsui and Collins proceeded methodically down the chromosome, double- and triple-checking at every point.*

Meanwhile, Williamson's story was coming apart. By this time he had obtained sequences of his "candidate" gene from two humans, one with CF and one without. There were no differences in the sequences that would explain the mutation. Further, the protein sequence coded for by the candidate gene did not look like anything he had expected. Because CF expresses itself as a defect in the transport of small molecules across the cell membrane, the CF protein was expected to be located in the membrane itself. Proteins found in the membrane, however, tend to have stretches of "waxy" amino acids marking the regions where they are embedded in the fatty membrane material, and Williamson's protein did not have such stretches.

Other groups, particularly White's, were furious that Williamson had published so hastily and caused them to give up the chase. Half-hearted efforts were made to catch up, but by this time Tsui and Collins were well

*Some of these poison sequences have still not been cloned in spite of all this intensive work, a fact that bodes ill for groups trying to make contig maps of entire human chromosomes.

out in front. They knew much more about this region of human DNA than anybody else. Tsui was taking each new piece of DNA from his walk or jump and hybridizing it to the DNA of many different animals—the same trick Kunkel had used to look for exons of the muscular dystrophy gene. Tsui was relying on the fact that exons tend to be conserved during evolution, while the much longer introns and regions between the genes do not.

Tsui and Collins marched and jumped across 250,000 bases, and in the process found four short regions that hybridized to the DNA of other animals. The first one, when tested with their set of families, was inherited slightly differently from the CF gene, so it could be ruled out. The second was Williamson's infamous gene. The third was an odd sequence found in many copies throughout the genome, so it too could not be CF. The fourth, located about 100,000 bases further along than Williamson's gene, might be it, they thought. Its inheritance was right, and it seemed to be expressed in all the appropriate tissues—although that, as they vividly remembered, had been the case with Williamson's gene too. Since they had as yet found only a tiny fragment of the gene, they soldiered on, looking for more pieces. At the same time they began the job of cloning a DNA copy of the messenger RNA. Just as with the muscular dystrophy gene, the messenger RNA was sure to be much shorter and easier to handle than the gene itself.

The promising new gene turned out to be only a tenth as long as the muscular dystrophy gene, but it was still very large, scattered in at least twenty-seven bits over the next quarter of a million bases. Finding the RNA message coded by these bits was frustrating; they were unable to find it in several different cDNA libraries from likely tissues. Luckily, John Riordan, a colleague of Tsui's, had managed to make cultures of cells from human sweat glands. It was in DNA made from the messenger RNA of these cultures that they first found a bit of the message—a very appropriate place when you remember that salty sweat is one symptom of CF. As more and more of the message became available, the group immediately began to sequence the gene itself, comparing normal and CF genes as they went. The protein was very large: 1,480 amino acids long.

Johanna Rommens, a young postdoc, was in charge of the walking process. She told me that the most exciting time of her five years in the lab occurred one long Friday evening in May 1989, as they painstakingly went through the results of the latest sequencing experiment. They found that in their CF mutant, the 508th amino acid from the beginning of the protein was missing. Three bases—one amino acid's worth of the code—had been snipped out of the mutant gene.

This was what they were looking for, the essential piece of the puzzle that Williamson had never found. But was it a fluke, a meaningless accident? In a gene this size, one would expect to see a few differences between any pair of copies chosen at random. To make sure, they had to sequence all the rest of the gene, which took several more months. Then, to lay any lingering fears to rest, they had to do a critical genetic experiment.

The new technique called PCR had recently become available. It enabled them to start with a few copies of a small stretch of DNA from any individual and multiply them millions of times. They applied the PCR technique to samples of DNA isolated from hundreds of members of their CF families, using it to magnify a short stretch of the gene that spanned the code for the 508th amino acid. They knew that if the gene was normal, then the piece they had multiplied would contain the critical three bases. If it was mutant, it should lack them. They then spotted these hundreds of samples of magnified DNA onto a nylon membrane and attached them firmly.

The next step was to carry out a kind of mini–Southern blot experiment, similar to Hans Lehrach's technique for mapping. They hybridized to the membrane two little specially built probes. One probe carried the critical three bases, while the other lacked them. Conditions were arranged so that the probes would only hybridize if they matched the gene perfectly.

They looked at 198 genes from the family members they knew had two normal genes. These hybridized only to the normal probe, meaning that none of them carried the three-base deletion. The probe with the deletion, on the other hand, hybridized to almost 70 percent of the 214 CF genes that were tested. They realized that the remaining 30 percent had to have some other mutational problem that their probe did not detect. This was the genetic proof that they needed. As soon as they had the entire sequence and the genetic data, they were ready to go. In September 1989 three massive papers, crammed with irrefutable results, were published in *Science.* Tsui's decade-long quest was over. The CF gene had, beyond a shadow of a doubt, been found.

DECOMPRESSION

Although both Tsui and Rommens felt decidedly let down after the frenzy of the chase, they knew that the task facing the CF research community was just beginning. To begin with, the sequenced CF protein turned out to resemble many other proteins that are known to reside in membranes and that are also known to transport molecules from one side of the membrane to the other. But does the CF protein itself transport something or not? The evidence is equivocal. You will recall that CF cells are unable to transport chloride ions across their outer membranes, but it turns out that this is not because they lack the protein that does the actual transporting. The channels for chloride transport still seem to be intact in the mutant cells, and can be activated by various tricks. Instead, the problem seems to lie with the process the cell normally uses to activate the chloride channels—a process triggered by a remarkable intracellular messenger molecule called *cyclic AMP.* Since cells from a CF patient are unable to transport chloride ions when stimulated with cyclic AMP, the CF protein might be involved in this triggering process rather than with the transport itself.

The *sequence* of the gene told Tsui's team only that the CF protein seems to reside in membranes and has the capability of binding to ATP, one of the building blocks of nucleic acids and, incidentally, a close relative of cyclic AMP. It told them nothing more about the protein's function. The *availability* of the gene would prove to be another story. Once it had been put together in the form of a clonable piece of DNA from which all the introns had been removed, it would be a flexible tool triggering an explosion of exciting research.

Rommens immediately began the task of making the gene clonable, starting with DNA that had been made from CF messenger RNA. Since all the introns had earlier been removed from this messenger RNA through the action of snurps, she could begin with fragments of DNA that were sufficiently shorter than the original gene that they could easily be inserted into plasmids. When she tried to piece these fragments together to make a functioning gene, however, the bacteria carrying the plasmids died. Repeated attempts also failed.

It appeared that this CF DNA was a poison sequence in bacteria. Perhaps, Tsui's group thought, part of the CF gene was being translated into a protein fragment that killed the bacterial cells. They used a clever fix: introducing mutations into the gene. These mutations did not change the amino acid sequence, but they suddenly and completely cured the problem with the bacteria.

The gene, now whole and functioning, was immediately made available to the scientific community. Results were not long in coming. Tsui and Collins themselves inserted the gene into a *retrovirus,* a virus that can carry genes into the nucleus of an animal cell. (We will learn more about these remarkable retroviruses in the next chapter.) They then mixed the virus with tissue culture cells from the pancreas of a CF patient, which, unlike normal cultured pancreas cells, were unable to transport chloride ions when they were stimulated with cyclic AMP. Although only a few of the cells successfully took up the virus carrying the normal CF gene, those that did became able to transport chloride and similar ions with the aid of cyclic AMP. In these cells the CF defect had been cured.

A large group of scientists from the University of Iowa, Tufts University, and the Genzyme Corporation of Framingham, Massachusetts, used their own copy of the CF gene to go a step further, using a retrovirus to express the gene in tissue culture cells taken from the inner lining of the lung of a CF patient. Again, in that small minority of cells that expressed the normal gene, the CF defect was repaired.

A long road still separates these early experiments from a cure for CF. There are great difficulties facing this kind of gene therapy, some of which we will explore in the next chapter. Still, it is an exciting beginning. As of this writing, a workable therapy seems far closer than it was as little as a year or two ago. CF is a strong candidate for such a gene-based cure because the tissues most drastically affected by the disease are directly accessible from outside the body. If a retrovirus can be made that infects a large fraction of the cells it contacts, a patient might be helped simply by breathing an aerosol of the virus. Of course, any such "cure" would only be temporary, for cells in the lungs' lining are constantly being sloughed off and replaced. And the question of how normal people or people with other lung diseases might be affected by accidentally breathing in these viruses has yet to be thought about, much less dealt with.

LOVE'S BLINKERS

What about the hope, entertained by Collaborative Research and many other companies, that a single test could be developed to detect CF heterozygotes in the population? That too has turned out to be elusive, for both scientific and political reasons. Unlike sickle-cell anemia, there is no one allele responsible for CF. As Tsui and Rommens discovered, about 70 percent of CF mutant alleles carry the three-base deletion. The remaining 30 percent consist of many different mutant alleles, certainly hundreds and perhaps thousands, none of which seem to be particularly common. These mutations are found all over the gene; thus, in order to make sure that a gene is completely free of mutations, one must survey it from one end to the other. Such a survey is very difficult to do, and has been accomplished so far in only a few cases.

Most of the mutations that have been tracked down involve tiny changes, but a few are more dramatic, like the deletions seen in the muscular dystrophy gene. Tsui and others around the world are starting a massive program to screen all the genes of all their patients, to look for correlations between the kind of mutation and the severity of the disease. Such tests may eventually explain the differences between the course of muscular dystrophy in Kevin and in Steve.

Because of this heterogeneity, however, the development of a simple diagnostic test for carriers will be much more difficult and expensive than the development of tests for sickle-cell anemia has been. A number of people, among them Arthur Beaudet of Baylor College of Medicine and Michael Kaback of the University of California at San Diego Medical School, quickly realized that a little knowledge about CF can be a dangerous thing.

Suppose a couple wishes to find out whether they are both heterozygous for CF, which would give them a 25 percent chance of having an affected child. They take a test designed to detect the three-base deletion. If both of them happen to be heterozygous for this mutant, which is the easiest one to find, then the test has done its job. Unfortunately, since only 70 percent of CF mutations are the easily detected deletion mutant, only half the couples at risk (70 percent times 70 percent) will be found. These couples will have been helped, but in the process far more couples might be harmed.

You will recall that about 1 Caucasian in 25 is a carrier of a CF mutant allele. This makes the chance that two such people will marry $1/25$ times $1/25$, or about 1 in 600. Thus, in about 160 out of each 100,000 couples tested, both partners will be heterozygous and at risk of having an affected child. Of these, however, only 80 can be detected by this early version of the test. In a way, they are the lucky ones, because they can be certain of their genetic makeup. Unfortunately, the test will cast far more couples into a limbo of uncertainty. In 2,800 of these 100,000 couples, only one partner will be found to have the easily detected mutant allele. The other partner *may or may not* have one of the many other mutants; at the moment, there is no easy way to tell. These couples now know only that their chance of having an affected child, which could have been as low as 1 in 2,500, has been revealed to be at least 1 in 300.* This is not high enough to be helpful; it is only high enough to cause endless anxiety. Would such couples be more likely to panic at any untoward event during their pregnancy, and choose a needless elective abortion?

The American and Canadian Cystic Fibrosis Foundations have refused to fund genetic surveys of the population at risk, both because of this uncertainty and because of the explosive political problems posed by widespread antenatal diagnosis and elective abortions. In spite of this, and of a call for a moratorium on CF testing by the American Society for Human Genetics in late 1989, a number of companies are anxious to enter the testing field. Some, indeed, have already sent letters to doctors pointing out that they might be liable if they do not offer testing.

Large bodies of information about CF carriers are already being gathered in several countries for good scientific reasons, particularly in order to determine the frequencies of the growing numbers of rare alleles that have been detected so far. What should be done with this information? Various complicated stopgap measures have been proposed, such as letting couples know the test result only if both partners test positive. Many unmarried individuals are also being tested, however. Should they be given the information from the test? In some cases, such as in a pilot program at Cal Tech, this is already being done.

Tsui feels that it will be just a matter of time before the entire gene can be surveyed for mutational changes. A series of probes that can detect most of the alterations in all of the exons of the gene will soon be constructed, though there may be too many such probes to use in a clinical situation. Eventually, if quick, easy, and accurate ways to sequence a quarter of a million bases emerge from the Human Genome Project, some of the ethical agonizing over these questions will disappear. In five or ten years, general screening of Caucasians for heterozygosity for these mutants may become as widespread as screening for the Tay-Sachs allele has become among Ashkenazi Jews.

A successful screening program for CF will immediately introduce

*One partner is certainly heterozygous. The other does not have a detectable allele but may have one of the alleles that make up the other 30 percent of CF mutants. He or she will have a 30 percent of a 1⁄25 chance of being heterozygous for such an allele, which means about 1.2 chances in 100. The chance that the couple will have an affected child will be one quarter of this, which is about 1 chance in 300.

further agonizing choices. CF is an "invisible disease" so far as society is concerned. In most cases, children with CF look like normal children, and they can, with effort and support, lead simulacrums of normal lives. In an illuminating study that has yet to be published, Michael Kaback interviewed the parents of 500 CF children, asking them how they rated the severity of their children's condition, the effect it had had on the lives of their families, and the financial impact of the disease. He asked the same set of questions of close relatives of the children, particularly uncles and aunts. In virtually every case, the parents had a decidedly more optimistic view about the effects of the disease than the more remote, and presumably more objective, relatives.

Because of these more positive attitudes of parents, the course of a CF screening program will be very different from the program for Tay-Sachs. The grim prognosis for Tay-Sachs confronts the prospective parents with the starkest possible choice. They are faced with the prospect of either terminating the pregnancy or of having a baby who will die, blind and paralyzed, in just a few years. As a result, Tay-Sachs testing programs have had an enormous impact. Among Ashkenazi Jews in North America, the number of babies born with the disease dropped by 90 percent from 1970 to 1990. Indeed, the programs have reached the point that only three or four babies with the disease are born each year to this group, compared with ten or twelve born to the non-Jewish population. While it is unclear exactly how much of this decline is due to the accelerating rate of intermarriage of Ashkenazi Jews with the rest of the population, the programs have certainly been responsible for most of it.

This has occurred in spite of the fact that many couples at risk never take the test, or may take it only when a baby is on the way, giving little time for diagnosis of the fetus before an elective abortion becomes too dangerous. There is also resistance among Orthodox Jews to the idea of prenatal diagnosis, though some rabbis have circumvented this problem by entering into simultaneous agreements with testing services and matchmakers. If the rabbi learns that a match is genetically unsuitable, a word to the matchmaker is enough to ensure that the potential couple, still innocent of their genotypes, can be steered towards different mates.

It might be argued that the Tay-Sachs program has succeeded because the Ashkenazi population is particularly well educated and very concerned with health-related matters, both of which are certainly true. One large determiner of the effectiveness of screening and testing programs, however, seems to be the severity of the disease. Consider Cooley's anemia, or beta-thalassemia, another devastating genetic disease. Children with this disease, which is common in the Mediterranean, lack an important component of hemoglobin. Most die in childhood or at the latest as teenagers from the accumulating consequences of severe anemia and iron overload due to the breakdown of their defective hemoglobin molecules. Thalassemia, like sickle-cell anemia, is a legacy of the malaria that was once widespread in these regions. In the past, heterozygotes for the beta-thalassemia mutant gene were protected against the severest ravages of that disease.

Beta-thalassemia is particularly common in the coastal regions of the island of Sardinia, where an astounding one-sixth of the populace are carriers of the mutant gene. One child in 150 was born with this disease in 1976, when widespread screening, testing, and education programs were first introduced under the leadership of Antonio Cao. A decade later the rate had been reduced to one in 1,500, a gain equaling that seen for Tay-Sachs, even though the population at risk was largely uneducated.

On the other hand, if a genetic disease is almost invisible to society at large, and is perceived by parents as being less severe than it actually is, and if on top of this various racial and societal factors also play a role, then the benefit of a screening program can be greatly reduced. Nothing illustrates this more vividly than the fate of screening programs for the sickle-cell allele.

Sickle-cell disease is an enormous problem in Africa, where an estimated 120,000 children are born with the condition each year. Because they are susceptible to a great variety of infectious diseases, including the malaria from which their heterozygous relatives are protected, few survive beyond the age of five. About 10 percent of U.S. blacks are heterozygous for the allele, and almost 4,000 children are diagnosed with the disease each year. In contrast to the situation in Africa, the immediate impact of sickle-cell disease can be greatly alleviated in the United States and other Western countries where childhood mortality results primarily from easily controlled pneumococcal infections. However, there are severe long-term problems caused by occasional massive sickling of the red blood cells. This leads to repeated circulatory crises, resulting in damage to many organs and tissues. These sickle-cell crises are often triggered by viral infections and can strike at any time. Their symptoms and consequences are diffuse and complex, far more difficult for the parents of these children to deal with or understand.

Widespread prenatal testing for sickle-cell disease was begun in 1975, some years *after* screening programs to detect heterozygotes had been started in a number of states. Up until that introduction, no sensible advice could be given to people who discovered through these programs that they were heterozygous. Many of them were confused and thought they had the disease rather than being asymptomatic carriers. Some carriers were actually advised not to have any children.

A decade and a half later, many prenatal testing programs for sickle-cell anemia are still spottily administered, expensive, and poorly explained. In 1987, only 272 prenatal diagnoses were carried out in the United States, even though the number of pregnancies at risk exceeded 15,000. Currently, only about a third of the pregnancies in which a sickle-cell fetus is diagnosed are terminated by therapeutic abortion, far lower than the 80 to 90 percent of fetuses terminated following a diagnosis of Down syndrome.* Tests to detect the disease among newborns have been more widespread and effective; most states now mandate this type of screening. Sickle-cell anemia requires intensive follow-up monitoring and therapy, however, and in many cases this has not been forthcoming.

America is not, of course, the only country in which political and racial factors have played a role in genetic screening and testing. In the Soviet Union, no Tay-Sachs screening program has yet been introduced, in spite of the very large Jewish population and in spite of repeated demands for it by the global genetics community. Sickle-cell anemia is widespread in the Middle East as well as Africa, but the first prenatal diagnosis for this disease in the Soviet Union was only made in 1989, as a result of screening a small number of pregnant women in Baku.

In view of the nature of the affected group, CF screening and testing programs are unlikely to be hampered by racial problems. They are, however, likely to fall into the group of programs hampered by parental and societal misapprehension about the severity of the disease, which will reduce their effectiveness. This can only be countered by strong educational efforts. As the programs grow in size, they are sure to spark intense debates about the quality of life, the role of therapeu-

*In one case I became familiar with, a couple had already had one daughter with sickle-cell disease, a child who had undergone repeated life-threatening crises. When their second baby was diagnosed *in utero* as having the disease, the mother wanted to terminate the pregnancy, but the father refused, accusing the doctors of racism and genocide. A reasoned discussion of the issues becomes very difficult in such an atmosphere of pervasive distrust.

tic abortions, and the role of physicians as advisers to prospective parents. In any case, screening and testing remain only palliatives. What is needed is a cure.

WHY CYSTIC FIBROSIS?

Johanna Rommens has become more and more sure that the little three-base deletion that accounts for 70 percent of the CF mutations arose just once at some point in the distant past, and for some reason spread through the Caucasian population. Has mere chance made it so frequent today, or is there some other explanation?

Whatever the cause of CF's high incidence, it is the genetic disease with the most impact on Caucasians, just as sickle-cell anemia is the genetic disease with the most impact on blacks. Indeed, CF has often been called the "Caucasian sickle-cell anemia." Of course, part of the reason that CF seems to be more frequent among Caucasians is that it is more likely to be diagnosed in Caucasians. The disease was first noticed in the 1930s. Both Kevin and Steve, you remember, were diagnosed only with difficulty. There are no firm numbers yet about the frequency of the disease in other racial groups, because most of them have only recently been exposed to Western medicine. There are many statements in the literature to the effect that the frequency of CF is vanishingly small among other racial groups, but this view is sure to change as more information becomes available.

Consider a simple model. Suppose that the frequency of CF in every racial group except for Caucasians is the same, and suppose further that the increase in Caucasians is entirely due to that mutation with the deleted amino acid that makes up 70 percent of Caucasian mutant alleles. In other racial groups, the CF alleles would be a heterogeneous collection, like the remaining 30 percent of Caucasian alleles.

Because the frequency of the gene is lower in other racial groups, the frequency of homozygotes will be lower too—dramatically lower. One twenty-fifth, or 4 percent, of Caucasians are heterozygous carriers of a mutant allele. In other racial groups, only 30 percent of 4 percent of people—about 1.2 percent—would be carriers. The chance that two members of a non-Caucasian group would marry and have an affected child would be only $(.012)^2$ times a quarter or 1 in about 28,000.

Should this simple model or something like it turn out to be true, the incidence of CF among non-Caucasians would be less than a tenth of the

Caucasian incidence. Such a small number of cases could easily be missed, particularly since death from untreated CF comes from lung infections or failure to thrive, both of which are common enough in the Third World to have many different causes.

In fact, this simple model may well be close to the real situation. The deletion mutation so common among Caucasians is very uncommon among other races. What has yet to be determined is whether the remaining CF alleles are as common among other races as they are among Caucasians; if they are, then we must ask why the allele with the deleted amino acid is so much more common in Europe than elsewhere. Perhaps this allele increased in frequency because in its heterozygous configuration it conferred resistance to some disease, in the way that the sickle-cell hemoglobin allele has increased in frequency among blacks because it confers resistance to malaria. Or perhaps it was a chance event at some distant time in Caucasian prehistory. Similar chance events have been observed to take place in historical times. A disease called porphyria variegata, in which victims have difficulty breaking down the red pigment in their blood, causes a variety of intermittent neurological symptoms. It is inherited as an autosomal dominant allele, which means that it can easily be followed from one generation to the next, unlike the recessive CF alleles that often disappear for many generations before they pop up again. Porphyria is ten times as common among the Boers of South Africa as among most other groups, and this high incidence can be traced back to a single woman who was one of the forty Dutch settlers who arrived in South Africa at the end of the seventeenth century. Did such a chance event increase the frequency of the CF allele in the distant past of Caucasians? If so, the increase must have been a dramatic one; the CF allele is so harmful that it must have fallen greatly in frequency since that time.

One might ask the question: Were the tens of millions of dollars devoted to this search worth it, or would they have been better spent developing the techniques needed to track down all the human genes so as to eliminate the necessity for such costly races in the future? A long-term cost-benefit analysis would undoubtedly conclude that the genome project should take precedence. But, as with most cost-benefit analyses, it totally ignores the fact that worldwide, perhaps thirty to fifty thousand children a year die slow, agonizing deaths from cystic fibrosis. It has proved possible, though difficult, to track down the gene now. A cure might lie five or ten years in the future. Had we waited for the gene to be turned up during a slow and logical progression through the genome, the cure might be as much as twenty or twenty-five years down the line.

Any cure will come too late for Kevin Riley, who died peacefully on January 3, 1991, still waiting for a heart-lung transplant. It cannot come too soon for the hundreds of thousands worldwide who are still suffering from the disease.

CHAPTER 10

The Early Days of Gene Therapy

The investigator reported that one-third of the rats were improved on the experimental medication, one-third remained the same, and the other third couldn't be reported on because that *rat got away.*

—Edwin Bidwell Wilson

Robert was a cynomolgus monkey, a small member of the macaque family. His immediate ancestors had lived for several generations in a primate colony at the National Institutes of Health in Bethesda, Maryland, but his more remote ancestors came from the Philippines. Many of his relatives had been involved in the development of a number of vaccines against bacterial and viral infections. Now he was one of a group of fourteen cynomolgus and rhesus monkeys chosen to be unwitting pioneers in a daring experiment. Researchers were going to attempt to alter the monkeys' genetic makeup. Even though this would take place in only a tiny fraction of their cells, and last only for a short time, any success would represent an important step toward the goal of human gene therapy. Robert and his small group of fellow primates were about to receive the undivided attention of two of the foremost laboratories in the United States that specialize in gene therapy. His fate, and that of his fellows, illustrates both the problems and the potential triumphs that await workers in this hotly competitive field.

The laboratories were those of W. French Anderson and Michael Blaese. The experiment being prepared for Robert and his companions was in principle straightforward, in practice very difficult. To begin with, a higher organism like Robert is a heterogeneous collection of thousands of different kinds of cells, organized into tissues in a very complex way. Each muscle in a complex structure such as the hand, for example, is made up of striated muscle cells, but the cells of each muscle are of distinctly different shape and orientation depending on the task the muscle must perform. Muscle cells are so highly specialized or *differentiated* to perform their task that they can no longer divide. Thus, they are unable to make new muscle cells to replace nearby ones should they be injured or die. This, you will remember, is the problem with muscular dystrophy—diseased muscle cells that are unable to make the protein dystrophin become disorganized and die. As one doctor described it to me, the muscles turn into hamburger. At least in humans, new muscle cells do not replace the old, though this can occur in some other animals.

Robert and the other monkeys in the group were about to donate a different but equally complex tissue to the experiment: a sample of their bone marrow. This soft and almost formless tissue is found in the inner hollows of all bones of the body. The surrounding bone gives it support and protection. It is responsible for producing the red and white cells found in an adult's blood, along with other structures such as the platelets. Maturing cells stream forth from the marrow via veins that pass through the surrounding bone. The marrow can be highly productive; within a few hours it can make up all the cells lost through the donation of a pint of blood.

The great majority of the cells in the marrow are also differentiated, or partly so. Some may go through at most a few more cell divisions before their descendants reach full maturity, at which point they stop dividing. But concealed among them are a few effectively immortal cells from which all the others are descended. These are the *pluripotent stem cells,* capable not only of dividing to reproduce themselves but also of giving rise to all the myriad kinds of cells in the blood. These versatile cells look and act like many other types of cell in the marrow, and it has not been possible to locate them with any certainty or to purify them. It is estimated that only about one in a hundred thousand cells in the bone marrow is one of these elusive pluripotent stem cells.

The stem cells of Robert and his companions were the ultimate targets of the experiment. If altered genetically, stem cells will pass the changes on to all the cells of the marrow throughout the lifetime of the organism. When cells that are more differentiated are altered, they and their descendants will eventually die out.

At the start of the experiment samples of marrow were removed and stored, and the monkeys were then subjected to intense blasts of radiation, more than twice the amount needed to kill them. Because marrow is one of the most rapidly multiplying tissues in the body, it is particularly susceptible to radiation damage. Death from radiation poisoning usually results from destruction of the marrow, because this interrupts the continuous stream of new blood cells that is essential for survival.

Robert and nine of his companions were saved from radiation-induced death by the reinjection into their long bones of the marrow samples that had been taken from them before the irradiation. The other four monkeys, in spite of the reinjection, died. This experiment would have had little point had the marrow they received been the same as the marrow that had been donated. In fact, though, something had been added to some of the cells.

That something was a bit of new DNA, containing genes that were not native to the monkeys' cells. One gene came from the bacterium *E. coli* and conferred resistance to the antibiotic neomycin. The other, from a human, coded for the enzyme adenosine deaminase (ADA). This enzyme catalyzes an important step in the breakdown of DNA and RNA that was discussed in chapter 3. When it is missing, this breakdown is blocked, causing deoxyadenosine to build up in the tissues. Other enzymes add a phosphate to this substance, which turn it into deoxy ATP, a compound essential for making DNA. Too great a quantity of it, however, will act as a metabolic poison.

Children missing the adenosine deaminase enzyme make up about a quarter of the victims of a disease called severe combined immune deficiency (SCID). (The causes of the remaining three-quarters of SCID cases have yet to be discovered.) They suffer problems with their immune system because T cells, an essential part of the immune system, are the most sensitive to the perils of excess deoxy ATP. Without help, SCID patients—like AIDS patients—will catch an appalling variety of bacterial, viral, and fungal diseases, their immune systems helpless to ward the infections off. A few of these children have been maintained in germ-free isolation, including "David," the famous "boy in the plastic bubble," who survived for several years before accidental infections finally broke through the defenses created by his doctors.

ADA deficiency, like gout and Lesch-Nyhan disease (which you will remember is another strong candidate for gene therapy), arises from an error in purine metabolism. Medicine has come a long way in the 150 years since Alfred Baring Garrod did his first simple experiments. As it turns out, however, the defect that can lead to gout and the defect that leads to ADA deficiency are separated by only one metabolic step.

About a third of SCID children are lucky enough to have siblings who can donate genetically similar bone marrow, marrow that they will not reject. A successful marrow transplant can effect a remarkable cure. If the child is ADA deficient, this cure can be traced entirely to the fact that the new bone marrow cells make ADA.

ADA deficiency is a very rare disease. At a recent meeting, Blaese showed a slide filled with the names of dozens of his collaborators and remarked ruefully that there were far more people currently working on the gene therapy for the disease than there were sufferers from it! Since the disease is so uncommon, no vast immediate benefits to humankind can be expected from curing ADA deficiency. The few victims will of course be helped, but the disease is primarily a kind of test case for new technologies. ADA deficiency is in many ways an ideal candidate for gene therapy. The exact cause of the disease is known, and work with cultured cells has shown that even a slight amount of ADA activity, as little as 5 percent of normal levels, should be enough to effect a cure. So even a partially successful attempt at gene therapy should have detectable and perhaps strikingly beneficial results.

Robert, like the other monkeys in the experiment, could make his own ADA. The experiment was only possible because his ADA could be distinguished from that made by the introduced human gene. After supplying the two genes to the cells of Robert and his companions, Blaese and Anderson planned to look for cells exhibiting two characteristics that Robert's original cells did not: they would make human as well as monkey ADA, and they would be resistant to neomycin.

The trick was to get the genes into the right cells. It would do little good to insert the genes into cells that were already differentiated, or that only had one or two cell divisions to go. Such cells would soon die, taking the new genes with them. Blaese and Anderson hoped to insert the human ADA gene into at least a few of Robert's immortal pluripotent stem cells. Since these cells make up only one-hundred-thousandth of the bone marrow tissue, a hefty fraction of the other cells would have to receive the genes as well. They had to find a highly efficient way to insert the genes into so many cells.

This is a tough problem. Many labs around the world are trying methods ranging from the ingenious to simple brute force to get foreign genes into mammalian cells. On the face of it, some of these methods have been remarkably successful. In one of the outstanding triumphs of gene therapy, functioning genes have been inserted into the nuclei of cells of mice and more recently of sheep and cattle, and have actually become integrated into the chromosomes and passed on to the next generation. The first of these experiments was carried out in 1984 by a group headed by Ralph Brinster of the University of Pennsylvania. The group was able to "cure" mice that suffered from a hereditary growth disorder by injecting a rat growth hormone gene into the eggs of mutant mice. Only about one in a hundred of the injection experiments worked, but in the few successful cases these *transgenic* mice grew to be one and a half times the size of normal mice. They were able to pass the gene, and their ability to grow, on to subsequent generations.

The potential of such an experiment is very great, for it would appear at first sight that if this could be done with the victims of ADA deficiency it would cure both the disease and the genotype. People with the disease would not only be cured, but they could have children secure in the knowledge that they would not be passing their damaged gene on to the next generation.

If this sounds too good to be true, it is. The mouse and sheep experiments are carried out with fertilized eggs—the most pluripotent cells of all. The genes are injected directly into the nuclei of the eggs immediately after fertilization by means of an ultrafine glass needle. The eggs are then implanted in the uterus of a surrogate mother. All the cells of the young that result from such a pregnancy are descended from the eggs, so that if the gene transplant "takes" in the egg cell it will be passed on to every cell in the young animal's body—including the sex cells that give rise to the next generation.

It is far more difficult to perform a gene transplant on a child or an adult. The affected tissues might be made up of billions of cells, far too many to inject individually. To get genes into a significant fraction of those billions of cells, a *vector* must be constructed, something that can be added to a cell suspension and then used to carry the genes directly to the nuclei of the cells. Ironically, the most successful vectors are also the most potentially hazardous. Many groups working on gene therapy have turned to the class of viruses called retroviruses, mentioned briefly in the last chapter. These include some of the most dangerous viruses known, such as the AIDS virus and many of the viruses that trigger cancer. When not inside cells, these viruses take the form of tiny balls of protein. If we again shrank ourselves to the submicroscopic size we attained in our tour of the nucleus, they would be about a yard across.

In the center of the protein ball, retroviruses carry a small number of genes on two identical copies of a short chromosome, which is made of RNA rather than DNA. They also carry a few molecules of a special enzyme, coded by one of those genes, that are released once the virus enters the cell. We have already met this versatile enzyme: It is reverse transcriptase, and once inside the host cell it can make DNA copies of the RNA chromosome and then make the DNA double-stranded. This double-stranded DNA, with the reverse transcriptase still attached, enters the nucleus of the cell. There the reverse transcriptase performs its final task, breaking the DNA of the host and inserting the DNA copies of the viral genome into the host's chromosomes.

While these viruses tend to have preferred settling places on the host chromosomes, they can settle, or *integrate,* at many different spots. Once integrated, the viral chromosome becomes a kind of fifth column, for it is now part of the host's genome. It may be passed on from one cellular generation to the next, sometimes for years, before it is activated and begins to make new copies of the virus. Or it may damage the genome in specific ways that greatly increase the risk of the cell's becoming cancerous.

Occasionally, a virus hidden on the host's chromosome is damaged by a mutation. When this happens it may remain stranded in the host's genome like the hulk of a ship mired in the Sargasso Sea, gradually dissolving away under the impact of further mutations until it disappears. At least a hundred thousand copies of old retroviruses and other retroposons in various stages of decay litter our genomes. Many of these must be the molecular scars of ancient infections, attesting to battles our ancestors waged with retroviruses millions of years ago. The scars show that retroviruses have played an important part in our evolutionary history.

In order to infect Robert's cells with human genes, Blaese and Anderson turned to a retrovirus that can cause leukemia in mice. Some critical viral genes were removed from the retroviral genome and replaced by the neomycin resistance gene and the human ADA gene. This disabled the virus, so that it could be grown in cells only with the aid of another "helper" virus. Because Blaese and Anderson did not want to risk giving Robert cancer, the "helper" virus they used was also disabled, but in a different way.

Both these viruses were added to Robert's bone marrow cells, and allowed to infect them. Then the cells were injected back into Robert's hip bone, to replace the marrow destroyed by the irradiation.

All of the ten surviving animals in Robert's group grew new bone marrow to replace the marrow destroyed by the radiation. When their blood was sampled, six of them showed slight traces of human ADA, but these traces soon vanished. Only Robert showed substantial amounts of the human enzyme, amounts that rose briefly to half a percent of his own ADA levels. Within six months, however, even this slight amount of human ADA had disappeared. Robert was his old self again, quite literally. What had happened? Somehow, it appeared, the foreign genes had been turned off. Even the genes themselves could not be detected in Robert's marrow cells, so either too few cells had been infected in the first place or the genes had somehow been cast out of the genome.

What was particularly frustrating about this experiment was that very similar ones, using the same viruses, had worked beautifully in mice. In one typical experiment, two out of nine mice showed remarkably high levels of human ADA, even more than the mouse ADA made by the same animals. And these huge amounts persisted for at least six months, the equivalent of twenty years or more in humans.

Blaese and Anderson found themselves in a thicket of questions. Suppose they had used a virus native to the monkeys, rather than a mouse virus? Would this have worked better? It would certainly have been more dangerous, for there was always the possibility that the disabled viruses could recombine with their helpers, giving rise to intact cancer viruses again. And what had really happened to the genes? Had they never gotten into most of the cells in the first place, or had monkeys evolved a way of chopping invading mouse virus chromosomes out of their DNA? This seemed possible, because when monkey cells carrying the mouse virus DNA are grown outside the animal, they too lose the DNA over time.

Monkeys are far closer to us than they are to mice, which was why it seemed logical to Blaese and Anderson to try in monkeys what had worked so well in mice. On the other hand, monkeys have far more sophisticated defenses against the invasion of viruses than mice do. They must have, because they live ten to twenty times as long, and we saw in chapter 1 that the ability to resist cancer-causing agents increases with increasing life span. So it is very likely that Robert's defenses were more sophisticated than Blaese and Anderson had anticipated. The defenses humans can marshal against the invasion of foreign genes may be even more highly developed.

HOW THE DYING HELP THE LIVING

To try to break through these defenses, Blaese and Anderson's groups have now joined forces with another laboratory at NIH, that of Steven Rosenberg. Rosenberg has been trying to use the body's own immune system to fight against cancer. In the process he has fortuitously developed a way to simplify the problems faced by the gene therapists.

Like the bone marrow from which many of its cells are derived, the immune system is made up of a bewildering variety of cell types. Some of these, the T-killer lymphocytes, are capable of finding and destroying other cells in the body that have been altered in such a way that they appear different from normal cells. Such cells include many cancer cells.

Rosenberg knew that a tumor taken from a cancer patient is swarming with cells from the immune system. These *tumor-infiltrating lymphocytes* are chiefly T-killer cells. A postdoc in his lab, Ilana Yron, had reasoned that this small subgroup of the body's lymphocytes must be the very best cells the immune system can produce to fight that particular cancer. Like the tiny fraction of a modern army that actually sees combat, they are on the front line of the body's defenses against the growing tumor. They may fail to halt or to destroy the tumor—indeed, they usually do fail—but this may simply be because there are not enough of them.

Rosenberg thought that if he could somehow increase the numbers of these particular cells, he might be able to augment the body's natural defenses. He cultured tumor-infiltrating lymphocytes out of a great variety of tumors that had been removed during surgery, but found to his frustration that they grew very slowly. Luckily, just at that time large quantities of a pure lymphocyte growth factor called interleukin-2, produced in genetically modified bacteria, were becoming available. When Rosenberg added interleukin-2 to his lymphocyte cultures, they went through a population explosion, multiplying several millionfold in a few weeks. Having started with a few cells from the original biopsy, he ended up with large refrigerators crammed with cell-filled plastic bags, holding as much as 450 liters of cloudy pink liquid.

When these hugely augmented forces of lymphocytes were transfused back into the patient from which the original tumor had come, they had a dramatic effect—about half the time. Tumors would shrink and in some cases disappear. In a few cases, less than 10 percent, they disappeared permanently. Some of these cases were particularly astonishing, since the patients had suffered from advanced malignant melanomas. These tumors, starting in the skin, usually spread rapidly throughout the body and are almost invariably fatal.

But most of the time, weeks or months after Rosenberg's treatment, the tumors returned. Lymphocytes taken from these new tumors now had little effect. Simply adding more cells to his original treatment only worked up to a point, for the sheer masses of cells transfused into the patient could damage blood vessels and muscle tissue. As with every other cancer therapy, it seemed to be necessary to wipe out every cancerous cell the first time. You will recall that one of the hallmarks of cancer is extreme genetic instability, snowballing into a huge variety of genetic alterations in the different cells of a tumor. This means that if even one tumor cell survives the treatment, it will be so genetically different from the original tumor cells that there is a good chance it will now be resistant to the therapy.

Rosenberg wanted to know what was actually happening to the cells he was transfusing back into the patient, and thought of a simple experiment. If he could label the cells with a reporter gene, one different from any that the cells normally possessed, he could see how many of them actually ended up in the tumors and could then follow their subsequent fate.

This was where his problems began. He was proposing a new and daring kind of experiment, in which a foreign gene was to be introduced into a human being for the first time. The gene he chose was the neomycin resistance gene from *E. coli.* It was to be inserted by means of a mouse cancer virus, but there was every expectation that neither the gene nor the virus would have any harmful effect. The patients themselves would all be terminally ill volunteers, and any altered cells would die with them. Nonetheless, the full gauntlet of regulatory agencies had to be run.

The saga took about two years, from June 1988 to the final decision handed down in March 1990. In the interim fifteen assorted subcommittees, committees, and agencies examined the protocol and modified it again and again. Beginning with various hospital oversight committees, Rosenberg continued through the human gene therapy subcommittee of NIH's Recombinant DNA Advisory Committee (RAC) and then through the full RAC committee. At the end of it all, approval had to be obtained from the Food and Drug Administration.

Preliminary approval for a small trial involving ten patients was given by James Wyngaarden, then the head of NIH, in January 1989. A few days later, however, a suit was filed by the environmentalist gadfly Jeremy Rifkin, head of a group called the Foundation on Economic Trends. This was one of a series of suits that had been brought by Rifkin in attempts to halt or delay various experiments that might release genetically altered organisms into the environment. Like most of Rifkin's suits, this one succeeded only in delaying the process and causing further review. The suit was settled quickly out of court, and the historic first treatment of humans with genetically modified cells began in May 1989. The number of patients permitted in the study was later increased by the RAC subcommittee.

Eight patients with metastasizing melanomas were treated with the genetically marked tumor-infiltrating lymphocytes, along with a hefty dose of unmarked lymphocytes to increase the likelihood that the therapy would succeed. One patient showed a striking remission over a period of five weeks, during which all the tumors seemed to disappear. After two months, however, new tumors appeared. Rosenberg found that very few of the lymphocytes from these new tumors carried the marker gene. The experiment told him what he had long suspected: The tumors had changed, and the population of lymphocytes fighting them had changed as well.

In the course of designing his experiments, Rosenberg joined forces with Blaese and Anderson. They hoped to use Rosenberg's system to fight both cancer and a more clear-cut foe: ADA deficiency. Since the monkey experiments using bone marrow had worked so poorly, Blaese and Anderson wondered whether lymphocytes, which would quickly colonize the body and perhaps replace the population of cells killed by deoxyATP, might be a better choice. Since Rosenberg's marker gene had had no influence on the success or failure of his cancer therapy, these would be the first real gene therapy experiments in humans. As Blaese put it, "We decided to put the ADA gene into lymphocytes, rather than beat our brains out trying to get it into stem cells."

This new protocol also had to go through lengthy hearings. There were two great difficulties. Perhaps the most awkward point was that disabled mouse leukemia viruses were still the vector of choice to get the ADA gene into the cells. The NIH workers would thus be infecting cancer-free patients with cancer viruses. True, the viruses were disabled and produced cancer in a very different organism, but it was possible that they could recombine with their helpers to produce a functioning cancer virus again. To reduce this problem, two different helpers, disabled in different ways, were used. This greatly reduced the risk that an intact cancer virus would somehow be reconstituted.

The second difficulty was that monitoring the course of the therapy could require that patients with ADA deficiency be denied the benefits of a current stopgap therapy that has recently become available. Adenosine deaminase can now be administered intravenously to patients with ADA deficiency who cannot be helped by marrow transplants. It is given in chemically modified form so that it is broken down slowly. Blaese, Anderson, and Rosenberg had to agree that this therapy would continue during the course of the experiment, even though this would make interpretation of their own results much more difficult.

On July 31, 1990, the NIH Recombinant DNA Advisory Committee voted approval, with only Richard Mulligan of the Whitehead Institute, a severe critic of the NIH group's approach, dissenting. In a statement to the press, Mulligan said, "The possible benefits of the experiment don't outweigh the risks. It should be scuttled." Certainly, there was no guarantee that the ADA-producing lymphocytes would survive once they were returned to the patient's body, or that they would lower the level of deoxyATP to the point that new T cells could appear. With so many unknowns, the experiment is as chancy as the one carried out on Robert and his companions.

The possibility of cancer, however, did seem small. No cancers traceable to the virus had been seen in monkeys infected with the same vector, and no sign of transformation into cancerlike cells had been seen in infected monkey or human cell cultures. Blaese's lab had accumulated data on fifty monkey-years of exposure to the virus, experiments in which a third of the monkeys' blood volumes have been repeatedly replaced by viral cultures with no sign of a cancer. Still, the critical question of what the viruses will do in a human remained to be answered—and unfortunately could be answered only by doing the experiment.

Approval from the full NIH committee and the FDA was obtained in early September 1990, and within a week and a half the first patient, a four-year-old girl, was being treated. Throughout the treatment she would continue the adenosine deaminase enzyme therapy. As of this writing the results of the gene therapy are not yet clear, but there have been some hopeful signs. When cultures of the girl's lymphocytes were infected with the ADA gene, the NIH group was greatly encouraged to find that the infected cells survived better and multiplied faster in tissue culture than had her original cells. This was a good indication that the new adenosine deaminase gene was carrying out its function in the cells, reversing the poisoning caused by the excess deoxyATP.

At the same time that approval was given for the ADA experiment, Rosenberg received the go-ahead for another gene transfer experiment, to be carried out on volunteers terminally ill with melanoma. Rather than simply infecting the patients with a reporter gene, he will attempt to infect them with a gene that might damage the tumors themselves. This gene produces a protein, called tissue necrosis factor, that kills cells. By using specifically targeted tumor-infiltrating lymphocytes, he hopes to destroy the tumors through the action of this gene, without doing too much damage to the rest of the body. One way or another, the NIH group hopes to achieve a success.

WHY IS GENE THERAPY SO DIFFICULT?

In the race to be the first laboratory to do successful gene therapy, the troika of Anderson, Blaese, and Rosenberg is out in front. At this point only a foolish punter would bet on anybody else. A breakthrough in ADA deficiency therapy would open the door to curing or alleviating many other genetic diseases, particularly diseases of the blood, such as sickle-cell anemia and hemophilia. Yet in spite of the excitement and media attention, some basic scientists, including Stuart Orkin of Harvard Medical School as well as Richard Mulligan, are critics of this method, considering it to be crude and clumsy.

Certainly, the pressures on clinical laboratories to obtain results are extreme. Every day the researchers treat patients whom they inevitably come to know well, and whose fears and hopes fuel the urgent search for a cure. Some cases are particularly poignant. While the majority of people who fall victim to cancer are middle-aged or older, with most of their lives behind them, some are young. Rosenberg has melanoma patients who are in their twenties, and all the victims of ADA deficiency are children. To prevent the waste of such young lives would be a wonderful thing.

There are also financial pressures. Anderson's research is partially supported by GTI Inc., which was founded by the Health Care Investment Corporation of New Jersey, a venture capital company with interests in biotechnology. Anderson has no direct financial interest in GTI and was at first reluctant to accept financial support, but the severe recent cutbacks in NIH funding helped pressure him into looking for outside support.

As a result of all these pressures, experiments are tried in a clinical setting that would not be tried were the subjects mice or rats rather than humans. Clinical researchers make their move before the system is well understood. This means that even if these early experiments in gene therapy succeed, it may take decades to figure out why they worked.

It is safe to predict that the outcome of the first ADA replacement experiments will be equivocal. Like Robert the monkey, the children may produce ADA at first, and then stop. Or the therapy may work in some children but not in others. Even if the attempt succeeds, the results will have to be followed for years. Suppose some of the recipients develop lymphomas, perhaps years down the line? Determining whether or not these cancers had been triggered by the viral vector will be very difficult, since the causes of cancer are still so uncertain.

How can we learn more about the basic biology involved? To begin with, it is striking that all the experiments conducted thus far work better, in some cases far better, in mice than in primates. The levels of human ADA or bacterial neomycin resistance expressed in both living mice and in isolated mouse cell lines are far higher, persist for longer, and are more consistent from experiment to experiment than they are in monkeys. This may simply be because laboratory mice are genetically very uniform while primates are not. As I suggested earlier, however, primates may also have developed new defenses against the invasion of foreign DNA. It is now known that part of the immune response involves special transport systems that move proteins around within cells. Perhaps in primates these systems are able to detect the invasion of a foreign virus even as it slips stealthily from the cytoplasm to the nucleus.

Just as with the effort to cure cancer, it will be necessary to understand the body's defenses against foreign DNA in great detail in order to perform effective gene therapy. One way to begin would be to ask whether genes are turned on in the cells of the primate immune system that are not turned on in the equivalent cells of mice. It may be possible to narrow the question still further, and ask whether different subsets of genes are turned on at the moment leukemia viruses invade primate and mouse cells. Tracking down such genes will be relatively easy. The hard part will be finding out what they do.

Another approach would be to develop new and more effective vectors to carry the genes into their target cells. There are many possibilities. Researchers at the University of Michigan have recently reported that it is possible to use common bacterial plasmids rather than the far more sophisticated and dangerous retroviruses to insert genes into human cells. The search for useful, specific vectors without side effects has only just begun.

Once the barriers to getting the gene into the cell are overcome, there will be other problems. Every genetic defect, large or small, is traceable to an alteration in the DNA. In principle, then, every defect is curable. What is required is a vector that does not trigger the body's numerous defense mechanisms, that invades all or most of the cells of the target tissue, and that adds a functioning gene to the cells it invades with a minimum of disruption to the genome.

Even more stringent requirements must be met if we are ever to cure the gene in such a way that it can be passed on to the next generation. Consider a recessive genetic condition such as ADA deficiency. A child suffering from this condition has two defective genes in each cell. At the moment, the retroviral vectors being used to try to cure ADA deficiency are very nonspecific; they can insert the functioning ADA gene they carry into many parts of the genome, sometimes in more than one place in a single cell. They are very unlikely to insert the new gene in place of an old one, so the original defective genes will still be present in their old positions even after the therapy.

Further, even after a successful gene transplant, different cells in the tissue will have different numbers of functioning genes inserted in different places, perhaps in the process disrupting other important genes. There is no way that such a scrambled set of genomes can or should be passed on to the next generation. Until far more precise *gene surgery* can be done, the possibility of curing defective genes permanently remains only a remote dream. Still, it is a truism that whatever we imagine we can accomplish. It is worth comparing the history of gene therapy to that of vaccination for smallpox.

The technique of trying to induce a mild case of smallpox in order to protect against a severe case later on goes back centuries—perhaps even millennia. The ancient practice of *variolation,* in which pus from a smallpox victim was rubbed into a scratch made in the arm of the person to be protected, seems to have been particularly common in the Middle East. A number of travelers brought the idea back to England and America at the beginning of the eighteenth century, and for the next hundred years many people, including members of the English royal family, were subject to variolation. Done carefully, with an amount of material tiny enough not to overwhelm the recipient's immune system, the treatment produced a milder disease and protected against subsequent attacks. But it was still incredibly dangerous. Between 1 and 5 percent of the people subjected to the procedure died of induced smallpox. Such were the ravages of the disease, and the fear it engendered, that this was considered an acceptable risk.

The idea of *vaccination* against smallpox can be traced back to Edward Jenner, a London physician who performed the first vaccination using virus from a case of cowpox in 1796. Cowpox is one of a family of viruses found in many different mammals that are closely related to the smallpox virus. In humans it produces a disease very like smallpox, but much milder. It is sufficiently like the smallpox virus that the immune system of a person infected with cowpox is stimulated to make antibodies that are effective against smallpox as well. (Milkmaids often contracted cowpox, which protected them against the disfiguring ravages of smallpox. Their perfect complexions explain why they figured so prominently in the bucolic fantasies and paintings of the eighteenth century.)

Jenner's idea of using a related disease rather than smallpox itself was a breakthrough, making the process of what later came to be called vaccination far safer. Controversy about Jenner's procedure raged during the nineteenth century, but as the practice of vaccination spread through the industrialized countries, the incidence of smallpox plummeted. In 1967 the World Health Organization announced a program to eradicate smallpox, which in that year had caused ten million deaths, chiefly of children in the Third World. The program was astonishingly successful. By 1977 the last spontaneous case of smallpox in the world was reported from Ethiopia. From its crude and dangerous beginnings almost three hundred years ago, the process of widespread immunization had brought about the eradication of the most devastating of human plagues.

The eighteenth-century doctors who cheerfully gave thousands of people doses of smallpox were working at a time when malpractice suits were unheard of and when anybody, high or low, could be struck down by that dread disease. Faced with a 50 percent mortality rate for those who contracted the disease, and the near certainty of dreadful disfigurement should the disease be survived, the public crowded to these doctors and paid huge sums for variolation.

The situation facing the gene therapists is similar in some ways, different in others. The diseases they are trying to cure are equally fatal, though not contagious. The cures currently being proposed may—if worst-case predictions are realized—lead to later problems. The thicket of restrictions and regulatory agencies that the NIH groups had to traverse bears witness to the fact that society now has the leisure to consider the implications of the proposed cures. As a result of this change in society's attitude, it has become essential to agonize over every step and to try to anticipate potential problems as much as possible. A sufficiently disastrous untoward result could delay progress in the field for years or even decades. It is not necessary to extrapolate three hundred years into the future to predict a remarkable impact of gene therapy, however—science moves far more quickly now than in Jenner's time. There is good reason to suppose that, once the problems with targeting and specificity are overcome, most single-gene defects will be curable at the gene level.

The damage caused by a defective gene before diagnosis will be much more difficult to deal with. By the time Down syndrome is diagnosed in a newborn, much of the developmental damage caused by the extra chromosome 21 has already been done, especially to the brain and heart. Simply turning off the chromosome will not be enough. Will it ever be possible, as I suggested in the introduction, to rebuild and cure such damage? Without the possibility of repair, gene therapy will be confined to a small number of diseases in which the damage is easily reversed by the healing power of the body itself.

Finally, we must ask the most difficult question of all: Just what constitutes a damaged gene? We have seen again and again in this book how clear-cut genetic diseases fade into a twilight zone of alleles with less and less significant effects, until they blend into the general genetic variability of the population. Steve Shepherd, though a victim of CF, can lead an almost normal life. Many other carriers of slightly damaged CF alleles no doubt have symptoms even less severe than Steve's. Should such alleles be "cured"? Are they entirely without benefit? If one penetrates far enough into the twilight zone, the boundary between normal and mutant alleles becomes completely blurred. By curing people of such alleles, we might be discarding an important part of our genetic heritage.

Are there any rules that will tell the gene therapists where to stop? I think there are. The last part of this book, though it may seem to wander very far afield from the topics we have dealt with up to now, will essentially be a search for these rules.

Part Three

EXTRAPOLATIONS

CHAPTER 11

Cancer Revisited

To conclude, the evil was so profound, the case so delicate, & the precautions necessary for preventing a return so numerous, that the operation including the treatment & the dressing, lasted 20 minutes! a time for sufferings so acute, that was hardly supportable— However, I bore it with all the courage I could exert. . . .

—Fanny Burney, recounting her mastectomy performed by Napoleon's surgeon, Baron Larrey, in 1811

When Renato Dulbecco proposed the daring idea for a Human Genome Project, he could not see any other way to make sense out of cancer, that infinitely complex disease. As we have seen, the project has become far more complex and diffuse than Dulbecco originally envisioned. It is not unreasonable to ask at this point whether exploring our genome in such detail *will* help in the fight against cancer. This chapter investigates that question by exploring a few of the more exciting recent discoveries in the field of cancer research. Each example illustrates some of the themes introduced in the course of this book. In each case, researchers are coming up against a wall of difficulties that can only be surmounted by ever more detailed explorations of the genome.

ANTICIPATING CANCER

Eldon Gardner was born in Logan, Utah, in 1909, and received his Ph.D. in genetics from Berkeley just before World War II. After the war he returned to his native state, briefly taking up a position as Assistant Professor of Biology at the University of Utah in Salt Lake City before moving to Utah State at Logan, where he spent the rest of his life.

The end of the fighting coincided with an explosion of funding for science on an unprecedented scale. This had been accomplished chiefly through the efforts of the remarkable Vannevar Bush, who headed up the U.S. scientific research effort during the war. One of the disbursers of this new largesse was the U.S. Public Health Service, which gave one of its earliest human genetics grants to the brand new medical school at the University of Utah. The grant was earmarked for the study of the genetics of cancer, particularly in Mormon families.

Utah is an ideal spot for such studies. The closely knit and hard-working Mormon families who form the backbone of the state and contribute most of the students to the university are a tremendous resource for human geneticists for a variety of reasons. First, most of them are from a rural background, which means that families are large and tend to stay in one place. Second, the Mormon Church is intensely interested in genealogy, and both the church and its constituent families keep careful records of who is related to whom. (There is a good reason for this, since the Church encourages rebaptism into the Mormon faith of deceased family members who had not been Mormons when they were alive. I have been assured by Mormon acquaintances that things are arranged so that startled ancestors are not suddenly yanked from their Episcopalian heaven and deposited willy-nilly into a Mormon one.) A third feature that makes these families so useful is their high average degree of education. Family members tend to be knowledgeable about science and enthusiastic about participating in studies, even when these involve multiple questionnaires, donations of blood, and sometimes even more invasive procedures.

Gardner was fascinated by the possibilities of studying the inheritance of cancer, and in his genetics course in the fall of 1947 he expounded at length to his students on this exciting area of research. After his lecture, an undergraduate named Eugene Robertson approached him with the news that in his home town there was an ideal family for such a study. The next weekend, Gardner visited the town with Robertson and began talking to family members and examining records. The grandmother of the family had died in 1909, and although the medical records were less than satisfactory the cause of death appeared to have been cancer of the colon. All three of her children had died of the same disease, as had three of her grandchildren by the time of Gardner's visit.

Gardner brought the resources of the new medical center into action. Over the next two summers, he tracked down fifty-one members of the family and drove them to Salt Lake City, where they were examined by endoscopy. This procedure could reveal the presence of *polyps,* small growths of the intestinal mucosa that often precede colon cancer. These come in a variety of types, the most worrisome of which are *adenomatous* polyps, meaning they are formed from glandular tissue. Six members of the family were found to have as many as several hundred adenomatous polyps. Only one treatment was available at the time: a full or partial colectomy. This drastic treatment saved their lives, for one or more of the polyps would certainly have become cancerous sooner or later.

In this small branch of what turned out to be an immense extended family, the gene for multiple polyposis could be traced back to the grandmother alone. Thousands of other family members were examined over the next few years, and none showed the condition. The gene must have arisen either in the grandmother or in one of her immediate ancestors (her mother had also died young). It was inherited in subsequent generations as an autosomal dominant, meaning that about half the children of a carrier and an unaffected person would be expected to develop the disease. The fact that all three of the grandmother's children died of colon cancer was simply an unlucky chance—though in a way it was lucky as well, since this unusual series of events had stuck in the memory of Gardner's student, setting in motion the train of circumstances that saved the lives of many of the victims' descendants.

Everybody concentrated on the multiple colon polyps at first, but Gardner noticed that the family members with these polyps often had other growths on their bodies—fibromas beneath the skin, osteomas that caused visible swellings of the jaw and other parts of the skeleton, and a high incidence of benign sebaceous cysts, particularly on the back. It was some time before he, a lowly Ph.D., could attract the doctors' attention to the existence of all these other growths, but he finally found a radiologist who was willing to examine X rays of the affected people in some detail and who discovered to his surprise that the osteomas could be found everywhere in their skeletons. Indeed, osteomas and other growths were sometimes found even in family members who showed no sign of polyps, indicating that the intestinal growths were only one manifestation of the disease, and not necessarily the most consistent one. Other abnormalities kept turning up; the victims of the disease often had extra teeth, for example, and patches of abnormal pigment in their retinas.

This particular variety of inherited cancer quickly acquired the unofficial name of Gardner syndrome, and this is how it is usually referred to in the literature. (It forms a fitting memorial to Eldon Gardner, who died in early 1989.) Doctors began to find cases of familial polyposis in many parts of the world. In some of these families the affected members had only the intestinal polyps and none of the many other manifestations. This variant of the disease was given a separate name: familial adenomatous polyposis. It too was an autosomal dominant, and so similar to Gardner syndrome in many ways that people began to suspect that the two diseases were caused by slightly different alleles at the same genetic locus.

Researchers knew that the gene was not on the X chromosome, but for a long time they had no other clue to its whereabouts. An important clue finally turned up in the form of a patient who, like the young boy B. B. we met in chapter 9, had an odd and apparently unrelated collection of symptoms. He was a forty-two-year-old man who was seen by physicians at the Roswell Park Memorial Institute in Buffalo in 1985. He clearly had Gardner syndrome, though no other members of his family had it. He was also mentally retarded, with an abnormal kidney, a missing lobe of his liver, and no gallbladder. When his chromosomes were examined very closely, it was found that one of his two copies of chromosome 5 was missing a tiny part of its long arm.

This was enough information for two different groups to begin an intense search for the gene. These were Ray White's human genetics group from Utah, and a group from the Imperial Cancer Research Fund under the direction of Walter Bodmer. Both groups began by collecting as many DNA probes as they could find that they knew were located in or near this region of chromosome 5. They then employed these probes in whole-genome Southerns, probing DNA from families in which Gardner syndrome or familial polyposis were common. Often the probes hybridized near informative RFLPs in these families, enabling the researchers to follow the RFLPs from one generation to the next. Both groups quickly found strong evidence of linkage between these RFLPs and the polyposis condition.

Now that the approximate location of this gene has been found, strenuous efforts are being made to walk to it and sequence it. There is intense curiosity about what the gene will turn out to be, for colon cancer is the second most common cause of cancer-related death in the industrialized world, just behind cancer of the lung. About 5 percent of the population can be expected to develop colon cancer at some point during their lives, and most of these cancers seem to develop from polyps. By contrast, the familial type appears to be rare; estimates of its frequency range from 1 in 20,000 to 1 in 5,000 people. This apparent rarity may be an illusion, however, for like so many of the genes we have met in the course of this book, the polyposis gene seems to represent just the tip of the iceberg.

The iceberg itself is gradually being revealed. One Utah family with the milder form of the familial disease has been investigated in great detail by Ray White and his group. Records of the family extend through seven generations; in five of these, virtually everyone at risk has been examined for colon cancer or polyps. At first the investigators used a fairly conservative criterion to pick out those family members who were affected; only those who actually developed cancer, or had at least five adenomatous polyps, or had some other very strong indication were placed in the affected category. When the symbols representing these people on the family tree were filled in, the condition seemed to follow the classic pattern of autosomal dominant inheritance through the generations.

Or at least it almost did. From the pedigree it became obvious that a number of family members should have had the disease but did not. Extensive analysis using RFLP markers showed that four of these were almost certainly carrying the gene. (The same analysis predicted that another sixteen family members were almost certainly *not* carrying the gene.) When the records of the proctological examinations were looked at again, it was found that all four of the suspected carriers did have polyps, but so few that they had not been included in the first screen. In contrast, fifteen of the sixteen predicted not to be carriers had no polyps at all, while the sixteenth had only one, unsurprising given that polyps occur occasionally in the general population.

This study showed that the effects of the gene could be detected even in virtually asymptomatic family members; the mutation appeared to be expressed in different people in enormously different ways. Limited as it was, this study raises complex and difficult questions about what should happen next in the study of this disease. When the gene is cloned and sequenced, as it soon will be, it will be possible to determine the exact mutational changes that are responsible for familial polyposis. There are likely to be many of them, just as there are many different CF mutations. If, however, the effects of even the most clear-cut of these mutations are hard to see in some of the family members and do not always lead to colon cancer, are doctors justified in sounding the alarm whenever a fetus is detected carrying one of these alleles?

Venturing further into the gray zone, what about all the mutations with even subtler effects? Are some of them associated with the overwhelming majority of cases of colon cancer that do not appear to run in families? What will happen when and if screening and testing programs are initiated? It seems certain that we will find slightly mutated alleles of the gene—alleles that might only occasionally trigger colon cancer—and that these alleles will turn out to be common in the population. Performing colectomies on the millions of people who carry such alleles would surely be an overreaction!

To add to the difficulty, it seems likely that many cases of colon cancer will be found in which this particular chromosome 5 gene turns out to be perfectly normal. Changes in the chromosomes, and many other more subtle genetic changes, have been observed in colon cancer cells. Any of these might be directly involved in many colon cancers.

Genetic screens for cancer alleles will soon become a possibility, but as soon as they do, problems such as these will surely dwarf the problems currently faced by screeners for CF or sickle-cell anemia. Since we cannot at the moment distinguish cause from symptom in cancers with any certainty, it is obvious that we will need to understand the genome and the changes that occur in it at a far more fundamental level if we are to make consistent predictions. Cancer is too subtle an enemy to give up its secrets easily.

THE VITAMIN CONNECTION

Leukemia, literally "white blood," is the name for a group of cancers in which certain kinds of bone marrow cells proliferate wildly. In some leukemias these cells are primitive, not too many cell divisions away from those pluripotent stem cells that gene therapists have had such difficulty finding. In others the cells are much further along the many branching differentiation pathways that lead to the bewildering variety of mature cell types normally present in the bone marrow. Whatever their history, as leukemic cells multiply they take over the circulation, interfering with and eventually totally preventing the normal functions of the blood.

In most leukemias the majority of the abnormal cells halt at a specific point. In each patient, however, a small proportion of the cells continue to differentiate for one or two more steps. It has been known for many years that leukemia cells taken out of the body and raised in tissue culture can be made to differentiate even further, through treatment with a strange and eclectic collection of chemical compounds. Dimethyl sulfoxide has this effect on some kinds of leukemia cells, perhaps by increasing the cells' permeability to unknown differentiation factors. Butyric acid works on other cell lines, for reasons that are not understood at all.

One of the most interesting of these compounds is vitamin A, along with its various derivatives. It was first noticed in 1925 that rats deprived of vitamin A developed precancerous lesions in the skin, which disappeared when the vitamin was reintroduced into the diet. In spite of these early observations, it was not until the late 1970s that a vitamin A–like compound, *retinoic acid,* was tried on tissue culture cells from human patients with various leukemias.

Retinoic acid had little or no effect on most of these cells, but it had a startling effect on one type. These were cells taken from patients with acute promyelocytic leukemia, a rare variety in which the multiplying cells are very primitive. (The disease is particularly dangerous because its victims can make only these primitive cells, and are unable to make any of the range of mature white cells that the body depends on to protect it against infection.) When in 1980 S. J. Collins and his colleagues from the Fred Hutchinson Cancer Center in Seattle treated cell cultures from this leukemia with tiny amounts of retinoic acid, they found that the cells stopped their mad proliferation and resumed normal differentiation. It was almost as if they had forgotten they were cancerous.

This finding immediately opened up the possibility of retinoic acid therapy for the disease. The first group to report a success was Wesley Miller's at the University of Minnesota. They treated a young man for whom conventional chemotherapy had failed. The effect on his blood was dramatic and immediate, with many classes of mature white cells appearing in the place of the useless and endlessly proliferating primitive promyelocytes. Unfortunately, the patient soon succumbed to a yeast infection from which his damaged immune system was unable to protect him.

Many other patients with this disease have been treated with retinoic acid since. The pattern has tended to be one of remarkable remission, usually (though not always) followed by relapse. A far-flung collaboration between French researchers in Paris and Lille and Chinese researchers at the University of Shanghai resulted in the finding that one particular form of retinoic acid—manufactured in quantity for the first time by the

Chinese group—is the most effective. Of twenty-two patients treated in France, for example, five were still in complete remission at the time the study was published. These remissions had lasted for periods of up to thirteen months. While this may seem a fairly high success rate for a disease that is almost invariably fatal, it must be remembered that the usual criterion for permanent remission is five years of freedom from the disease. None of the French or Chinese patients have achieved that yet.

So far, then, this story seems very similar to that of many other cancer therapies: initial striking success followed, after some variable period of remission, by eventual reemergence of the disease. There are some new clues to be found in this particular tale, however. One is provided by retinoic acid, which has turned out to be very important in cellular development and differentiation—the very things that are disturbed in the cells of leukemia patients.

The importance of retinoic acid in these processes was discovered by embryologists and cell biologists working quite independently of cancer researchers. It was found that normal embryonic development in experimental animals could be profoundly disturbed by the application of extra retinoic acid to various parts of the developing embryos. This discovery led to the detection of gradients of different concentrations of retinoic acid in the embryos. The gradients were apparently maintained and generated by specific proteins found in all animal cells, which bind retinoic acid very tightly. The genes coding for these proteins are called retinoic acid receptor genes, and several of them have been cloned by Pierre Chambon and his group at the University of Strasbourg. Chambon's group has recently succeeded in mapping one of these retinoic acid receptor genes to the long arm of human chromosome 17.

The second clue is provided by chromosome 17 itself. Most of the primitive proliferating cells in patients with acute promyelocytic leukemia have a particular type of chromosomal defect, in which parts of chromosomes 15 and 17 have been exchanged or *translocated.* This translocation is not a normal genetic recombination event, but rather a highly abnormal one. In a normal recombination event, the pieces of chromosome that interchange are matching pieces exchanged between copies of the same chromosome. Such an exchange might involve a swap of corresponding pieces between a cell's two copies of chromosome 15. Alleles might be switched in this process, but the order of the genes on the two chromosomes would be unchanged. In a translocation, by contrast, the abnormal exchange results in the production of grotesque new hybrid chromosomes made up of parts of two very different chromosomes.

When you consider the almost infinite number of different translocations that could occur between different parts of the various human chromosomes, it is remarkable that this specific 15–17 translocation is so common in promyelocytic leukemia. It is much rarer in other types of cancers, and is almost never seen in normal cells. The translocation seems likely to be either a cause or a consequence of the cancer itself; it is not found in the patients' other tissues, and could not have been handed down from the patients' parents, for such a translocation would adversely affect the critical cell division called meiosis. Since meiosis leads to gamete formation, the result would be highly abnormal eggs or sperm.

In 1990, the Pasteur group cloned and sequenced the chromosomal region where chromosomes 15 and 17 are joined together. In patient after patient with the disease, they found that a mysterious gene from chromosome 15 had been fused directly to the retinoic acid receptor gene they had earlier tracked to chromosome 17. The result was a hybrid gene.

The discovery of this hybrid gene was a very exciting breakthrough. Somehow, by processes that have yet to be understood, this gene must disturb normal retinoic acid metabolism, perhaps making the cells less sensitive to retinoic acid. Dosing the patient with very high levels of retinoic acid appears to be enough to overcome this insensitivity, permitting near-normal differentiation to resume.

Of course, you know by this time that no story about cancer is as simple as it first appears. The operative word in this case is *near-normal.* Even when patients respond well to retinoic acid therapy, their new mature white blood cells differ slightly from normal white cells, retaining traces of their past lives as wildly proliferating promyelocytes. This means that the fusion of the retinoic acid receptor gene with the unexplored gene from chromosome 15 cannot be the only genetic change that has happened in the generation of these cancers. Further, since not all patients with the disease have this fused gene, the fusion cannot be the sole genetic event causing this leukemia. Still, the fact that retinoic acid can almost completely reverse the course of the disease suggests that the fusion must be one of the most important genetic changes involved. In the case of this cancer, we seem to be getting very warm indeed.

Before you assume that acute promyelocytic leukemia is about to be cured, however, consider the following. When the mature, healthy white cells of patients who had been successfully treated with retinoic acid were examined, they did not show the translocation. Their chromosomes were perfectly normal, even though the cells still showed those slight traces of the days when they had been part of the teeming horde of cancer cells in the bone marrow, and even though the immature white cells of those same patients examined before retinoic acid treatment had apparently all carried the translocation.

One possible interpretation of this confusing result is that there is a mixture of cancer cells in the marrow. Somehow, the cells with the translocation are affecting other cells that do not have it, causing them to behave cancerously too, like potentially good kids trapped among bad companions. When retinoic acid is added, it is apparently the cells that do not carry the translocation that can differentiate most rapidly into mature white blood cells.

Should this turn out to be true, it will open up whole new levels of complexity in cancer research. The possibility that cancer cells can influence the behavior of normal cells, causing them to act as if they themselves are cancerous, has already been raised by researchers investigating some unusual tumors of the gonadal tissue called teratomas. At least some of the cells extracted from teratomas are perfectly normal, and can give rise to normal animals in the next generation. Perhaps this is because they have left behind those genetically damaged companions that caused them to behave cancerously.

The many possible interactions between normal and abnormal cells imprisoned together in the same tissue, and the ways that abnormal cells might be able to alter the genotypes of normal ones, are only beginning to be explored. It will be essential to find many signposts in the genome in order to distinguish all these various cells from each other and follow their interactions.

GENE ANTI-THERAPY

Acute promyelocytic leukemia is not alone in being associated with a defined genetic event. Many other cancers have been found to be accompanied by specific swaps between different chromosome arms. One of these is a less severe form of leukemia called chronic myelogenous leukemia, in which the cells that multiply uncontrollably are fairly mature white cells. These cells are differentiated enough to be able to function properly, but they are produced in such huge numbers that they too eventually interfere with other normal functions of the blood. This leukemia can be kept at bay, sometimes for years, with chemotherapy, but sooner or later some unknown event will cause the patient's system to be flooded with masses of even more primitive cells, now too primitive to do their job. Untreated, this second phase of the disease will soon lead to death.

Cells taken from most (but again not all) people with chronic myelogenous leukemia show a particular chromosomal translocation. The translocation was first noticed in a patient with the disease who happened to come from Philadelphia: Thus, scientists now refer to the so-called Philadelphia chromosome, in which parts of chromosomes 9 and 22 have been juxtaposed. Cloning and sequencing of the junction of these two chromosomes revealed that once again two genes had been fused together. The one from chromosome 9 is an oncogene called *abl.* The other, just as with the 15–17 gene fusion we looked at a moment ago, is a gene of unknown function.

As soon as this fused gene was discovered, a number of different groups of investigators asked an obvious question about it, one answerable through what might be called gene anti-therapy. Suppose this gene were inserted into normal tissue culture cells, or even normal animals. Would the gene by itself be enough to cause cancer?

The answer, remarkably, was yes. But it was the wrong kind of cancer! Repeated experiments using a retrovirus to carry the fused gene into mouse or human tissue culture cells resulted in transformation of these cells into cancerlike cells. The fused gene was also inserted into cells that were then mixed with normal embryonic cells to make *chimeric* mice: mice in which the tissues were a mixture of the two types of cells. Cancers developed in some of these mice. In all these cancerous cells, however, differentiation was halted at a very late stage, resulting in a proliferation of cells of the kind found in the lymph nodes rather than in the bone marrow. These were the kinds of cancers that the *abl* gene was capable of causing by itself, so the fact that it was fused to the unknown gene from chromosome 22 seemed to have had no effect.

Finally, after much experimentation, a group in David Baltimore's lab at MIT managed to force the fused gene to produce the "right" kind of cancer. They inserted the gene into a retrovirus, infected mouse bone marrow cells with it, then finally injected the infected cells into mice whose bone marrow had previously been destroyed by irradiation. Even so, over half the mice did not develop a cancer at all, and something that looked like chronic myelogenous leukemia developed by itself in only a tenth of them. The rest of the mice developed a variety of different solid tumors and leukemias, often more than one in a single mouse.

There is, of course, no way of knowing what the results would have been had the same experiment been carried out on human beings. I have already suggested that we and our close relatives have acquired in the course of our evolution many sophisticated defenses against the various genetic changes that lead to cancers. Mice and other short-lived animals might have fewer of these defenses. Perhaps this is why the fused gene readily causes such a variety of cancers in mice.

These gene anti-therapy experiments raise many questions. When the retroviruses infected tissue culture cells they seemed to produce the "wrong" kind of cancer cells; infecting whole mice, they sometimes produced the "right" kind. What aspect of using whole mice creates such a different result? Is there something about the environment within the bone marrow that is different from that experienced by tissue culture cells in a petri dish, and if so, what is it? Perhaps the retrovirus inserts the fused gene into different parts of the genome in the cells of whole mice and in tissue culture cells. If so, might these different insertion processes cause changes in other genes, changes that help push the cells toward different types of cancer? Can these various genes be cloned, and can their original function be discovered?

All these questions will be answered in greater and greater detail as knowledge about the mouse genome grows. Indeed, you can see that it is possible to make as strong an argument for a Mouse Genome Project as for a human one.

VNTRS AND CANCER?

Changes in the chromosomes are the hallmark of many different kinds of cancer. Translocations are particularly common in cancers that affect the bone marrow and the lymph nodes, the tissues that produce the cells of the immune system. Specific translocations tend to be associated with each type of tumor.

The litany of such tumor-translocation associations grows longer each year. In the immune system, receptor molecules found on the surface of T cells play a central role in recognizing foreign proteins and cells. Different tumors involving T cells are associated with different chromosomal translocations. These various translocations have one thing in common: They cause the gene for the T-cell receptor to be fused to an oncogene. This fusion has the effect of turning the oncogene on permanently, and presumably inappropriately. When tumors involving B cells, the cells that actually make antibodies, are examined, translocations are found that have a different result. These translocations fuse the genes for the antibodies themselves to oncogenes.

Burkitt's lymphoma, a tumor of the lymph nodes that often starts in the neck and jaw, is the result of a complex interaction involving the state of the body, the state of the genome, and the environment. In Africa, this frightful cancer seems to spread through populations like an infectious disease. The process that leads to the lymphoma starts with a severe infection by Epstein-Barr virus, the same virus that causes mononucleosis. This produces an explosion of antibody-producing B cells. When the viral infection is accompanied or followed by a malarial infection, the B-cell population increases even further. In most cases of Burkitt's lymphoma, somewhere in this hugely expanded population of B cells a critical chromosomal translocation occurs. The result is a fusion of an antibody gene, usually one on chromosome 14, with an oncogene called *myc,* which is found on chromosome 8. It is not known, of course, whether this is the event that sets the course of these cells irreversibly toward cancer, or whether it is some earlier or later event.

In these lymphomas, called "endemic" lymphomas because they seem to be largely confined to certain African populations, the fusion occurs at a specific point in the antibody gene. Other Burkitt's lymphomas turn up occasionally in people elsewhere in the world who have not been subjected to these horrendous infections. When these "sporadic" cases are examined, the fusion to the *myc* oncogene is found to have taken place at a different, but equally specific, part of the antibody gene.

Why should these highly precise translocations take place again and again in different people? One could argue that it is just chance; after all, there are a lot of cells in a lymph gland, and all sorts of chance rearrangements of the chromosomes could take place in them. However, the chance that such specific juxtapositions would occur again and again in a genome consisting of three billion bases is vanishingly small.

The fact that antibody genes and their relatives are so often involved turns out not to be a coincidence. Similar though less dramatic translocations are old stuff to antibody genes. The genes that make our antibodies are initially pieces of gene scattered along the chromosome, separated sometimes by millions of bases. In the genomes of B cells, and only there, they are pasted together by a series of translocation-like processes to make a complete gene. This happens through the action of specific enzymes, active only during the period when the B cells are completing their differentiation in the bone marrow and the lymph nodes. The pasting process is also aided by short stretches of DNA, only seven or nine bases long, that are called recombination signal sequences. When these tiny stretches are removed, the pieces of antibody gene cannot come together properly.

Some of these recombination signal sequences are found near various pieces of the antibody gene on chromosome 14. A group at the Hôpital de Saint-Louis in Paris has recently shown that similar sequences are also found near the *myc* gene on chromosome 8. The evidence is now strong that the translocations found in so many immune system cancers arise because the normal pasting process that makes antibodies somehow goes awry, bringing the antibody gene and the oncogene together in a fatal combination.

Many recombination signal sequences of various kinds are scattered throughout the human genome. For example, you will remember the VNTRs we met in chapter 8, which are so useful in DNA fingerprinting. Up until recently VNTRs had been widely dismissed as uninteresting "Simpson Desert" regions, of no particular importance to the genome. When Alec Jeffreys discovered the first VNTRs, however, he noticed that parts of them resembled bacterial recombination signal sequences that had been found in *E. coli* several years earlier. It seems likely that the high rate of mutation that causes VNTRs to increase and decrease in size may be due to the activity of these signal sequences. Perhaps VNTRs will turn out to have an important role after all. They might indicate major points of weakness in the genome that can result in cancers, and they might also have other functions that have yet to be hinted at. If there is any moral to this story, it is that nothing about our genome can be taken for granted and that no part of it can be safely ignored.

These tales about cancer are only a sampling, but enough is enough. If this chapter has confused you with a deluge of information, I hope it has also served to demonstrate that Dulbecco was quite right: We must learn far more about our genomes and how they are put together before we can understand and finally defeat this subtle enemy.

There are, however, even more complex secrets locked in our genomes.

CHAPTER 12

Voices Heard in a Tiny Piece of Chromosome

For the Snark's a peculiar creature, that won't
Be caught in a commonplace way.
Do all that you know, and try all that you don't;
Not a chance must be wasted today! —Lewis Carroll

The young Asian man was a twenty-year-old student at the University of British Columbia. He had been having bizarre auditory hallucinations and delusions since age fourteen, but had not told his family about them until a month earlier. Now his worried family had brought him to the Health Sciences Center Hospital on the UBC campus to be examined by a psychiatrist. His mother told the psychiatrist that his school work had been suffering for a long time, and that he seemed to be becoming progressively more and more divorced from reality.

The psychiatrist, Dr. Anne Bassett, examined him further. He had great difficulty concentrating, was easily distracted, and showed no particular interest in the psychiatrist's questions. His short-term memory was impaired. It was very hard for him to switch from one task to another. He answered her questions in monosyllables, and attempts to provoke an emotional response from him failed.

As she talked to him he would often make inexplicable jumps in his reasoning, moving from topic to topic in ways that at first made no sense, though gradually the tenuous connections between the jumps became clearer to her. She recognized this pattern as one often mentioned in the psychiatric literature, where it is sometimes called "knight's move thinking." In the unavoidably fuzzy world of diagnosis of mental illness, then, his case seemed to be remarkably clear-cut. He appeared to be a classic schizophrenic.

About 1 percent of people in Western societies are hospitalized at one time or another during their lives for what is diagnosed as schizophrenia. The term itself dates from 1911, when it was coined by the Swiss psychiatrist Paul Eugen Bleuler to cover what appeared to be a variety of paranoid and delusional states, the onsets of which often coincided with puberty.

The disease often appears with little warning, and it can, after repeated episodes that incapacitate its victims for varying periods of time, disappear as mysteriously as it came—though it often leaves a persistent neurological deficit behind.

Although the layperson often confuses schizophrenia with split or schizoid personalities, the conditions appear to be quite different. Schizophrenia has such a variety of manifestations, however, that there has been endless argument about whether or not it should itself be split into a number of different subcategories. Hebephrenic or early-onset schizophrenia is usually differentiated from paranoid or delusional schizophrenia, and these in turn have been distinguished from a variety of "schizophreniform" disorders that contribute to what is called the schizophrenia spectrum. Over the years, as many as twenty different subcategories have been assigned.

To make things even fuzzier, psychiatrists could always fall back on the category of "borderline schizophrenia," which can encompass a great variety of slight deviations from accepted behavior. The danger of this approach has been vividly illustrated by the revelation of the abuses practiced by the Soviet mental health system. Under the direction of politically motivated doctors such as A. V. Snezhevnevsky, Soviet psychiatrists diagnosed political dissidents and members of various religious sects as having "sluggish schizophrenia" because they opposed aspects of the Soviet system. Under pressure from the World Psychiatric Association this system of political diagnosis is slowly being reformed, but its consequences are sure to plague Soviet psychiatry for years.

Despite the excesses of diagnostic zeal evident in the psychiatric establishments of both communist and capitalist societies, clear-cut cases of schizophrenia are easily recognized and distinguished from other mental disorders. Attention span and reaction time are often impaired in schizophrenics. In cases involving delusions, inner voices and visions apparently distract the attention of the patient.* In other cases, the patient simply has great difficulty in concentrating on a task and relating to his or her surroundings.

As is typical of schizophrenics, the student examined by Dr. Bassett responded readily to low doses of the tranquilizer haloperidol. The drug alleviated the more obvious symptoms and enabled him to resume his studies successfully. Yet odd features of the case continued to bother Dr. Bassett. Her young patient was unusually short, even for an Asian, and clearly overweight. And his face looked very different from those of other members of his family. His eyes were exceptionally widely spaced; his forehead bulged; and the back of his head was flattened. There was more than the usual webbing between his fingers and toes, and an ultrasound exam revealed that his left kidney was an odd shape and in the wrong position. Indeed, the patient's mother volunteered to Dr. Bassett that she had been concerned that there was something physically as well as mentally wrong with her son for a long time.

Partly, this was because of his worrisome behavior, but it was also because the mother's fifty-two-year-old brother had also been diagnosed as having schizophrenia. As with his nephew, symptoms of the disease had become pronounced at age twenty. He too had experienced auditory hallucinations and disorientation, and had responded readily to low doses of neuroleptic drugs. Yet her son and her brother had even more in common: They also shared a remarkable physical resemblance. She showed Dr. Bassett dozens of family pictures, pointing out the similarities as they appeared again and again.

Dr. Bassett, who was in the midst of a schizophrenia research rotation at the time, also had a strong background in human genetics. She knew that a syndrome of this type, encompassing a variety of defects and affecting physical appearance, can often indicate chromosomal abnormalities. Indeed, people from different families who share the same chromosomal abnormality may resemble each other more than they do their own siblings. This is particularly obvious in the case of Down syndrome, but it is common with other chromosomal problems as well. An experienced investigator can sometimes guess, just by looking at the patient, what the chromosomal problem is likely to be. But no one at Dr. Bassett's hospital had seen this particular complex of symptoms before.

The next step was to examine the uncle and other members of the

*Remarkably, schizophrenics who have been deaf from birth claim to "hear" voices, providing strong evidence that there really is a brain malfunction involved.

family. The uncle proved to have a set of physical abnormalities very similar to those of the son. In his case the left kidney was completely absent. The mother's other son and her other three brothers showed no signs of schizophrenia or any other mental illness, nor did they have the short stature and other physical abnormalities shared by the student and his uncle.

Blood samples were taken from both uncle and nephew, along with samples from the young man's mother and father and his other three maternal uncles. White blood cells from the samples were treated so as to catch the chromosomes at the point where any abnormalities would be most likely to be detected. Both the student and his uncle had the normal number of chromosomes: forty-six. On first examination the chromosomes looked normal. When they were examined carefully band by band, however, uncle and nephew were both found to have the same small abnormality. One member of the largest pair of chromosomes, the pair called chromosome 1, was a little longer than usual. It had a small amount of material inserted into the longer of its two arms. None of the other chromosomes appeared unusual, and the insertion was too short for the cytologists who examined the chromosomes to be able to determine where it had come from.

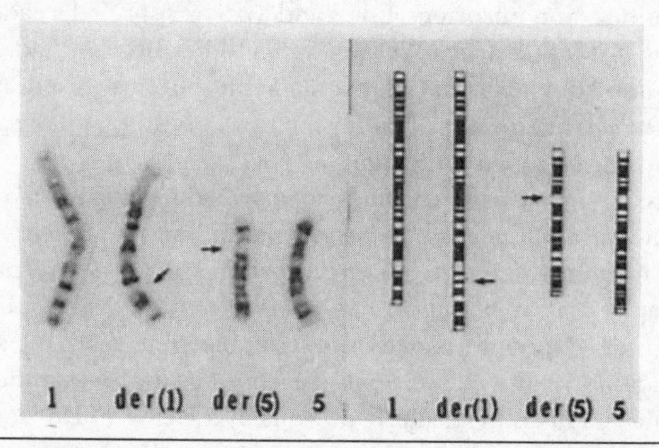

Figure 12–1. Chromosomes from the mother of the student with schizophrenia. One of her chromosomes 1 is normal, while the other has a small insertion of chromosome 5 material at the point of the arrow. One of her chromosomes 5 is also normal, while the other has a deletion of exactly the material that was inserted into chromosome 1. These two abnormal chromosomes balance each other, so that she shows no trace of the abnormalities that affect her brother and her son.

The student's father's chromosomes were completely normal, as would be expected; so were those of the three uncles. The chromosomes of the mother provided the needed clue. One of her copies of chromosome 1 also had the tiny insertion, but compensating for it was a tiny deletion near the base of the long arm of a smaller chromosome, chromosome 5. Because the gain and loss of genetic material canceled each other out, she showed no physical abnormalities.

How could this have happened? Neither of the mother's parents were alive, so their chromosomes could not be examined, but in one of them, or in one of their immediate ancestors, a piece of a chromosome 5 must have broken out and been inserted into a chromosome 1. Further, this must have occurred in a cell destined to form sex cells. The result was that the abnormal chromosomes were passed on to some progeny and not to others. (Unlike the much larger translocations discussed in the last chapter, this small change was not enough to interfere with the process of meiosis.)

Because the two abnormal chromosomes were found together in the mother, the mutational event had to have occurred quite recently. Had it happened more than one or two generations ago, the different chromosomes carrying the tiny insertion and deletion would have been passed to different progeny and would be very unlikely to turn up in the same offspring. Indeed, by the mother's generation the two altered chromosomes were already beginning to part company. The abnormal chromosome 1 had been passed on to the mother's brother and son, but the abnormal chromosome 5 had not.

Both schizophrenic members of the family had received from their parents two normal copies of chromosome 5 and one normal chromosome 1, but also an abnormal chromosome 1 with a small piece of chromosome 5 inserted into it. As a result, they had three copies of that small piece of chromosome 5, instead of the usual two.

Small as it is, this tiny extra piece of chromosome consists of about thirty million base pairs and probably carries a thousand or so genes. Three copies of all these genes seem to be enough to disturb normal development slightly, although it is unlikely that there are specific genes on this segment that direct the formation of extra webbing between the fingers and toes, or of a malformed kidney. Is there likely to be a gene for schizophrenia on the segment? Perhaps not; perhaps the association of schizophrenia with this chromosomal aberration is simply another slight developmental disturbance that cannot be attributed to a single gene. Nonetheless, the discovery of this association opened up the exciting possibility that one of those thousand genes is a gene for schizophrenia.

SEARCH FOR THE SNARK

Middlesex Hospital is a soot-blackened Victorian stack of bricks squatting a few yards from London's bustling Tottenham Court Road. It is a teaching hospital for the University of London, whose main colleges lie scattered throughout Bloomsbury to the east. The Department of Psychiatry is housed in a much less unappetizing building down a quiet nearby street. Here, a young psychiatrist named Hugh Gurling has been trying to track down the genes for schizophrenia and other mental illnesses.

Gurling is convinced that there really are genes for these conditions, and that finding them will be a great step forward. He remarked to me that when he tells his schizophrenic patients a mutant gene may be responsible for their disease, their relief is palpable. It removes a great weight of guilt, since schizophrenics commonly feel their condition is somehow their fault.

About 5 percent of cases of schizophrenia are, like the ones in British Columbia, clustered in families. The rest seem to be isolated instances, in which the parents and siblings are unaffected. Gurling and his associates have spent years tracking down families in which schizophrenia is common, working out their pedigrees, looking for patterns. They are not alone: Several other groups, in Edinburgh, Stockholm, and Bethesda, Maryland, are also studying apparently familial cases of schizophrenia. The families being studied are widely scattered, in England, Scotland, Iceland, Sweden, and the United States. In all of them, a very large proportion of family members have been affected by schizophrenia or allied mental disorders.

Gurling's group has access to seven unrelated families, five Icelandic and two British, with a total of 104 members. Of these, an astonishing 39 have been diagnosed as schizophrenic by one or more criteria. Other members of the families exhibit a variety of problems, ranging from severe clinical depression to excessive alcohol and drug use. Indeed, among the living family members, only 56, or slightly over half, seem free of problems. And 13 of these have married into the families from outside. This lends support to the idea that there might be a single dominant gene for the disease in these families, passed down from generation to generation but sometimes, for unknown reasons, skipping whole generations.

Gurling's families had been selected from a large pool in which schizophrenia is common. In most families where mental illness seems to be clustered, some members are schizophrenic, but others have a very different mental illness: manic depression. Some cases of manic depression are also strongly suspected to be inherited, and Gurling was afraid this would confuse the genetic picture. He was after "pure" schizophrenia.

Unfortunately, Gurling had no guarantee that all the cases of schizophrenia in his families were the result of mutations in a single gene; a quite different gene could have been passed down in each family. This was made more likely by the fact that schizophrenia is inherited so irregularly, sometimes behaving as though the gene were dominant, but sometimes skipping generations. The severity of the disease and the way it is expressed also vary greatly from one family member to another. Gurling's group had no idea where to begin.

Then came the news from British Columbia about the possible involvement of a small piece of chromosome 5. At last, the group was able to find its bearings in the vastness of the human genome, and begin to do meaningful linkage studies.

Such studies depend on the ability to follow two or more genes through families, to see if they are inherited together more often than would be the case if they were not linked together. Unfortunately, as we saw in chapter 8, human mating habits and family sizes are very inconvenient from the geneticist's point of view. With fruit flies, the geneticist can be confident that males and females placed in the same bottle will eventually mate, no matter what strange mutations their partners may carry. Humans are a good deal choosier, and cannot be mated to order. Further, they produce distressingly few children compared with the fecund fruit fly or the even more fecund bacterium. Even if the human geneticist finds an ideal family for a linkage study, there may be too few children to say anything firm about whether the genes are linked.

Even so, such advanced tricks as RFLP mapping can be used to track down the position on the chromosomes of "yes or no" genes that, like the gene for CF, always behave the same way as they are passed from generation to generation. The problem becomes much more difficult if one is dealing with "yes-no-maybe-sometimes" genes, like the ones that might be responsible for schizophrenia.

Consider the two pedigrees for schizophrenia shown in figure 12–2, which have been redrawn from a paper from Gurling's group.

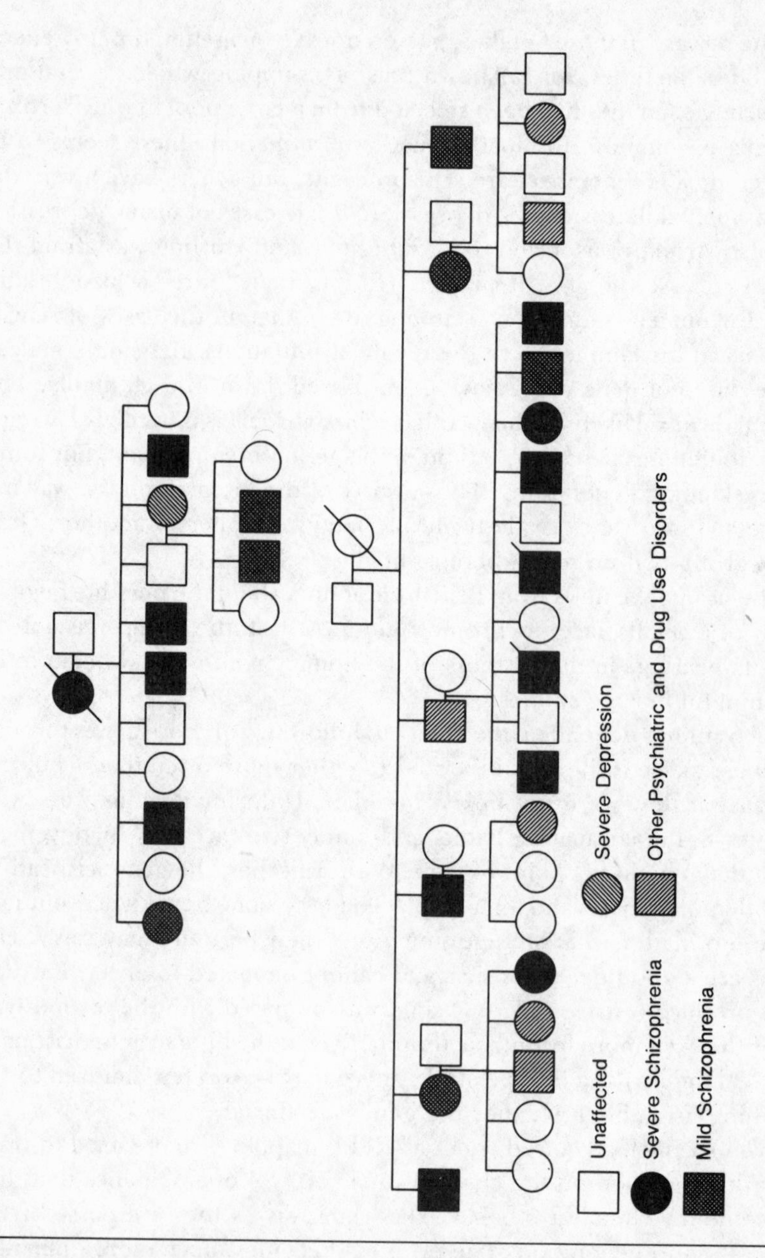

Figure 12–2. Familial schizophrenia pedigrees. Clear symbols represent unaffected individuals. Solid symbols represent individuals with fairly clear-cut schizophrenia, while the other symbols represent a variety of conditions ranging from depression through phobias to alcoholism and drug use disorder.

The conventions for drawing human pedigrees are simple. Circles are females, squares males. A diagonal line through the symbol indicates that the person was dead at the time the pedigree was drawn up. Shaded symbols are people with the condition in question. To indicate the range of different mental problems found in these families, Gurling has used various patterns, with well-diagnosed cases of schizophrenia represented by solid black symbols.

The first pedigree appears to follow the classic pattern of a dominant gene. The character is expressed in every generation, though as you can see it varies considerably in its effects. As you would expect, roughly half the progeny of a heterozygous carrier and a normal person are affected.

The second pedigree presents a very different picture. The one set of grandparents shown in the family tree appear to be unaffected. Of their six children, two are schizophrenic, one is an alcoholic, and three display schizophrenia-like symptoms. In turn, the children of these children have an even greater variety of mental disorders. Remarkably, most of the schizophrenic grandchildren are concentrated in the family of the non-schizophrenic alcoholic. In contrast, the one second-generation schizophrenic who reproduced has three children, none of whom have schizophrenia (though one is an alcoholic). In short, the pattern of inheritance seems to be completely different from that of the first family tree. The gene acts more like a recessive one, skipping generations.

A geneticist would argue that the gene is unlikely to be recessive, however, because that would mean two out of the four people who married into the family would have to be carriers of the gene. That would seem to be stretching coincidence too far—unless, of course, some of the people who married into the family were related to it by earlier family ties, which is not impossible.

Suppose we assume that the gene really is dominant, as suggested by the first family tree. To make this assumption fit the second tree, we would have to assume also that its expression in this family is extremely variable—so variable that its effects sometimes do not appear at all, like those of the gene for Gardner syndrome. The different patterns seen in the two trees reinforce the uncomfortable possibility that the two families are actually carrying different genes for schizophrenia, one with clear-cut and the other with variable expression. These are, after all, unrelated families, so there is no particular reason why the schizophrenia genes found in them should be the same.

Other models are possible. For instance, schizophrenia may be caused not by one gene alone but by the accumulated effects of several. In this model, a person might develop schizophrenia if he or she has a sufficient number of schizophrenia alleles at various genetic loci—if, in other words, a kind of genetic threshold has been reached. This is called a *polygenic model.*

If the polygenic model were correct, then the families picked by Gurling's group to study would be simply one extreme of the continuum of families found in the human population as a whole. The genes for schizophrenia would happen to be concentrated in these families, just as genes for tallness might be concentrated in some families and genes for shortness in others. After all, these families were picked out of thousands simply because of their high incidence of schizophrenia. Perhaps it is quite erroneous to conclude, on the basis of these families, that a mutant allele at a single gene is enough to cause schizophrenia. It would be like picking out that small fraction of families who happen to have only a mixture of very tall and very short members, and concluding from them that there must be a single gene for height, alternative alleles of which specify very tall or very short people.

An even more extreme view is that schizophrenia is entirely a result of the environment. If so, the disease could again be expected to cluster in families, particularly in those who happen to live in an environment conducive to its development. Environment certainly plays a role, for there are many cases of identical twins in which one is schizophrenic and the other is not. But it is not the only factor. Strong evidence points to a genetic component to schizophrenia as well. For example, the siblings of schizophrenics show an increased likelihood of having the disease. While this alone is not enough to show a genetic effect, since siblings also share an environment, careful studies have tried to disentangle the two. One of the best was a Danish study that examined children who had been adopted at an early age and thus separated completely from their biological relatives. The children with schizophrenic biological relatives showed a significantly higher incidence of schizophrenia than their carefully matched equivalents with no schizophrenic relatives.

What this genetic component might consist of is far more difficult to understand. Perhaps the data can best be explained by assuming that some cases of schizophrenia are caused by a single gene and others are polygenic—with a large environmental component in both situations. This possibility encourages Gurling and the others involved in the hunt, because it suggests that at least some cases of schizophrenia may be caused by a "major" gene.

When the evidence from the Canadian schizophrenics suggested that such a gene might reside on chromosome 5, Gurling's group was able immediately to conduct a simple test. They turned to the restriction fragment length polymorphism markers that we first met in chapter 8. A number of probes were available in the general area of the critical segment on chromosome 5, but not all of them could be used in any given family. You have already seen how it is necessary to find a gene or RFLP marker in which two or more alleles are present in a given family in order for the marker to be useful in linkage studies. As the number of markers grows, it will become easier to find ones that are useful, but in the meantime the human genetic map remains highly imperfect, with whole chunks missing.

Gurling and his group went to their freezer, where samples of all the DNA from their families were kept, and did whole-genome Southern blots using all the probes that were available. Two of the probes revealed RFLPs that were *segregating* (that is, two or more alleles were present, able to be passed to different family members) in each of the seven families. Given these data, the group could begin its linkage analysis.

The model they decided to use was that of a dominant gene with variable degrees of expression. Having settled on this model, they were faced with two major decisions. The first was to decide how often the gene expressed itself. If this number were set too low, then too many people who really did not carry the schizophrenia gene would be included in the analysis, weakening the power of the test. The second issue was how to decide which affected family members to include. Should they include people with mild schizophrenia-like symptoms, for example, or those with mental problems unrelated to schizophrenia that might be due to varying expression of the gene? The group solved this problem by running several different tests for linkage, first using only the family members who were clearly schizophrenic, then gradually including more and more of the "fringe" phenotypes.

The computer program moved the imaginary schizophrenia gene along the chromosome, calculating the likelihood of linkage to the two markers each time it was moved. They found to their excitement that the likelihood peaked near one of the markers. Both markers showed linkage to the presumed schizophrenia gene: They did not even have to postulate a low level of expression.

The data showed one worrying feature, however: As more and more fringe phenotypes were added, the linkage got stronger and stronger. This should not have happened. As people with a less obvious phenotype were included, the gene should have been expressed more and more erratically, increasing the proportion of mistakes and decreasing the likelihood of linkage being detected. Indeed, when only people with absolutely certain cases of schizophrenia were included in the analysis, there was just a slight indication of linkage, but not a significant one. For linkage to become really apparent, some of the fringe phenotypes had to be included.

On the other hand, it was remarkable that both the English and the Icelandic families showed linkage. This would seem to indicate that the schizophrenia gene is widespread, and that the same gene was present in all seven families they looked at.

WHERE ARE THE GENES FOR SCHIZOPHRENIA?

Then the story began to fall apart. Three other groups in different parts of the world began looking for the same kind of linkage in their own collections of families prone to schizophrenia. A far-flung group of scientists coordinated by Kenneth Kidd at Yale University examined a very thoroughly documented set of seven related families from the far north of Sweden. They used the same markers as Gurling's group, and also found some others that were useful. And they detected no linkage at all to the suspect region of chromosome 5.

The schizophrenia in these Swedish families seemed to be different from that in Gurling's families—more severe, and very resistant to treatment with drugs such as haloperidol. Because members of the Swedish families could all be traced back to a small group of people who had moved into that remote area some three hundred years before, it seemed likely that the schizophrenia gene (if it existed) was the same in all these families. Moreover, because it showed no linkage to chromosome 5, it appeared to be different from the gene that had been looked at by Gurling's group.

More data was soon forthcoming. Fifteen families from the southern part of Scotland were examined by a group at the University of Edinburgh. As in Gurling's study, these unrelated families showed a high incidence of schizophrenia, along with a great variety of other psychiatric problems. Unlike Gurling's families, these included manic depression. Once again, the gene or genes for schizophrenia in the various families seemed to be inherited as dominants, but would often skip generations, showing that they were not always expressed. As with the Swedish families, no linkage to the markers on chromosome 5 could be detected.

Finally, and more recently, a set of five North American families was examined by members of Gurling's group in collaboration with a group at NIH, and again no linkage was found. What did all these discouraging results mean? The most depressing possibility was that the Gurling group's first analysis was some kind of statistical artifact, a fluke that might happen once in a thousand or so cases and that just by irritating coincidence seemed to indicate linkage of schizophrenia to chromosome 5. Through such a fluke, apparent linkage could have been detected even in the absence of major genes involved with schizophrenia on that part of the chromosome.

A less dismal alternative is that there really are a number of major genes for schizophrenia in the human population. Any group of unrelated families, like those in the first Gurling study, might easily carry more than one of these genes. Even if a gene on chromosome 5 happened to be one of them, it might be carried by only one or two families in any given study, and any evidence for linkage would tend to be swamped by the effects of other families carrying different genes.

There may be a way to test this possibility. One intriguing approach that has not been tried would be to pool together all the groups of families in the various studies (excluding the Swedish group, in which all the families were related and none showed any linkage to chromosome 5). One could then pick out of this pooled collection only those families that seemed to show a slight hint of linkage to chromosome 5. Would these families have anything in common? Perhaps they might have an unusual number of individuals with a certain type of schizophrenia. Perhaps they might, when gathered together, appear to share other details of linkage and inheritance suggesting they all carried the same gene. I discussed this idea with Gurling, who has been looking into it.

Perhaps, though, the Gurling group was wrong to choose a dominant gene model. As I suggested a moment ago, schizophrenia could easily be polygenic—caused by many genes. Would this alternate model be able to explain the Canadian family with the translocation? Maybe—if an extra dose of a certain allele of a gene carried on that little piece of chromosome 5 were enough to trigger the symptoms of the disease in an otherwise normal person. But when that allele was present in the normal two doses rather than three, it would simply be one of a number of genes that could be involved in schizophrenia.

One powerful argument for why most schizophrenia should be caused by polygenes is that the disease is worldwide in its distribution and very common—twenty-five times as common in Caucasians as is CF. This alone suggests it is unlikely to spring from only one or two major genes.

If a genetic disease is common, the alleles that cause it must occur very frequently in the population. If we were to assume that mutant alleles at one or two genes could be responsible for all cases of schizophrenia, we would have to ask what would cause such disastrous alleles to be so plentiful. This question is especially pertinent because schizophrenics have fewer children than the average, so such alleles should be removed quickly from the population.

By contrast, the polygenic model assumes that alleles at many different genes, each with a small individual impact and a tendency to be strongly affected by the environment, are responsible for most cases of schizophrenia. It is easier to see how such alleles could be common. Under other genetic and environmental circumstances, they might play other, less damaging or perhaps even beneficial roles. We have already seen such a situation with gout. Like most cases of that disease, the majority of schizophrenia cases may occupy a twilight zone, in which many of the genes involved can play both positive and negative roles.

A number of variations of the single-gene and the polygenic models have been proposed by many workers in the field. I suspect that both models will turn out to be partly right. Single genes might be responsible for some of the familial cases, with polygenes contributing to many of the cases that occur in isolation.

Despite our limited current understanding, enormous advances have been made in therapies for schizophrenia, and slower but substantial advances have been made in the attitude of society and families towards mental illness. Anne Bassett's patient responded well to drug treatment and has since returned to school. The favorable outcome of his case had much to do with the strong support of his family throughout his ordeal. Schizophrenics who both do not respond well to drug treatment and lack such familial and societal support often face a far bleaker fate.

Such cases lend urgency to the search for an understanding of the genetics of schizophrenia. Should most schizophrenia turn out to be polygenic, researchers will face a very difficult problem. Ordinary genetic models are woefully inadequate at handling cases in which a number of genes are involved in the expression of a character. Because the genes can interact with each other and the environment in so many different ways, a geneticist trying to tally them is in the position of someone who can only count, "One, two, three, many." In the case of schizophrenia, there could be as few as four or five genes involved, or as many as hundreds.

Still, this does not mean that we cannot eventually track down at least some of the genes involved. Once a sufficiently complete genetic map for the human genome is available, some of these polygenes will surely be located. To see how this will eventually be accomplished, and to get a glimpse of the problems that will arise on the way, let us turn to a group of organisms very different from ourselves, in which this tracking-down process is much more advanced.

While we have focused in this book on the human genome, exactly the same techniques and approaches can be taken with many other organisms. In some cases other creatures are far better suited than humans to particular experimental challenges. One of these can be found in what Dr. Johnson called the vegetable part of creation.

STALKING THE WILD TOMATO

Tomatoes are native to the Andes, which are the original source of an astonishing variety of edible plants—potatoes and sweet potatoes, beans, squash, cocoa, and peppers among others. (This area is also, as we have become only too aware, the source of the infamous coca bush.) In their native Andean lowlands and highlands, wild tomatoes of the genus *Lycopersicon* are found in the form of various species that differ greatly from each other, ranging from fifty-foot vines to small bushes with fruit that is almost invisible. Edible tomatoes used today in Peru attain a great range of colors when ripe—red, green, white, yellow, orange, and even purple.

By the time the Conquistadors arrived in Mexico, a few of these varieties had passed from one Mesoamerican culture to the next, diffusing up the Isthmus of Panama to meet them. These tomatoes had already been selected over thousands of plant generations for size, plumpness, and flavor. The Conquistadors brought some of them back to Spain, and the plant quickly became a staple of Mediterranean cuisine. The cherry tomato is one type that has come down to us almost unchanged from that time.

This vegetable booty of the New World now represents far more wealth than all the gold that Spain plundered from the ruined empires of the Aztecs and the Incas. Even so, the plants the Conquistadors brought back carried only a tiny fraction of the available genetic variation in this cluster of species. This was particularly so because the plants the Indians were growing had already been selected to breed true. This involved selection for the facility of mating with themselves, something most wild tomatoes do not do.*

Tomatoes were slow to be accepted in much of the Old World. Because they are members of the deadly nightshade family, northern Europeans assumed they were probably poisonous. They could be grown as ornamentals, but never eaten. This superstition was publicly challenged in 1820 by one Robert Johnson, a world traveler who had no fear of tomatoes. He stood on the courthouse steps in Salem, New Jersey, and ate several in front of a horrified crowd without any ill effects.

Even after this convincing demonstration, the myth that tomatoes were deadly took decades to fade. They are now the eleventh most important agricultural crop in the United States, with a total harvest in 1989 of over twelve million tons. Although this represented a hundred pounds of tomatoes for every man, woman, and child in the country, it seemed not to be enough to satisfy our insatiable demand; in that year we still had to import another 700,000 tons.

Most tomatoes do not end up in salads, but instead contribute their inimitable flavor to ketchup, pizza, spaghetti sauce, tomato soup, and innumerable other foods. The tomatoes that are used for these foods are not plump and luscious, tenderly plucked from the vines at the peak of ripeness by workers in the fields. On the contrary, they are harvested by machines as soon as they begin to turn red, hurled by conveyor belts into trucks where they are piled hundreds deep, and dumped into huge warehouses to await reduction to tomato paste. These field tomatoes are the vegetable equivalent of battery chickens. Ideally, they should be hard in order to withstand the rigors of mechanical picking, have a high density of the soluble sugars and other solids that contribute to the paste, and have a sufficiently acid pH to discourage bacterial growth during processing. These properties are very different from those that we find desirable in a garden tomato.

Indeed, any increase in these properties can have an enormous economic effect. For each tenth of a percent increase in the soluble materials that can be converted to paste, processors will save seven million dollars a year in energy costs alone, not to mention the great increase in yield of tomato paste per acre. As a result, agribusiness has belatedly realized that the wild strains of tomato growing in their native Peru are an im-

*One notable exception is *L. cheesemanii,* confined to the Galapagos Islands, which produces tiny orange fruit and which must pollinate itself because there are no insect species on these remote islands capable of cross-pollinating it.

mensely valuable resource. These strains are sure to differ from our common domestic strains of tomato in all these respects, as well as in many others, such as disease resistance. Some of the alleles carried by the wild strains will be of great potential economic value, provided that they can be extracted from the domesticated tomato's wild relatives without bringing undesirable characteristics along with them. As we explore these problems, we will see that they reflect in miniature the problems we will face as we try to understand, and perhaps eventually to modify, the genes responsible for polygenic traits in our own species.

Charles Rick, considered by his peers to be the guru of tomato genetics, is a large, gentle man with a white beard who inhabits an office in the heart of the University of California campus at Davis. Unstable-looking shelves bulging with books and documents line the office from floor to ceiling. I expressed to Rick some relief that the epicenter of the recent San Francisco earthquake had not been any closer to his domain, and he replied that he was relieved as well, since he would otherwise surely have been buried in his own accumulated records of thousands of crosses between plants of the tomato family.

Rick took me around his greenhouses on the edge of campus, where many of the wild relatives of our domesticated plants were thriving. One species, *Lycopersicon hirsutum,* was particularly striking, producing small green fruit covered with long white hairs. Its pungent odor was striking as well, far stronger and more unpleasant than that associated with most tomato plants. Yet because this variety is completely resistant to the common insect pests that afflict the domestic tomato, *L. esculentum,* it represents a valuable source of resistance genes. These genes must be handled with care; were they simply to be introduced into the domestic tomato *en masse,* they would make the fruit far more toxic to consumers than any current pesticide treatment.*

Some of the other species populating Rick's greenhouse were once found in the coastal plains of Peru, but had been wiped out as a result of intensive agriculture there. In one of the ironies that abound in recent agricultural history, the only tomatoes now grown commercially in Peru

*This happened recently with potatoes. A disease-resistant variety called Lenape was released to the farming community by the USDA, and shortly afterward a Canadian investigator growing some in a greenhouse became ill from eating them. The strain was hastily withdrawn.

are those that have been imported from North America. Unfortunately, the North American species is very susceptible to infestations of the roots by nematode worms that are native to Peru and that have multiplied through the widespread use of irrigation. As a result, for a time, disaster threatened the Peruvian tomato industry. The reader will no doubt guess where genes resistant to the Peruvian nematodes were found—in the native plants, of course, driven to near-extinction in their usual haunts but still luckily to be found in Charles Rick's greenhouses. Rick and his coworkers were able to save the Peruvian tomato industry by reintroducing these genes.

Although the genetic potential of many species remains to be explored, Rick no longer travels to the interior of Peru to replenish his supply of native plants. The camp of his son, an anthropologist, was briefly invaded by Marxist guerrillas in 1986. Along with most other scientists, the Rick family took this rather broad hint and abandoned the Peruvian highlands to the guerrillas.

Figure 12–3. Some of the tomato species Charles Rick works with. The one in the center is of course *L. esculentum,* the familiar garden tomato. *L. chmielewskii* is to its right, *L. hirsutum* is at the lower left, and *L. cheesmanii* from the Galapagos Islands is at the lower right.

In another greenhouse just south of Davis, I was shown some rather unhappy-looking representatives of a wild tomato with the tongue-twisting name *Lycopersicon chmielewskii.* This plant, found only in Peru, is about as large as a domestic tomato plant, but it produces tiny hard green fruit about a fifteenth the size of those of *esculentum.* These fruit, inedible as they are, are nonetheless the proud possessors of 10 percent soluble solids, compared with only 5 percent for *esculentum.* A moment's calculation will tell you that if *chmielewskii* genes could transfer into *esculentum* and double the latter's soluble solids without harming any of its valuable properties, the tomato industry could stand to gain at least a third of a billion dollars a year. Particularly since these would be direct energy savings, this is a goal well worth striving for.

Over a decade ago, Charles Rick appeared to have succeeded in producing plants with large red fruit and almost twice the soluble solids of the domestic strain. He did it by bringing in genes from yet another wild Peruvian species, *L. minutum.* Unfortunately, few fruit were produced per plant, and the fruit were soft, spoiling rapidly and often cracking as they ripened. In spite of the care taken with the crosses, undesirable characters from *L. minutum* had been dragged in along with the desirable ones.

Rick's student Steven Tanksley, now at Cornell, wondered whether it would be possible to combine the old-fashioned methods of plant breeding with the new methods of molecular biology, to locate on the chromosomes the genes responsible for such characters as soluble solids and see which of them could be most easily pried free from undesirable traits that might be linked to them on the chromosomes. He knew that more than one gene must be responsible for determining the levels of soluble solids, for the same reason that more than one gene must cause schizophrenia. Characters controlled by various alleles of more than one gene are called *quantitative characters,* and they are the most slippery and difficult kinds of characters that the geneticist has to deal with.

Tanksley began with *L. chmielewskii,* the sad-looking species I had seen in the greenhouse. This wild tomato grows well in its native habitat of Peru but very badly in California. When it is mated to the domesticated tomato *esculentum,* the yields of the hybrids are poor. Indeed, many of these hybrid plants will not even set fruit. Some of the quantitative alleles that produce this problem are presumably the same ones that adapt the plant to growth in the cool, mid-altitude valleys of Peru. They make it very maladapted to the hot, low-altitude valleys of California.

These alleles may or may not have anything to do with the amount of soluble solids in the fruit, but it seems quite possible that some of them do affect this character. One obvious adaptation to a high-altitude existence would be an increase in the amount of soluble solids, making it more difficult for the fruit to freeze. Alleles producing this effect, however, might easily turn out to have other effects making it very difficult for the plant to survive at low altitudes.

In addition to these alleles with their multiple influences, the wild species brings with it a huge freight of other quantitative genes for which the plant breeder has no use—genes for small fruit size, for slow ripening, for thick skin, and so on. In a few places even the order of the genes on the chromosomes is different in the two species, leading to problems of fertility in the hybrid. The result of all this complexity is that it would take many generations of patient backcrossing and selection before the economically useful genes could be sorted out from the useless or harmful ones.

This is where Tanksley hoped that the tools of molecular genetics would come into play, allowing him to locate the genes for these quantitative characters on the chromosomes. His goal was to plan out a rational strategy for transferring the genes that the plant breeder wanted and leaving the unwanted ones behind.

The first thing required to accomplish this was, as you might suspect, a good genetic map. By the time the molecular biologists appeared on the scene, the genetic map of the tomato was already excellent, with over 300 mutant genes pinned down to loci scattered over the 12 chromosomes. However, this map suffered from a great drawback that made it useless for locating these quantitative genes. The mutant alleles at the various mapped loci usually made the plants ill or malformed, so that no individual plant could carry all these alleles at once; it would die first. And without such a plant, it was not feasible to map all the many genes that might contribute to a single quantitative character. This is because the quantitative polygenes might be anywhere on the chromosomes, so that tracking them down would require endless crosses to tester plants that each carried just a few of the mutants.

Tanksley and his colleagues thus were forced to construct an almost completely new map for the tomato, one much more useful for this kind of hunt. Like similar human maps, the map used RFLPs and a few visible markers. It also achieved the completeness of the earlier map, with over 300 markers. The great advantage over the earlier one was that any number of the markers could be combined in a single plant. Like humans carrying various combinations of RFLPs, plants with different RFLPs are quite healthy, since most RFLP markers have no obvious effect on the plants themselves.

At the time of this writing about 3,000 RFLPs and other markers have been mapped in the human genome. Although these are far more thickly scattered than in the tomato map, a great deal of tedious work was involved, and the map's construction depended largely on a few good human pedigrees with many family members. As we saw with the CF story, even with all these RFLPs, tracking a human gene down to a particular chromosome is still an enormous task.

Plant breeders have a far easier time of it. They can start with inbred lines of plants that differ at many RFLP markers, and mate them together. The plants that result are heterozygous for all these RFLPs. When *these* plants are mated with each other (or, as is possible with many complaisant plants, with themselves), the RFLPs will distribute themselves in innumerable combinations in the following generation. If two RFLPs are always, or nearly always, inherited as a unit, this shows they are tightly linked.

This is just the kind of cross that Gregor Mendel, the founder of genetics, carried out almost a century and a half ago with his famous pea plants. Because Mendel relied on visible characters, however, he was only able to follow one or two genes at a time. Because he followed so few genes, he never detected any genetic linkage. The modern molecular biologist can follow hundreds of RFLPs in a single cross. Indeed, Tanksley's group did just this with tomatoes. An inbred line of *L. esculentum,* the domesticated species, was crossed to an inbred line of a close relative, *L. pennellii.* The hybrid was allowed to fertilize itself, producing plants with many different combinations of *esculentum* and *pennellii* genes in the next generation. Random pieces of DNA from *L. esculentum* were used to probe whole-genome Southerns of these varied progeny, in order to see which RFLPs were inherited together. A beautiful map was quickly constructed.

Good as this map was, each marker still had to act as a signpost for a large chunk of DNA. As the size of the tomato genome is about 700 million base pairs, probably carrying about 40,000 genes, each marker stood for a piece of the genome about 2 million base pairs long, in which some 125 genes were embedded.

The next step in the hunt for quantitative genes was to search for a tomato relative that was dramatically different from *L. esculentum,* preferably one differing from it in some quantitative character that might be economically useful. An obvious choice was the Peruvian species, *L. chmielewskii,* with its fruits full of all those sugars and other valuable solids.

Although the hybrids between *esculentum* and *chmielewskii* were miserable creatures, pollen could be obtained from them. This pollen was used to fertilize *esculentum* flowers. Since each of the hybrid pollen grains had some different combination of *esculentum* and *chmielewskii* genes, the plants in the next generation were a wildly eclectic bunch, with all kinds of combinations of fruit size, pH, color, soluble solids, and so on.

Next it could be asked which of the known RFLP and other gene markers that had already been mapped were associated with which of these characteristics. The DNA from 237 of these diverse progeny was probed with the original set of probes, and the data were analyzed using an ingenious computer program developed by Eric Lander of Harvard University. Unfortunately, not all the 300 RFLPs and other markers were present in this hybrid; only about 70 of them could be used. Still, this was enough to learn a lot about where the genes were that made *esculentum* and *chmielewskii* fruit so different.

Consider size. Some of the hybrids had very large fruit, others very small fruit. Most had fruit that was intermediate in size. When Tanksley and his coworkers looked at the plants with very large fruit, they found that these almost always carried *esculentum* RFLPs that they knew were located on certain regions of chromosomes 1, 4, 6, 9, and 11. When they looked at the plants with very small fruit, they found that those almost always had *chmielewskii* RFLPs in those same regions. In addition, fruit with lots of soluble solids tended to have *chmielewskii* RFLPs on certain regions of chromosomes 3, 6, 7, 8 and 10, while those with lower levels of soluble solids usually had *esculentum* RFLPs at these same locations.

It must be remembered that the RFLP markers are almost certainly not located in the genes that actually affect fruit size, soluble solids, and pH. Like buoys that mark sunken treasure, the RFLP markers give only a rough idea of where these genes are located. It would take just as much work to walk along a tomato chromosome from the nearest RFLP marker to the gene that actually affects the character as it did to track down the gene for cystic fibrosis! Furthermore, only about half of the variation in fruit size, soluble solids, and pH could be accounted for by these RFLP-marked regions. The rest of the variation must have come from the environment, from genes with small effects, and from interactions between the genes and the environment that were too complicated to be unscrambled by these crude methods.

There were regions on several of the chromosomes that contributed substantially to both fruit size and soluble solids. It was striking that in every case, a region that increased fruit size decreased soluble solids, and vice versa. This confirmed the worst fears of the plant breeder, for it appeared that these characters were negatively correlated. The result would be that if one tried to select for large fruit size starting with these hybrids, one would inevitably select for low soluble solids.

Yet even when such characters are negatively correlated, molecular biology provides hope for the breeder. The genetic map of tomatoes is so crude at this point that it is quite possible that the genes for soluble solids and for fruit size will turn out to be different, even though they are closely linked on the chromosome. In the old days before the advent of molecular biology, it would be necessary for the plant breeder to produce a huge number of plants and hope for a rare recombination event that would separate the gene that was wanted from the gene that was not. This was both expensive and risky. Today, with the aid of the RFLP map, the breeders now know the approximate locations of the genes they are interested in. In principle, it should be possible to find the genes, clone them, and eventually determine how they work. And it should eventually be feasible to do gene therapy on the domestic species, producing tomatoes of any size, pH, and soluble solid content, and launching a new era of designer tomatoes.

Unfortunately, problems are already beginning to arise. I spent much of an afternoon with Joseph de Verna, one of Tanksley's coworkers, going over in detail the results of further crosses he had carried out to try to track down more precisely just where the genes for soluble solids are relative to the RFLPs on the chromosomes. The effects of the soluble solids genes could still be detected in a few of these new crosses, but in most of them the effects tended to vanish like the victims of Lewis Carroll's Boojum.

There were four reasons for this. First, de Verna was only working with one out of perhaps half a dozen genes that contributed to the character in the original species, and the effect of the single gene in isolation was slight and difficult to measure. Second, in each cross there is always some unavoidable environmental variation that contributes to uncertainty in the results. This effect loomed as large in each of his later crosses as it did in the original one, tending to mask the already small effects of the isolated individual genes. Third, each cross shuffles *all* the genes of the plants, dealing them from the genetic deck in different orders and combinations. As a result, these pioneer molecular plant breeders are rediscovering a truism plant and animal breeders have known for decades: A gene may behave one way when it is embedded in a background of many other genes from one strain, but in quite a different way when it is embedded in a background of genes from another strain. Finally, it is difficult to keep the environment constant from one generation to the next. You will remember that the *chmielewskii* plants grow far better in Peru than in California, illustrating the fact that what happens in one environment may be quite different from what happens in another. Indeed, in early 1991 Tanksley's group published an extensive study of this very effect, showing that many of the genetic regions they had previously tracked down in the hybrids gave quite different results when plants carrying them were grown in different parts of California and in Israel.

What relevance does all this fuss about tomatoes have for humans? Put simply, it is likely that the genetic properties of human quantitative characters will be found to resemble those of tomatoes. We already know from classical genetic analysis that human characters such as height and skin color are inherited in a quantitative way. Half or more of the variation in these characters appears to be controlled by a relatively small number of genes, just as the quantitative characters in tomatoes are. But will this apply to schizophrenia? Susceptibility to alcoholism? And even intelligence? Is the variation we see in these characteristics, or at least a good fraction of it, actually controlled by a relatively small number of genes that in principle could be tracked down and understood?

There are many roadblocks in the way of doing this. One is that the sorts of crosses the tomato breeders can carry out with relative ease cannot be done with humans. Even though many more RFLP markers are known for the human genome than for the tomato, we do not have access to highly inbred lines of humans or—luckily—the power to make them mate with each other. We certainly could not produce huge numbers of progeny from such crosses even if we could make them. A second roadblock is that the effects of human quantitative genes, like those of such genes in the tomato, are likely to vanish the more we try to isolate them, for exactly the same reasons that caused the effects of de Verna's tomato genes to fade away even as he tried to follow them through a series of crosses.

A third difficulty has to do with the complexity of the quantitative character being examined. A quantitative character such as height is straightforward: You measure it with a ruler, and few major genes seem to contribute to it. Perhaps it will turn out that there are only a few major genes contributing to schizophrenia as well. But a quantitative character such as intelligence is sure to be far more complicated. Trying to measure it with a ruler such as an IQ test has given rise to endless controversy in the past. Any genetic search based on the assumption that half a dozen genes control intelligence is sure to fail in the future. We will see in the next chapter why this statement can be made with confidence.

CHAPTER 13

Fanfare for the Common Person

Think not that the nobility of your ancestors doth free you to do all that you list; contrarywise, it bindeth you more to follow virtue.

—Pierre Erondell

Hospitals have played a pivotal role in the exploration of the human genome, and we have visited quite a few of them in the course of this book. The last one I will introduce you to is the Hôpital de Saint-Louis in Paris, not far from the Gare de l'Est. This hospital was founded by Henri IV in 1607, the year after a devastating plague decimated the city. He named it after its most logical patron, the sainted Louis IX, who had died of the plague in Tunis in 1270 on the way to his second crusade.

A few of the original buildings have managed to survive fires over the years. Steep-roofed, gray-walled, and bedecked with flowers, they form a peaceful oasis in the center of the hospital complex, quite dwarfed by the modern buildings that surround them. Some little distance away lies a quiet street of offices and administration buildings, the rue Juliette Dodu, named after a forgotten heroine of the Franco-Prussian war. Here another war hero presides benignly over one of the front lines in the exploration of the human genome: the Centre d'Études du Polymorphisme Humaine, commonly known as CEPH.

He is Jean Dausset, tall and distinguished, with a Gallic profile. When the Germans invaded France in 1940, he fled to North Africa and joined the free French army. With his doctor's degree he found himself in a medical division, in charge of blood transfusions. This introduced him, literally under fire, to the problems of matching the blood types of donor and recipient.

These puzzles of immune compatibility and incompatibility continued to fascinate him on his return to France after the Liberation. He knew that most of the difficulties with transfusions lay in reactions between antibodies of the recipient and certain *antigens,* specific molecules found on the surface of the red blood cells given by the donor. Not all the problems are due to this interaction, however, and he wondered whether some of them could be traced to the much rarer white blood cells. If they had antigens that differed from those of the red cells, this could be the source of further complications.

In fact, this turned out to be the case. Dausset's pioneering work was the first glimpse in humans of a new and complex maze of antigens, now known as the HLA (human lymphocyte antigen) system. These antigens float in the membranes on the surface of white blood cells, in the same way that the better-known antigens of the ABO and Rh systems float on the surface of red blood cells. The HLA gene complex has turned out to be the second most *polymorphic* set of genes yet found in the human species—meaning that many genes in the HLA complex come in a great variety of allelic forms. (We met the most polymorphic genes, the VNTRs, in chapter 8.)

Dausset received the Nobel Prize in 1980 for his pioneering work on HLA. He has remained entranced by the implications of this and other polymorphic genetic systems such as RFLPs in the study of human genetics and the exploration of the human genome. This lifelong interest led him to found CEPH in 1984 for the further study of human genetic polymorphism. The group now has some seventy-seven collaborators worldwide (Collaborative Research was initially part of CEPH), and together they are developing their own independent RFLP map of the genome. The principle on which CEPH was founded was to establish cell lines from large human families, and then make the DNA from these cell lines freely available to workers around the world to help them track down genes.

Howard Cann, an expatriate American who has been with Dausset for ten years, showed me around the spotless facility and treated me to a cup of French coffee which was—to put it mildly—unlike anything I had ever tasted. He drank calmly from his cup while I looked in vain for a potted plant to pour mine into.

The CEPH collection consists of frozen white blood cells from some sixty-one large families. About half are from Ray White's collection of large Utah families; the rest are from France, Venezuela, and several other countries. In forty-four of these families cells have been obtained from all of the grandparents, making them particularly valuable for genetic studies. Cann showed me how DNA extracted from these cells is packaged sterilely and shipped without charge to qualified investigators. Quality control is essential in work of this kind. Before any DNA is sent out, it is hybridized to a known set of VNTR probes to check for contamination. In effect, the genetic fingerprint of each DNA sample is taken.

The other half of the CEPH enterprise involves the accumulation of a growing set of genetic probes that can be used to track RFLPs through the various generations of CEPH's families. CEPH is engaged in a long-range collaboration with the American Type Culture Collection to put together "kits" of probes from particular chromosomes or parts of chromosomes. These kits will also be made available to the scientific community without strings.

CEPH, like the computer data bank at Los Alamos and the sorted gene libraries from Los Alamos and Livermore, is an essential part of the exploration of the human genome.

On my visit I chatted with Dausset about the consequences of the discovery of all this polymorphism in the human species. I commented that repeated studies had shown how genetic differences between individuals chosen at random were far greater than any slight differences between races, and wondered what conclusions he drew from that. He gave an amused and very French response. "Ah, you Americans," he said. "Always concerned about race. I am reminded of a comment Napoleon made after one of his battles. When he had surveyed the heaps of dead, he remarked, 'This is nothing that one night in France will not put right.' Let me add that this matter of race is nothing that one night of random mating would not put right!"

Dausset is quite correct in principle, though it would take more than one night of random mating to remove completely the phenotypic extremes among humans that feed racial prejudice. Still, his reply had avoided my question about individual differences, a question that lies at the heart of concerns that many people have expressed about the Human Genome Project. As I remarked to Dausset, racial differences have repeatedly been shown to account for only some 3 to 10 percent of the genetic diversity harbored by our species. It is possible to predict that as the Genome Project advances and more is learned about our genes, claims of a genetic basis for racial prejudice will begin to sound more and more hollow.

The reverse of this argument, however, is less comforting to liberal sensibilities. Over 90 percent of genetic differences are found between individuals within races. Alleles of many different genes are sure to be discovered that have a measurable impact on personality, behavior, and aspects of intelligence. There is no reason to suppose that these alleles will not be found in all races approximately equally. While it is to be hoped that prejudice against races will be lessened by these discoveries, will prejudice against the individuals who carry certain of these alleles be increased? We are now ready to examine that question.

THE GENEALOGICAL TRAP

My father was inordinately proud of the fact that a remote ancestor of his, also with the name of Wills, had been a privateer. He felt that this added a touch of the romantic to what was otherwise an extremely humdrum genealogy.

As a child, I was agog for tales of this dashing figure, whom I imagined to be rather like Errol Flynn in *Captain Blood.* Alas, my father informed me, the reality was not so exotic. The ancestor, who was my father's great-great-great-grandfather and flourished during the time of the Napoleonic wars, had received a charter from George III permitting him to rob the ships of the French and their allies on the high seas. With a group of other investors, he bought a couple of ships and sent them out. (I like to think that his ships were under the command of captains who *did* resemble Errol Flynn. Childhood illusions die hard.) Then he sat in his London office, patted his large stomach in anticipation, and waited for the money to roll in.

Roll in it certainly did. The resulting fortune lasted all the way down to the time of my grandfather, who managed to lose the remainder of it in 1920 in a scheme involving patented lifeboat davits. This immediately plunged the family from the upper-upper to the lower-lower middle class.

The legend of the privateer, however, lived on. As both the fortune and the name had come down to my father's time, he imagined that his ancestor's business acumen and propensity for making fortunes had too.

As my genetic expertise grew, I was able to point out to my father that in fact he shared only about 1⁄32 of his exotic ancestor's genes, or roughly three percent. If all my father's ancestors were followed back five generations, with the number of forebears doubling each generation, his family tree would be made up of sixty-one people in addition to the privateer. While most of these were probably upstanding citizens, a few were likely to be lunatics, habitual drunkards, embezzlers, and the like. And all of these would have contributed at least as many genes to my father's chromosomes as had the privateer. Figure 13–1 shows a sketch contrasting my father's view of his genealogy with mine.

Predictably, these arguments had no effect on my father. Just as a gambler never remembers the horses that lose, the person who is proud of his or her ancestry never mentions or thinks about the overwhelming majority of those ancestors who were run-of-the-mill, much less the small percentage who might have been criminal, insane, or retarded. Genealogical trees do tell us some interesting things—they just happen not to be the things that the genealogists think are important. In figure 13–1, you will notice several of them.

The first is that it is much easier to trace back a paternal line than a maternal one, because of the name. The maternal lineage is usually followed out for only one or two generations and then abandoned in most genealogies, thus immediately ignoring most of the maternal half of a person's ancestry. And much of the paternal ancestry is ignored as well. As one travels back in time, the paternal lineage receives a genetic contribution at each generation from a different maternal lineage with a different family name. Going back as few as one or two generations, these multiplying maternal lineages become progressively harder to find and to trace. Few of us know our paternal great-grandmother's maiden name, or even how to track it down. Within a few generations there are dozens of such lines to be traced back, only one or two of which the genealogist concentrates on.

Something else that is apparent from the figure is that if all our ancestors were unrelated to each other, their numbers would exactly double as we went back through successive generations. I would only have to follow my family tree back to A.D. 1000 or so to find my ancestors numbering in the billions—far more than the total number of people living on the planet at that time. This is obviously impossible, but the difficulty can be resolved if we take inbreeding into account. It turns out that all of us have a great deal of inbreeding in our ancestries, though very little of it was close enough to upset the theological establishment at the time it took place. The cumulative effect of this inbreeding is to reduce the number of our ancestors, since the same person may turn up in many places in a family tree. This is particularly the case when the tree is traced

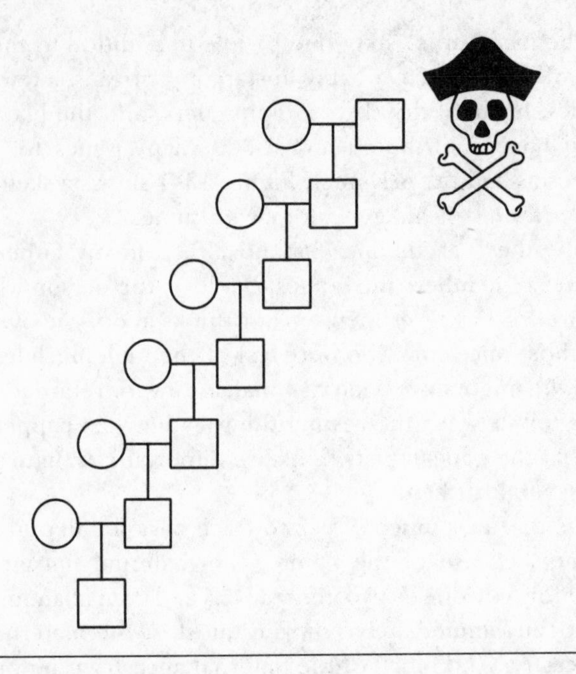

Figure 13–1. Above, a five-generation family tree in which only the patrilineal line of descent has been followed back, and, on facing page, the same tree when all the ancestors are traced back.

back to a time when our ancestors lived in small genetically isolated villages.

All this inbreeding does not reduce the diversity of our ancestors, however. Were we patient enough, most of us could trace our ancestries back through multiple lines, not to one village but to dozens or hundreds, scattered across entire countries or even continents. Because of this, each of us represents a small sampling of the genetic heritage of a large fraction of the human species.

None of these fine points bothered my father. He simply assumed that somehow his ancestor's desirable characters had been passed down to him via the paternal line through the intervening four generations. There is some small possibility that this might have happened if all of his ancestor's rather raffish entrepreneurial skills had been wrapped up into a single gene. That gene would then merely have had to make its way through five successive generations down to my father. The chance that a gene will be passed from a parent to a particular offspring is one half.

Figure 13–1. Continued

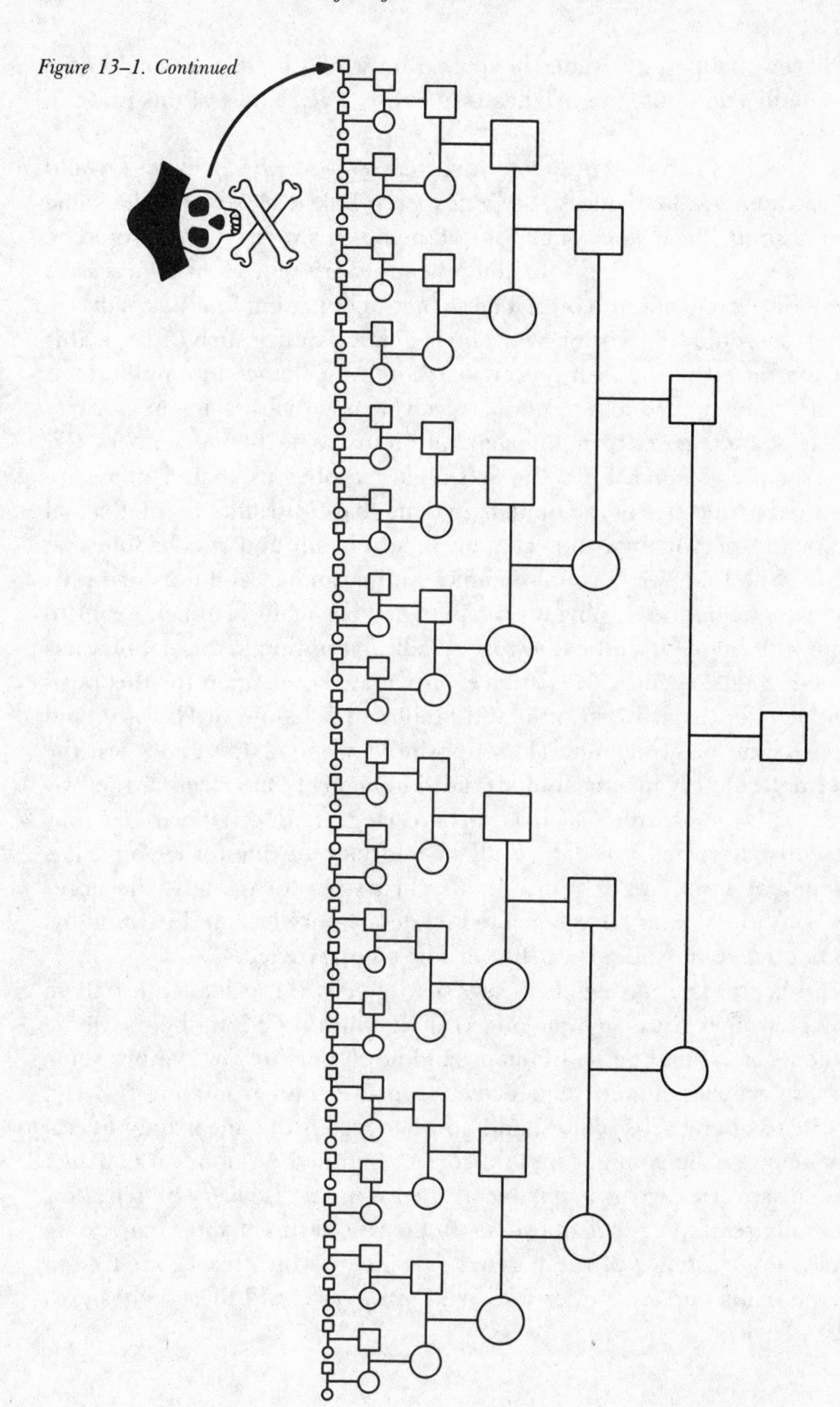

To have it happen five times in succession would be like tossing a coin repeatedly and obtaining five heads in a row. The chance of this is $(1/2)^5$, or $1/32$.

If as many as four genes were involved, however, the chances would lessen dramatically. Unless the genes were linked closely on the same chromosome, the chance of all four of them reaching my father would be the same as the chance of obtaining heads every time in five successive tosses of a group of four coins. The chance of obtaining four heads in one such toss would be $(1/2)^4$ or $1/16$. The likelihood of five such tosses is this fraction raised to the fifth power, or about one chance in a million.

Such a genetic model works very nicely if you are following one or two clearly defined genes from one generation to the next—the gene for CF, for example, or an RFLP. But it should be obvious that if there are inherited properties of the human genome that contribute to intellectual ability and personality, they cannot be so blatant and predictable that they can be followed through families in the same straightforward way.

What may not be so obvious is whether there really are such genes to begin with. Indeed, a long liberal and radical tradition in this country has argued against their very existence. Articulate spokesmen for this position include Richard Lewontin and Stephen Jay Gould of Harvard, and Leon Kamin of Princeton. The opposite viewpoint, that genes are the chief determiners of our abilities and intelligence, has been argued by conservative spokesmen such as Richard Herrnstein of Harvard, Arthur Jensen of Berkeley, the late William Shockley of Stanford, and Hans Eysenck of London's Institute of Psychiatry. Unfortunately, the arguments on both sides have become hopelessly mired in, and colored by, the political philosophies of the various men involved.

The brutal fact is that there are such genes; the evidence for them grows stronger with each passing year. It will not be long before some of them are located on RFLP maps, and not long after that before some of these are cloned and sequenced. No matter how wounding this fact may be to liberal sensibilities, and how dangerous the unearthing of real genetic inequalities might be to a society founded on political equality, these discoveries are as inevitable as the eventual discovery of genes for manic depression or schizophrenia. In the rest of this chapter I am going to take the existence of these genes as a given, which leaves me free to ask the really pertinent question. What properties will these genes turn out to have?

THE RANDOMNESS OF GENIUS

Francis Galton, Charles Darwin's cousin, lived solidly within the Victorian era, enmeshed there in both time and outlook. An adventurous and observant traveler in the Middle East and southwest Africa as a youth, on his return to England he quickly rose in the scientific establishment. His work in psychology, anthropology, and biomathematics was seminal. His invention of the idea of correlation led to the foundation of the field of statistics. He was the first to suggest the use of twins to measure the roles of heredity and environment in the expression of different characters. One outcome of his lifelong interest in anthropometrics was the development of fingerprints (real, not DNA fingerprints) as a means of identification.

Galton loved to collect data, and did so with abandon. Like the Count on "Sesame Street" he counted everything, though unlike the Count he put the data to good use. He collected data on phenomena ranging from the proportion of pretty girls on the streets of British cities (London won) to the number of fidgets per minute in attentive and bored audiences at Royal Geographic Society lectures.

One enduring passion, in a lifetime devoted to quantification, was the measurement of genius. Galton defined a genius as did Dr. Johnson: as "a man endowed with superior faculties." As this was long before the concept of an intelligence test or intelligence quotient, he could only roughly quantify levels of genius on the basis of the degree of eminence reached by the men (and the few women) from many walks of life whom he studied. The questions he asked form the core of the present chapter. Is genius inherited, and if so, how?

The first question was answered resoundingly in the affirmative in Galton's 1869 book, "Hereditary Genius." The book is a remarkable farrago of shrewd observation, blatant editorializing, historical revisionism, anticlericalism, racism, and sexism. In short, it is wonderful fun to read now that his more outrageous statements have been softened by the patina of time.

Galton divided up both historical figures of the Western world and eminent Europeans of his time into those who excelled in jurisprudence, politics, military prowess, literature, science, music, and so forth. He then determined, as far as he was able, the degree of eminence achieved by close relatives of these people. Finally, he used the ratio of eminent people to the general population to ask whether the relatives of eminent individuals were eminent more often that one would expect by chance.

Mirabile dictu, they were. The sons of judges were far more likely than the general population to be judges, the sons of churchmen to be churchmen, and so on. The likelihood dropped off rapidly with more remote degrees of relationship. Galton, however, was not so naive as to suppose that there were hereditary factors for jurisprudence or religiosity. He knew, though in the book he said surprisingly little about it, that the environment, particularly parental example and influence, must play a role in determining the sorts of careers that sons embark on. His contention was simply that the characteristics of intelligence, determination, political skills, and so forth that set eminent people above the ordinary were likely to be shared by their close relatives, something that in his view the data amply bore out.

In a few cases the pattern did not hold. Mathematicians seemed to stand alone, except in such rare instances as the Bernoulli family. Galton shrewdly explained this by pointing out that eminence in mathematics is very difficult to achieve. Only a tiny fraction of people skilled in thinking in the abstract can become the mathematicians whom the world remembers. It is perhaps less demanding to become eminent in other fields.

These early studies of hereditary genius shaped Galton's thinking profoundly, and led him in later years to propose the idea of eugenics. His concept of eugenics has since fallen into disfavor, but his observations about genius remain. What are we to make of them now?

Remarkable family trees do exist, though it is very difficult to disentangle the effects of genotype from those of the environment. Perhaps the most famous of these trees is that of the Bach family.

We know a lot about the early generations of this tree because Johann Ambrosius Bach (1645–1695), the father of Johann Sebastian (1685–1750), was very proud of his distinguished musical ancestry. Johann Ambrosius was a court musician in the Thuringian town of Erfurt. He began to collect music written by his family, a collection that over time became known as the *Alt-Bachisches Archiv.* His famous son, and later his grandson Carl Philipp Emanuel, added to the compendium. The earliest works included were those of a great uncle, Johann Bach (1604–1673), and there were contributions from many collateral members with various degrees of composing skill.

The Bachs traced their ancestry to one Veit (Vitus) Bach, who had arrived in Thuringia, probably from Hungary, around the year 1550. He may actually only have been visiting Hungary temporarily, since other members of the Bach clan were apparently living in Thuringia at that time.

Veit Bach arrived in Thuringia at a propitious time. Both during and after the Thirty Years' War, this part of Germany was an island of musical activity, both religious and secular. It was somehow able to preserve the music so loved by its native son Martin Luther against the grim encircling tide of Calvinists. In Thuringia every cathedral or church of consequence needed an organist, and most of the tiny secular and ecclesiastical courts in the region required resident conductors, composers, and musicians. Indefatigable genealogical researchers have tracked down over fifty Bachs who held positions of this sort, including many whose works are still performed.

Figure 13–2 shows an abbreviated family tree of the Bachs, taken from a book about the family. In view of what it leaves out, this tree has more than a little resemblance to my father's selective view of his ancestry. Almost all the members listed are musicians, which leaves out the numerous Bachs who were not. No women appear in the tree, and the descendants of the sisters of the various Bachs are also invisible because their names were not Bach.

Without a doubt, there was an unusual amount of musical talent in the Bach family tree. But conditions had to be just right for it to be expressed. Had Veit Bach stayed in Hungary instead of returning to Thuringia, the Bachs might have disappeared from history. Further, Bach's life spanned a period when polyphony was giving way to harmony, when the invention of new instruments and the introduction of the equal-interval scale opened up wonderful opportunities for new combinations of sounds. In short, the musical environment was ready for a Bach, and a Bach appeared.

One cannot help but wonder whether some of the less distinguished Bachs would have obtained the posts they did had they not carried the Bach family name. Although all five of Johann Sebastian's sons who survived childhood became musicians, their attainments varied widely. For example, he had difficulty in placing Johann Gottfried Bernhard, his third son, in a position as organist; despite his undoubted talent, the young man's tendency to run up gambling debts and insult his patrons did nothing for his career.

The presumption that genes for musical talent run with unusual plenitude in the Bach family is open to test, though the test itself would try the patience and ingenuity of any historian who attempted it. The trick would be to seek out those invisible members of the family tree—the male offspring of women of the Bach clan whose names were not Bach. These offspring carried the same genes as the Bachs. Would they be as rich in musicians as those who carried the name as well as the genes? I am willing

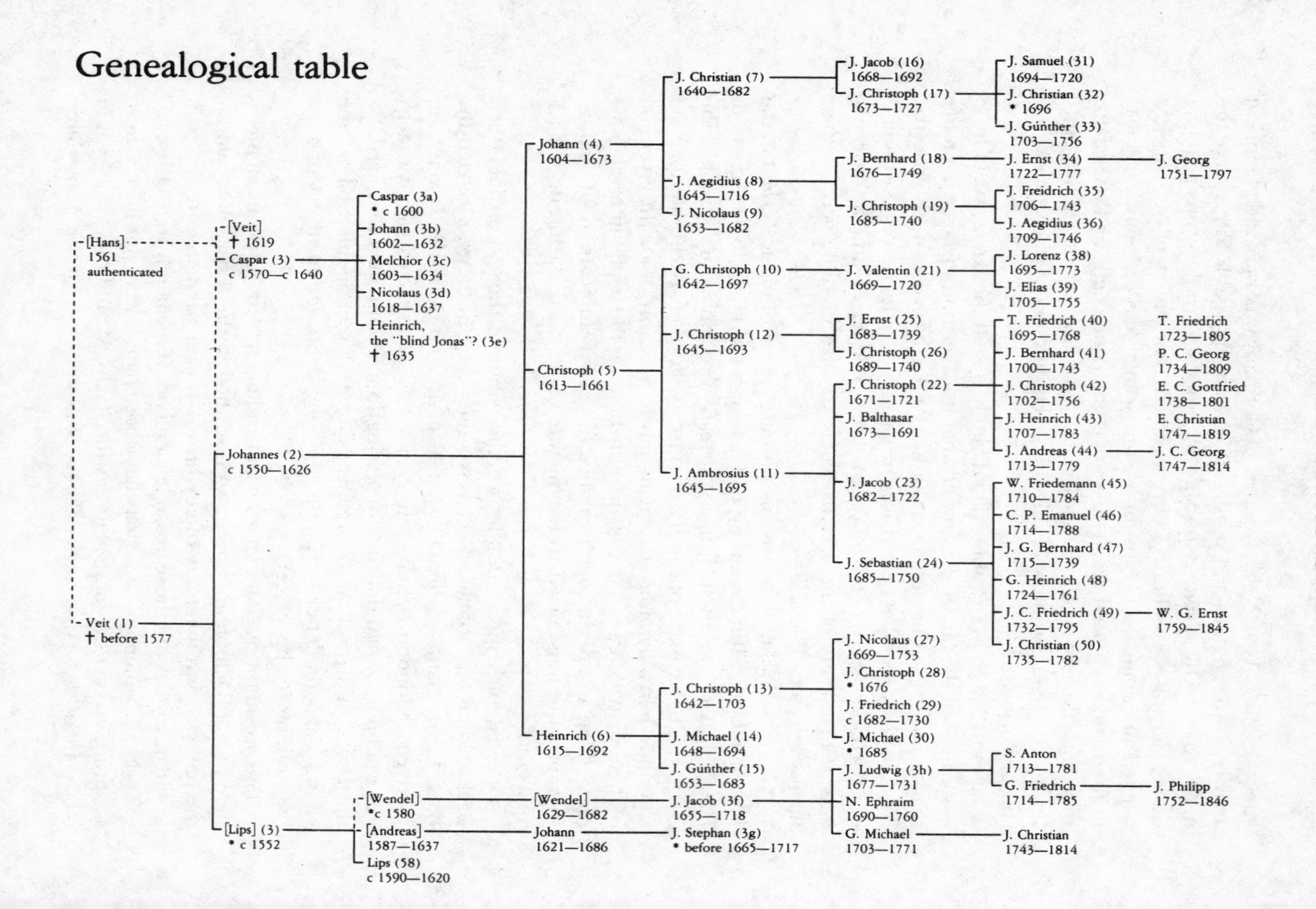
Genealogical table
[Hans] 1561 authenticated
Veit (1) † before 1577
[Veit] † 1619
Caspar (3) c 1570—c 1640
Caspar (3a) * c 1600
Johann (3b) 1602—1632
Melchior (3c) 1603—1634
Nicolaus (3d) 1618—1637
Heinrich, the "blind Jonas"? (3e) † 1635
Johannes (2) c 1550—1626
[Lips] (3) * c 1552
[Wendel] *c 1580
[Andreas] 1587—1637
Lips (58) c 1590—1620
Johann (4) 1604—1673
Christoph (5) 1613—1661
Heinrich (6) 1615—1692
[Wendel] 1629—1682
Johann 1621—1686
J. Christian (7) 1640—1682
J. Aegidius (8) 1645—1716
J. Nicolaus (9) 1653—1682
G. Christoph (10) 1642—1697
J. Christoph (12) 1645—1693
J. Ambrosius (11) 1645—1695
J. Christoph (13) 1642—1703
J. Michael (14) 1648—1694
J. Günther (15) 1653—1683
J. Jacob (3f) 1655—1718
J. Stephan (3g) * before 1665—1717
J. Jacob (16) 1668—1692
J. Christoph (17) 1673—1727
J. Bernhard (18) 1676—1749
J. Christoph (19) 1685—1740
J. Valentin (21) 1669—1720
J. Ernst (25) 1683—1739
J. Christoph (26) 1689—1740
J. Christoph (22) 1671—1721
J. Balthasar 1673—1691
J. Jacob (23) 1682—1722
J. Sebastian (24) 1685—1750
J. Nicolaus (27) 1669—1753
J. Christoph (28) * 1676
J. Friedrich (29) c 1682—1730
J. Michael (30) * 1685
J. Ludwig (3h) 1677—1731
N. Ephraim 1690—1760
G. Michael 1703—1771
J. Samuel (31) 1694—1720
J. Christian (32) * 1696
J. Günther (33) 1703—1756
J. Ernst (34) 1722—1777
J. Freidrich (35) 1706—1743
J. Aegidius (36) 1709—1746
J. Lorenz (38) 1695—1773
J. Elias (39) 1705—1755
T. Friedrich (40) 1695—1768
J. Bernhard (41) 1700—1743
J. Christoph (42) 1702—1756
J. Heinrich (43) 1707—1783
J. Andreas (44) 1713—1779
W. Friedemann (45) 1710—1784
C. P. Emanuel (46) 1714—1788
J. G. Bernhard (47) 1715—1739
G. Heinrich (48) 1724—1761
J. C. Friedrich (49) 1732—1795
J. Christian (50) 1735—1782
S. Anton 1713—1781
G. Friedrich 1714—1785
J. Christian 1743—1814
J. Georg 1751—1797
T. Friedrich 1723—1805
P. C. Georg 1734—1809
E. C. Gottfried 1738—1801
E. Christian 1747—1819
J. C. Georg 1747—1814
W. G. Ernst 1759—1845
J. Philipp 1752—1846

to make a small wager, say ten Maria Theresa thalers, that they would not.

Of course, only the rankest environmentalist would argue that the talent exhibited by this multitude of Bachs was entirely due to environmental circumstance. The Bach family tree must have been rich in genes that, in the right combinations and under the right circumstances, helped to produce this remarkable musical flowering. Nevertheless, there must have been many other families then, and there must be even more now (since the world's population has increased tenfold), that are just as rich in genetic potential, but whose unfavorable environment renders them unable to express it fully.

It is ironic that the salient fact about Galton's analysis of such families turns out not to be the one that he considered important at the time. He was struck by the observation that eminence tends to be grouped in families, but we have seen that environment must play a very large role in such grouping. Meanwhile, he ignored the equally striking fact that over two-thirds of his eminent people stood quite alone, springing from undistinguished ancestors and giving rise to equally undistinguished descendants. Eminent people, whether they sprang from distinguished families or not, appeared in every social class, and in many different times and circumstances. So fixed was he on the idea that genius was inherited in some simple fashion that he never followed out the implications of his other observation that *eminent people can appear anywhere, anytime.*

This fact must be accounted for. It can be explained easily if the environment is the sole determinant of genius, for this is just the pattern one would predict. If, on the other hand, there really is a hereditary aspect to genius, it must be very different from the one Galton imagined. He assumed that characteristics such as intelligence were passed in discrete fashion from one generation to the next, though of course in 1869 (long before the founding of the science of genetics) he had no idea how this might be accomplished. This, coupled with his cousin Charles Darwin's proposal of the idea of natural selection, led him to the supposition that such characters can be selected for, and then to the idea of eugenics.

Yet if Galton were right, genius should have quickly become concentrated in the most successful classes of society, since bright people tend

---

Figure 13–2. Family tree of the Bachs. Note the absence of females and paucity of nonmusicians.

---

to be more successful than the average and to mate with other bright people. The English House of Lords should have been bursting with genius. Clearly, things must be more complicated than he imagined. For example, we can provide a very different genetic explanation of the observed pattern by postulating a rich substratum of genetic complexity underlying every family and group of the human species. Normally unremarkable in its expression, it is occasionally able to give rise to eminence if the genes are shuffled in the right way and nurtured in the right environment. Galton himself may have dimly perceived such an idea (though he was probably only waxing lyrical, as authors tend to do at the ends of books) when he wrote at the end of *Hereditary Genius,* "We may look upon each individual as something not wholly detached from its parent source,—as a wave that has been lifted and shaped by normal conditions in an unknown, illimitable ocean. There is decidedly a solidarity as well as a separateness in all human, and probably in all lives whatsoever."

THE CAVE OF GENIUSES

Let us imagine that some calamity is about to overtake the planet, such that the only chance for survival of the human species is to build a very deep cave and stock it with food and other supplies sufficient for a small band of people. A thousand might be a reasonable number. This would be enough people to ensure survival in case of further accidents, and also to ensure an adequate sampling of the diversity of the human gene pool. It is your job to pick the people. How would you go about it?

You have been given an opportunity to mold the human gene pool for many generations to come. Some of your decisions will be easy. All the people you pick should still be well within their reproductive years, since they will be required to have lots of children. And they should, of course, be tested for fertility. You want every member of this Noah's Ark to contribute to the next generation, if possible. On top of this, they should be free of obvious physical or mental defects.

These decisions are fairly easy. You are now faced with a much harder one. What kinds of people should you pick for this founding population? Let me suggest three possible ways in which you could fill the places.

First, you could use a single criterion to pick the lucky few: their score on an IQ test, perhaps.

Second, you could try to gather together as brilliant a group of diversely talented people as possible. You could pick scholars, teachers, scientists, engineers, artists, dancers, athletes of various kinds, musicians, entrepreneurs, explorers, actors, writers, architects, reporters, even (to make the list of human talents complete) lawyers and politicians. Your list would undoubtedly be eclectic, but with a thousand places available it should be possible to fill them with a remarkable range of accomplishments.

There would be problems. As I pointed out earlier, your candidates would all have to be young, so you could not include the sort of genius that develops only with age and accumulated wisdom. Statesmen are a small subset of politicians, for example, but at age twenty they may not be distinguishable from the more venal majority. Further, your sample might be heavily weighted with people whose talents fade early and disappear before cementing any real record of accomplishment. And on top of this, many of your talented crowd will be intensely competitive and egotistical, perhaps not the best characteristics for survival together in an underground cave.

The third possibility, and certainly the most democratic, would be to pick candidates by lottery. Even restricting yourself somewhat to people endowed with youth, health, and freedom from obvious defects, you would have plenty of people to pick from. And you might even, by the luck of the draw, end up with some very talented individuals. Most, however, would be quite ordinary by whatever standards you chose to apply to them. And some would certainly be decidedly subnormal in certain areas—intelligence, for example, to the very vocal distress of all the geniuses about to be left behind to perish.

However you decide to pick them, this group will be the new founders of the human race. The question is, which of these three methods of selecting them will preserve sufficient genetic variability to ensure that, in the future, the whole spectrum of talents present in the human race can reappear? Remember that your goal is a long-term one: You are interested not so much in the properties of the small group you are trying to preserve, as in the properties of future generations. Bearing this in mind, let us construct several models for the genetic variability in the population, and see what the consequences of your different methods of picking the survivors will be.

THE MENSA MODEL

Our first method for picking survivors turns out to be very like the scheme used to recruit members to the Mensa Society. To attain membership in this society, one must score in the top 2 percent of the population on an IQ test. The idea for Mensa arose in the course of a series of talks given on the BBC in 1945 by the late Cyril Burt, in which he explored various utopias and their consequences. He suggested that one way to achieve a utopia would be to start a society of people with high IQs who could, like the ruling elite of Plato's *Republic,* help to shape and direct society toward its eventual improvement. The question of what such a society might be like has intrigued thinkers and philosophers for centuries. One possibility that particularly fascinated Burt was that people with high intelligence might exhibit other favorable qualities if some of the genes controlling intelligence also happened to control other aspects of personality or physiology.

Some suggestion that this might be true came from a study begun in California in 1921 by the psychologist Lewis Terman. Terman selected 1,528 children who scored 140 or higher on the Stanford-Binet IQ test, then followed them for decades in a project that continues to this day. The children were tested repeatedly as they grew up; they were given batteries of personality tests; and every aspect of their lives was followed. When they married, their spouses were tested. The fates of their marriages were recorded. Children who resulted from the marriages were also tested. These children also turned out to have high IQs on average, though not nearly so high as those of their parents.

Most of the group grew up to be far more successful than the average in both their personal and their professional lives. They tended to end up in the professions, particularly in teaching, in large numbers. By age thirty-five, thirty-one of them had appeared in *Who's Who in America,* and three had already been elected to the National Academy of Sciences. Their unemployment rate was a twentieth of the average. Their marriages were more stable as well. And as if this were not enough, they were both physically and mentally healthier than the majority of their contemporaries. Some, of course, had physical and mental problems, but these were decidedly the minority.

Burt and others used Terman's data to suggest that genes conferring intelligence also conferred other desirable properties. Of course, another and equally plausible interpretation is that physically and mentally healthy children are more able to express their potential through successful test-taking than are children who have problems, no matter how intelligent the latter.* Regardless of whether genetics might be involved, Terman's very thorough study would appear to suggest that selecting for high IQ tends also to select for physical health. The question of whether the reverse is true is one I will address in a moment.

So, let us begin with the Mensa approach. I will use it to construct a very simple-minded genetic model for intelligence, one that will soon need to be made more complicated. How many genes might be involved in intelligence? Your first inclination, I am sure, is to reply: A lot. Consider all that these genes would have to do. They must contribute in various ways to brain structure, function, and development, to the biochemistry of the body, and to the biochemical interface between brain and body. Some of them may be concerned with the structure and development of the peripheral nerves, while others may influence the acuity and function of sense organs. Any and all of these genes could interact strongly with other genes and with the environment. You can see right away that any model for intelligence has the potential to become dauntingly complex.

There is also direct molecular evidence that many genes have at least the potential to be involved. It is now possible to determine the fraction of our hundred thousand genes that are switched on and actively making messenger RNA in the brain. As it turns out, about a third of them are. Of these, about half—some fifteen to twenty thousand—are turned on *only* in the brain.

The genes that interest us here are the polymorphic ones. If every human being has the same version of a gene, then that gene obviously cannot contribute to variations in intelligence. At the genetic loci we know about, about a quarter to a third are detectably polymorphic. Assuming that this proportion holds true for the genes that are active only in the brain, then about five thousand of these genes will be polymorphic and have the potential to contribute to variations in intellect. These genes will be scattered throughout the chromosomes.

Obviously, some of these genes will have large effects, others small, and most perhaps none at all. Some will be highly polymorphic like the

*As an illustration of how potential can sometimes shine through the most severe masking, consider the case of Nigel Hunt, an English child with Down syndrome. He was raised in a highly stimulating environment by his parents, both of whom were schoolteachers. Young Nigel succeeded in writing a charming little book that was later published, making him the only child with Down syndrome to have succeeded as an author. One can only speculate on what his potential might have been, were it not for that extra chromosome 21.

genes in the HLA complex, others much less so. But if our simple Mensa model is correct, variation in many of these genes should have some impact on intelligence. If you select the inhabitants of your cave for high IQ, in the same way that Terman did, then you should be selecting for alleles that confer greater intelligence. Put crudely, this means that there must be specific alleles at these loci that make certain people smarter and different alleles that make others dumber. Smart people have smart alleles at more loci than dumb people do.

If, in order to stock your cave, you were to give an IQ test to a large group of people, you would find the mean score to be about 100, with 95 percent of your group scoring between 70 and 130. The scores would fall along the familiar bell-shaped curve, with the "tails" of the distribution made up of those few who scored very high or very low.

Suppose there are five thousand polymorphic loci that are turned on in the brain and that contribute to intelligence. Suppose further that there are two alleles, a "smart" allele and a "dumb" allele, at each locus, and that each occurs equally frequently in the population.* A person of average intelligence should then carry about five thousand smart alleles and five thousand dumb alleles. Different people of average intelligence will of course carry different permutations of these alleles; because there are so many genes involved, the number of possible permutations will be very large. There are lots of ways for somebody in this population to be of "average" intelligence.

Thus, although the average is five thousand, most individuals will carry more or fewer smart alleles. If we draw a graph plotting the numbers of people against the numbers of smart alleles carried by each, we will get another normal or bell-shaped curve that looks remarkably like the curve obtained when an IQ test is given to a large population (figure 13–3).

For the IQ distribution, as we saw, 95 percent of the people fall between an IQ of 70 and an IQ of 130. This is the actual distribution that is observed. Suppose it were caused by our hypothetical collection of smart and dumb alleles. It is easy to calculate the expected distribution of smart alleles given our assumptions about them, and it turns out that 95 percent of the population should carry between 4,900 and 5,100 smart alleles. (See the notes to this chapter if you want to follow how this is done.)

Now that we have charted out its consequences, our simple Mensa

*This is a simplification, of course, because there could be different numbers of alleles at different frequencies at the various loci. But the simplification does not affect the subsequent argument very much.

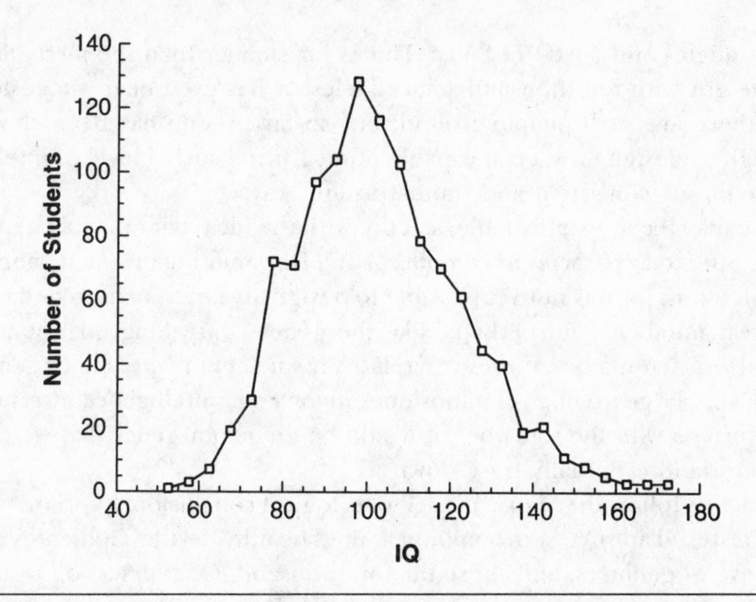

Figure 13–3. Distribution of IQ scores for 1,225 Scottish school children, all aged eleven. The distribution falls approximately along a bell-shaped curve.

model starts to come apart at the seams. To begin with, why should the average person have five thousand smart alleles? Why not four thousand, or six thousand, or ten thousand?

A lower limit would quickly be reached, of course, since somebody with only four thousand smart alleles would have an IQ of minus 200. If our population lost too many smart alleles, it would be in big trouble! But there is no obvious reason why the population should not, with time, move the other way and gain smart alleles while losing dumb ones. At the extreme upper limit, somebody with ten thousand smart alleles and no dumb ones would have an IQ of 3,100, quite sufficient to become master of the universe. Obviously, this has not happened. The upper limit of human intelligence, while it cannot be measured by ordinary IQ tests, shows no sign of being at such a godlike level.

One way to get around this problem and still retain the simple Mensa model is to arbitrarily reduce the number of genes involved in intelligence. Suppose there were only fifty instead of five thousand. Then 95 percent of the population would have between 40 and 60 smart alleles, and the smartest possible person in such a population would have 100 smart alleles and an IQ of 250.* This is far smaller than the incredible IQ we got with ten thousand smart alleles. It has even been suggested that there are such people around, but so far no one has been clever enough to design an IQ test capable of measuring such a level of intelligence in any consistent and standardizable way.

You may feel uncomfortable, as I do, with the idea that only fifty genes contribute to a character as complex as IQ. Discomfiting or not, it cannot be ruled out, for it is not yet possible to design an experiment to find out the real number. And perhaps, like the genes controlling quantitative characters in tomatoes, there are a relatively small number of major genes and a very large number of minor ones involved in intelligence. Presumably, just as with the tomatoes, it would be the major genes that we can measure and eventually track down.

So let us follow the Mensa model to its logical conclusion. Suppose you administered an IQ test to a million young, healthy, fertile candidates for the cave of geniuses, and chose the top thousand. Of course, you would have to be sure that your test was free of culture and language bias.

Figure 13–4 shows how far out you would be able to go along the tail of the distribution. I have needed to put the tail under a magnifying glass, because the top thousandth is such a tiny fraction of the whole distribution. The people you chose would have IQs of at least 145, with an average of about 150. In fact, your sample would be very much like Terman's, though you would have made greater efforts to remove sources of bias than he did in the less socially and culturally conscious 1920s.

Now we can ask how effective this selection has been. How many smart alleles will the selected sample carry if fifty genes control IQ? This question turns out to be a bit difficult, because to calculate that number we have to consider the effects of the environment. An individual may have a high or a low IQ for two reasons, genotype and environment, and on top of this different genotypes and different environments may interact in complicated ways.

The proportion of the variation in a trait that can be attributed to variation among genotypes is known as the heritability of the trait. Again to simplify, if genes contribute substantially to intelligence it should be very easy to select for them; the heritability of intelligence should be high. If, on the other hand, environment is primarily responsible, then it

*These numbers may not seem to add up, but I have taken into account in my calculations the fact that if there are a smaller number of genes each gene will have a larger impact on IQ.

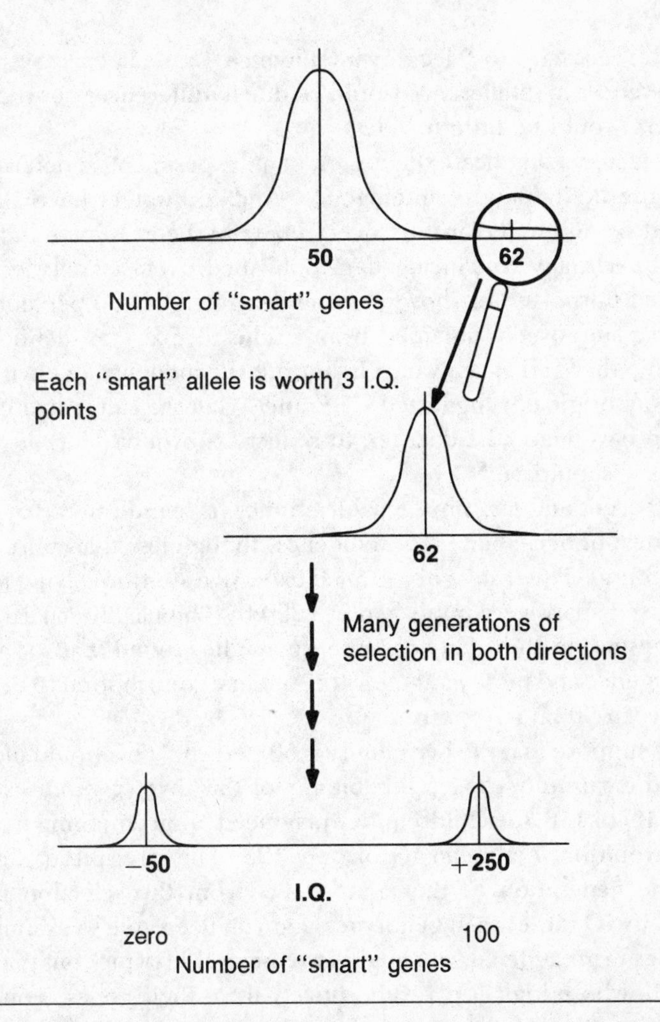

Figure 13–4. The result of selection for IQ on the assumption that fifty loci control this character and that there are equal numbers of "smart" and "dumb" alleles at each locus in the population. It is assumed that genotype and environment contribute equally to the character, which has the result of slowing down the selection process. If you continued the selection in both directions for long enough, however, you would eventually reach the result shown at the bottom of the figure.

should be very hard to select for intelligence. In the latter case so few of the differences in intelligence would be due to differences between genes that there would be little to select.

Many factors complicate the design of an experiment to determine the size of the heritability of intelligence. And the waters have also been muddied by an unpleasant discovery made by Leon Kamin and others. Some superficially convincing data published over several decades by Cyril Burt purported to show a heritability for IQ of 80 percent. These data were supposedly obtained from studies of pairs of identical twins raised together and apart, which factor out the influence of environment in the way originally suggested by Francis Galton. Unfortunately, Burt seems to have made the data up, to reflect his own opinion of what the heritability should be.*

More recent and less impeachable studies have indicated that there is a good deal of heritability to intelligence, though less than Burt claimed to have found. The results of the most extensive study involving identical twins raised apart were published in 1990 by Thomas Bouchard and his colleagues at the University of Minnesota. They found that for a variety of intelligence and personality tests the genetic contribution to the scores fell between 50 and 70 percent.

Let us suppose that the heritability is 50 percent. This would mean that even if the carefully chosen inhabitants of the cave of geniuses had an average IQ of 150, the children they produced upon emerging from their shelter would have a mean IQ of only 125. This precipitous fall in IQ from one generation to the next results from the selection of cave-dwellers by IQ rather than genotype. Had you been able to examine their genotypes in great detail, you might have been able to pick out that subset of people whose high IQs result entirely from their genes. You cannot do this with your IQ test; you are forced to guess the genotype from the phenotype.

Once the average IQ has dropped to 125, however, it should drop no further. The people you selected, and their children, will have on the average fifty-eight smart alleles and forty-two dumb ones, compared with fifty and fifty for the population from which they were chosen. The surviving population would certainly have a higher IQ than the average, and would have gained accordingly. Would anything have been lost in the process?

*There is some possibility that he really did collect some of these data, lost them, and then reconstructed them from faulty memory.

If you have obtained no other impression from this book so far, you should be convinced that the human genome is a huge and complicated affair. Are we justified in applying such a simple model to it, and proceeding on this perhaps dangerous assumption? Suppose that our initial hunch was correct and there really are a lot of genes contributing to intelligence. We have already seen that it is difficult for the Mensa model to account for that. Might there not be better ways than an IQ test to select our population of survivors in order to preserve all the many manifestations of human talent?

THE RENAISSANCE PERSON

In 1860, the Swiss historian Jacob Burckhardt published a seminal essay on the Italian Renaissance in which he attempted to explain why *l'uomo universale,* the Renaissance Man, had appeared at that time and in that place. He pointed out that mankind's freedom from the intellectual fetters of the Middle Ages allowed people to develop for the first time as individuals rather than as members of the state or the church or the town. Spurred by discoveries about the civilizations of antiquity, the collective knowledge of Western man began to grow. Just before the knowledge explosion gathered full force, it briefly became possible for a single man to span all of human knowledge and excel in it.

This ultimate universal man was Leonardo da Vinci, who deliberately hid away in his notebooks the full extent of his explorations. His contemporaries knew him mostly as a remarkable but excessively experimental artist and architect whose chief creations tended to fade or peel, or to be so grandiose that they were not even begun. The men of the Renaissance themselves awarded the accolade of most remarkable of the universal men to the architect Leon Battista Alberti (1404–1472), whose modest motto was, "A man can do all things if he will."

Alberti was born in Genoa, at a time when his father was in exile from Florence. He came from a family of powerful merchant bankers, but because he was born under the bar sinister his legitimate cousins managed to ensure that he obtained none of his father's estate. Forced briefly into beggary, he finally managed to obtain a position as a minor papal secretary. His first architectural commissions were probably obtained because church officials were not required to pay him a fee, and indeed, he moved all his life from one vague and undefined position to another. Primarily, he was an architect, and much of his other work reveals the precision and logic of an architect's eye. His constructions were marked by a resolute sense of classical proportion, which was particularly striking in his most famous work, the facade of the church of Santa Maria Novella in Florence. But he went far beyond this, writing treatises on an astonishing variety of subjects and exploring every avenue of science, art, music, poetry, politics, and philosophy that was available to him.

At age twenty, he published a comedy in Latin, which he passed off as the work of a mythical Roman author and which fooled everybody for ten years until he admitted authorship. He followed this with a stream of works that influenced his contemporaries immensely. Some were written in his native Tuscan, which he showed was as regular as Latin and therefore just as valid a language in which to write prose. He did a great deal to popularize the vernacular, such as establishing an annual poetry competition at which Latin verse was actually banned.

De Picturis, written in Latin and then translated into Tuscan, was the first formalization of the rules of perspective that were just beginning to be explored by Brunelleschi and Masaccio. Alberti also tossed off books of maxims, jokes, and mathematical games, a valuable treatise on horsebreeding, and a guide to legal practice. In the *Descriptio Urbis Romae* he demonstrated surveying techniques that made it possible to draw an accurate map of a city from a particular perspective; this opened up the entire field of surveying. He was also responsible for founding the science of cryptography, inventing the cipher wheel, double encryption, and many other refinements that did not come into general use until centuries later. Cardinal Colonna commissioned him to raise an ancient Roman wreck from the bottom of Lake Nemi near Castel Gandolfo. Though the ship broke in half, enough remained intact for him to write a treatise (unfortunately lost) on Roman shipbuilding. Alberti's ten-volume work on architecture, published in 1485 some years after his death, was translated into nearly every European language. It continued to be read and studied widely until well into the last century.

His physical as well as his mental achievements were extraordinary. An anonymous biographical fragment, which may have been written by Alberti himself, tells how he could tame the wildest horse and ride it for hours without fatigue, fling a spear through a breastplate of the thickest armor, and jump the height of a man from a standing position. Further: "Once he threw a little silver coin up against the roof of a high church with such power that his companions heard it strike the vault."

Figure 13–5. The facade of Santa Maria Novella in Florence, designed by Leon Battista Alberti.

Alberti was remarkable, but even he excelled far more in some areas than others. None of his musical compositions have come down to us, though the biographical fragment tells us that they were "highly esteemed by the music masters." His poetry is pretty heavy going when matched with that of Dante and Ariosto. Even his architecture is quite standard compared with the incredible soaring dome with which his friend Brunelleschi crowned Florence's Duomo. Still, he obviously was tremendously good at a great variety of things. Let us speculate about what his genotype might have been like.

Suppose you could partition up intelligence, skill, and talent into some finite number of categories—say ten. Suppose further that a set of fifty genes is responsible for the genetic variation in each of these ten categories, with half the alleles at each genetic locus being "good" and half "bad." And suppose that, using these tests, you would like to fill the cave of geniuses with polymaths like Alberti. The task would be daunting. You would have to construct ten fair, balanced, and culture-free tests to distinguish people who excelled in all ten of the different categories. Constructing a test for artistic ability might be very difficult if you had to gain the agreement of people like Jesse Helms! Then you would have to administer the tests to a million or so people and pick the thousand who had the highest average in all ten categories.

Many people would do well in some tests but badly in others. Very few would do well in all. One way to find polymaths would be to pick those who scored in the top half in the first test, then pick from these the ones who scored in the top half of the second test, and so on. If you started with a million people, by the time you repeated this process ten times you would end up with the thousand who scored in the top half in all ten tests.

The genetic impact of this whole elaborate procedure would be minimal. If we assume, as we did with IQ, that the heritability for these ten different characters is fifty percent, then among our polymaths we will have an average of 51.7 good genes and 48.3 bad genes in each category, compared with an average of 50 good and 50 bad in the original population. Is it worth incurring the wrath of Senator Helms for such a tiny gain?

You will be equally frustrated if you try to fill the cave with geniuses rather than polymaths. Suppose you pick the hundred top scorers in each of the ten tests. You will be picking specialists rather than generalists, so your selection will appear to have a greater impact. The mean number of good genes carried by the hundred people who score at the top *in a particular category* is about 62. But of course your selection process will have had no impact on the genes in the other nine categories that are carried by your selectees. The result is that when these people marry and have children the average of good genes in all ten categories among the children will drop to only 51.2. Any large gain you manage to make in a particular category will be diluted as these genes are mixed through the whole population. You will have taken the dangerous course of filling your cave with high-strung geniuses for no appreciable genetic gain.

So it is very hard to select for a Renaissance person. If our crude model has any relation to reality, it is apparent that even Alberti could not have excelled in all ten of our categories. Given the size of the population of Renaissance Italy, the chance of such an extreme polymath appearing among them is vanishingly small. Still, the fact remains that he was a remarkable polymath. Does this mean that our Renaissance Person model is unrealistic? Of course it does.

THE CAVE OF PLEBES

Lewis Terman found the health and strength of his high-IQ children to be better than the average, contradicting the public's image of the smart child as a bespectacled wimp. He did not carry out the reverse experiment of picking children with good health and determining whether their IQs were higher than the average, but he would almost certainly have found that they were. These characters are positively correlated, though the correlation is probably a weak one.*

If such positive correlations exist among our ten categories it should be much easier to select for a person who excels in several of them. Some alleles for mathematical ability might easily have a positive effect on certain kinds of musical ability or certain categories of general intelligence. Alberti's talents seem to have been heavily loaded towards spatial perception—consider his art, architecture, engineering, and map-making skills. If genes were in part responsible for his talent for perceiving things in three dimensions, then it seems that they produced a number of correlated effects.

However, correlations are of two kinds, positive and negative, and the latter are *a priori* equally probable. We have just seen how negative correlations can hinder the process of artificial selection in tomatoes. Negative correlations due to genetic differences might be produced in three general ways. First, alleles at certain loci might have a positive effect on one trait, but a negative effect on another. Second, alleles at a locus might produce one effect, while alleles at a closely linked locus might produce the opposite effect. Third, alleles at a single locus might have

*The depressing academic performance of many college athletes would seem to argue against this, but it must be remembered that in most cases their coaches give them little chance to study. Further, athletes are not necessarily healthy; they are people with the determination needed to push their bodies beyond the normal bounds of strength and endurance. Distressingly often, they abuse their bodies in the process.

positive but opposing effects, so that selecting for one desirable character would inevitably select against another, perhaps equally desirable one.

One might imagine that the alleles that lent Alberti his highly effective way of looking at the universe prevented him from perceiving the world in other, equally valid ways. His art has a cold, remote, and precise feel to it—not surprising in view of his preoccupation with problems of perspective. Could he have become a warm and passionate artist? Perhaps, given a different environment, he could have, but perhaps not. If not, then the possession of one set of alleles for artistic ability might actually have ruled out possession of other alleles that influence different kinds of artistic ability.

Negative correlations pose other dangers to your cave of geniuses as well. Talent might correlate negatively with emotional stability, or the skills valued by civilized society might correlate negatively with the abilities required for survival in a post-cataclysmic world. In the process of selecting for genes for talent in your little population of survivors, you will want to avoid simultaneously selecting for others with negative effects of unknown magnitude. You could of course try to avoid this problem by applying additional tests for emotional stability and survival skills, but this would reduce the effectiveness of your selection for talent still further.

A moment ago I calculated that there are five thousand polymorphic genes expressed only in the brain. Yet none of our crude models have managed to deal with more than five hundred of them. The possible interactions, both positive and negative, among the alleles at all these five thousand loci are numberless, particularly when different environments are added into the equation. Who knows what fraction of these alleles are important when we may have no way of detecting their effects at present?

Faced with such unknowns, and with the fact that your most rigorous selection would have such a tiny effect on the genetic makeup of your band of survivors, I think you will agree that the safest and fairest way to choose them is through a lottery.

By drawing your group of a thousand at random, you will be sure to preserve virtually all the genetic variability of the original population. Members of this democratic group will certainly vary more in their abilities than a highly selected group might have done, but leaders are sure to emerge to help guide them through the disaster. You will not end up with a Graustarkian army consisting entirely of field marshals, as might have happened had you filled your cave with geniuses. Indeed, while you have traded your cave of geniuses for a cave of plebes, it should still contain all the seeds of genius for future generations.

After exploring all these models, we have come back to the astonishing and diverse reservoir of genetic variability that is concealed in the chromosomes of the average man and woman. We can now provide, as Galton could not, a genetic explanation for why geniuses seem to appear at random. If we cannot deliberately select for genius because of the complexity of the human gene pool, it is hardly surprising that natural forces cannot select for genius either. It must happen in an almost infinite number of ways, through fortuitous and unpredictable combinations of genes and environment. These combinations arise occasionally from the complex substratum of genetic variation that we all possess, genius or no.

NO BAD GENES, ONLY BAD ALLELES?

Given our present state of knowledge, there is no way we can rationally apply selection to such ill-understood qualities as intelligence and talent. This understanding is usually the end point of discussions about eugenics. Now, however, there is a new dimension to the problem, one being opened up by our exploration of the genome.

I suggested a moment ago that a time will soon come when alleles will be discovered that can be associated with various mental abilities. These will probably first be detected by linkage studies, in which certain RFLPs will be found to be more common among groups of people exhibiting certain talents—musicians, physicists, architects, and so on—than among the general population. All or nearly all of these associations are likely to be weak, like the association of IQ with physical health that Terman found. We will not find alleles *for* architectural or musical ability, but we will find alleles that may *influence* the development of these abilities.

The RFLPs will serve as signposts, pointing to nearby alleles of genes that are important in the development of the brain and the intellect. Unlike the alleles in my simple-minded Mensa model, however, the common alleles at these loci cannot be characterized as "good" or "bad." They will all do important tasks that have been shaped by their evolutionary history, and they will all be carried by millions of people. Determining the function of each of these genes with their various alleles will pose an enormous challenge to our understanding, for these genes will provide important clues to why the brains of different individuals function differently.

Once these genes have been cloned, it will be possible to ask a different question. In most people, these genes will function properly. Different people will have different alleles that function in slightly different ways, but the alleles will all do their job. Will we find rare alleles at these loci that do *not* function properly? What will their effects be?

In our exploration of the genome we have seen such rare mutant alleles at a number of loci, many of which give rise to mild or severe genetic illnesses. There is no reason to suppose that the genes that control the development of the brain will be any different. We are sure to find mutants among these genes that are the equivalent of Lesch-Nyhan or ADA deficiency. Their effects may be much subtler, but most of them will also be unequivocally harmful.

At any one locus, such harmful mutant alleles will be uncommon. But recall that fifteen thousand of our genes are expressed only in our brains. This is such a large number that most of us probably carry several of these harmful mutant alleles. Because they are individually rare, however, the numbers of these alleles in any individual will be small. We know this because of data collected on the children of severely retarded people. About two-thirds of these children possess normal intelligence. There is still much argument about what proportion of people classified as feebleminded owe their mental deficit to genetic reasons. If this proportion is substantial, then these data suggest that damage to the genome in such people seems to be removed easily by the normal genetic shuffling that goes on between generations.

It will be much more difficult to track down the rare harmful alleles at other genetic loci important to brain function if they happen not to be polymorphic, because there will be no RFLPs to point the way. Even when we know the whole sequence of the human genome, these genes will just be part of the anonymous crowd unless mutant versions of them are discovered that contribute substantially to mental illness or deficit.

Now we can reconsider the question posed at the start of this chapter. In sum, it should not distress us that there are large differences in individual talents and abilities that can be attributed to genetics. Much of this variation, which has built up slowly over the course of our evolution, contributes to the remarkable tapestry of genetic polymorphism in our species. What can and should concern us is that some of this variation consists of actual damage to these functional genes. This damage can in principle be repaired. While selecting for geniuses is beyond our capacity at the moment (and perhaps will be forever), it may be possible to pick out and heal rare mutant alleles that adversely affect mental functioning.

There is no way that bad genes can survive millions or billions of years of the rigors of natural selection. Bad alleles of these genes, however, are arising all the time. We may not be able to select for geniuses, but perhaps we can increase their number, and raise the general level of intellectual attainment of the population as well, by repairing these truly bad alleles. This could be accomplished without altering the complex and delicate balance of genetic variation in our population that has been built up over evolutionary time. The mapping of mental abilities in the human genome will point the way to this spectacular possibility.

CHAPTER 14

Ethical Hurdles and Opposing World Views

Mother, mother, let us come home!
The door is bolted, we can't get in!
We are in the dark and frightened!
Mother, mother! Alas, alas!

—Hugo von Hofmansthal, *Chorus of Unborn Children,* from Richard Strauss's opera *Die Frau ohne Schatten*

In this book I have done my poor best to squeeze the complexities of three billion bases into three hundred pages. The small beachheads of understanding that scientists have established here and there in the genome are dwarfed by the huge stretches of our DNA that we simply do not comprehend at all. I think you will have been struck in reading this book, as I was in writing it, how certain themes emerge again and again.

One is that the human genome is an organic whole, molded by billions of years of evolution. Exons, introns, retroposons and retroviruses, regulatory regions and the "Simpson Desert" regions have all been shaped by natural selection and by chance events to produce a construct of enormous complexity. It is far too soon for us to say that one part of the genome is more important than another, or that we should concentrate on understanding certain parts of it to the exclusion of others. Even the lowly VNTRs, superficially as meaningless as the whorls and arches found at the ends of our fingers, might provide important clues to cancers. Ten percent of our genomes have been built through the activities of retroviruses and retroposons. Why? We have much to learn about the processes that have shaped our genes.

Another recurrent theme is that every genetic disease and other genetic character that we have looked at in the course of this book, no matter how straightforward it might first appear, shades off into endless complexities. Except when the effect of one allele overwhelms everything else, as in Tay-Sachs, innumerable combinations of alleles at many loci interact with the environment to influence how each character is finally expressed.

It should now be clear to you that the Human Genome Project, originally touted as a way to solve these problems, is nothing more than a jumping-off point. It is a device for concentrating the minds of scientists on ways to solve immediate problems—how to cure cancer and mental illness, how to screen the population for genetic defects, and indeed how to draw the line between genetic defects and genetic normality. Beyond the Project, larger problems loom.

THE FOUNTAIN OF YOUTH

At the close of the last chapter I suggested that there are no bad genes, only bad alleles. In making that statement I could be accused of a kind of evolutionary chauvinism. The statement implies that any gene that has been shaped during the course of evolution for hundreds of millions or billions of years must be "good." As a result, we alter such genes at our peril. Is such chauvinism justified?

In most cases, I think it is. Yet, we are certain very soon to find genes that look as if they could be improved. To take one example, in the course of human evolution our life spans have lengthened enormously. Our nearest living relatives, the chimpanzees, have a maximum life span of only about thirty to thirty-five years. Genetic events during the past five million years of evolution have thus resulted in at least a doubling of our life spans. If we could understand those genetic events, it seems quite likely we could double our life spans again.

What untoward biological and societal effects would this have? I suggested a few of them in the introduction. There could be many more, including totally unexpected ones. After all, since our life spans have taken millions of years to lengthen, time has been provided for many other genetic and accompanying societal adjustments to take place.

In spite of these, extremes of human behavior can still result in aberrations, such as the emergence of merciless and unbalanced leaders of tribes or nations. Perhaps such aberrations have not been selected against because our brief life spans allow new leaders quickly to replace the especially savage ones. If we unleash immortality on the human species without somehow tempering these extremes at the same time, the result might be a terrible disaster. Consider that if such technology were available now, Stalin and Franco might still be at the height of their powers! In that sense, my evolutionary chauvinism is justified.

Leaving these qualms aside, there is no doubt that the temptation to try to lengthen our life spans will be overwhelming. Who would not trade the chance of some genetic or social maladjustment further down the line for extra decades of productive life? Faced with such a prospect, ethical qualms about tampering with the essence of humanity will surely be swept aside. Few people will want to serve as controls, condemned to a life of threescore and ten when everyone around them is living sevenscore. One can only hope that the inevitable effects of any genetic maladjustment, when they finally become apparent, will not be too serious.

DESIGNER GENES OR DESIGNER ENVIRONMENTS?

I closed the previous chapter with a glimpse of a utopian world in which legions of gene therapists might track down harmful alleles of all types and cure them. Yet such a utopia is unlikely to come true during our lifetimes or during those of our children. It would be too expensive, for one thing. Each of us acquires in the course of our lifetimes two or three new severely harmful mutations, some of which, particularly if recessive, are passed on to our children. Perhaps ten billion harmful recessive mutations enter the human gene pool each generation to join those already present. A truly effective gene therapy program would require that these harmful alleles be searched out from among the quadrillion or so genes currently possessed by the entire human species and then cured. We are many centuries—millennia, perhaps—from finding and curing even a tiny fraction of these harmful alleles.

I contend, however, that we do not have to wait for millennia to achieve benefits from our growing understanding of the human genome. To begin with, consider that the attitude of Western society toward people with genetic handicaps like Down syndrome has changed dramatically in the space of a few decades. The nature of the chromosomal abnormality leading to Down syndrome was only discovered in 1959, though its existence had been suspected since the 1930s. The discovery of the specific problem removed the mystery from the condition; these people turned out to be no different from you or me, barring a single genetic error with very large consequences. The syndrome became no fault of theirs, nor of their parents. The categorization of Down syndrome in the 1970 Encyclopedia Britannica under "monster" shocks us now. We have come a long way in twenty years.

Exploration of the human genome will inevitably extend our understanding further, into regions of human behavior where the geneticist has not yet dared to venture. Pinned on my wall is a wonderful cartoon by Mankoff from the *New Yorker,* in which an accused felon stands before a judge and pleads, "Not guilty, Your Honor, by reason of genetic determinism!" We laugh, in part because we know his plea is wrong. Genetically speaking, the accused felon is just like you and me. The eighteenth-century criminals who were shipped in chains to Australia produced descendants who are among the most law-abiding on earth, as Robert Hughes has superbly shown in his book *The Fatal Shore.* The downtrodden of America traditionally swelled the criminal classes as they first arrived. When and if they were permitted to join the mainstream, their overt criminal activities largely ceased.

During a talk with Jo Kiernan, the assistant district attorney responsible for prosecuting some of the most repellent criminals in San Diego, I asked whether she thought there could be some genetic reason for their behaviors. She dismissed the idea contemptuously. "They know exactly what they are doing," she said. "They are responsible for their conduct."

Still, there is a hint of something genetic here. It has often been observed that the police and the criminals they deal with come from the same social stratum. Why should one group be law-abiding and the other not, if criminal behavior is engendered entirely by the environment? Is it possible to ask, as we did with intelligence in the last chapter, whether there could be a genetic component to antisocial behavior? And if so, what would that genetic component consist of?

Without doubt there are no genes for criminal behavior per se, for criminality is far too diffuse a problem to be traceable to a few alleles at a few loci. This does not, however, rule out the possibility of associations of certain alleles with criminality. If such associations are found, they will appear first from linkage studies with RFLPs. The associations will surely be very weak. This means, you remember, that while there might be a slightly higher frequency of a particular allele among criminals, the majority of criminals will not carry the allele and the majority of people who do carry the allele will not be criminals.

What could such alleles actually *do?* Criminal behavior is often characterized by short-sighted and impulsive actions. One could imagine alleles affecting hormone levels or levels of neurotransmitters that could trigger such behavior under some circumstances. In addition, criminals are almost by definition insensitive to the rights of others. Could alleles at some loci affect attention span in such a way that it is more difficult for children carrying them to interact normally with other people as they are growing up? Different, more severe alleles at such loci could be contributors to schizophrenia, in which isolation from everyday stimuli is often pronounced.

Yet each of these alleles, if they exist, must have only marginal effects. One can imagine many circumstances in which they might be beneficial rather than harmful. A rapid response to stimuli, for example, is a very good thing to have if one's environment is a dangerous one. Further, if these alleles turn out to be common enough in the population to be detectable by RFLP mapping, then they must form part of our evolutionary heritage. It would be as difficult, and as senseless, to purge the population of them as it would be to purge it of all the alleles for blood type B.

The immediate benefit of such studies will be the growing realization that, as with Down Syndrome children, we should not view criminals as monsters somehow marked off from the rest of humankind. Physiological reasons may contribute to some of their problems, but aside from these minor difficulties they are genetically just like anybody else. Jeremy Bentham (1748–1832), the first real champion of the idea of rehabilitation for criminals, was hundreds of years ahead of his time in this regard. While humanitarian views of physical disabilities and mental illness have begun to permeate society, most of us have yet to adopt Bentham's compassionate view of criminal behavior. Eventually we surely will. As Madame de Staël remarked, "*Tout comprendre c'est tout pardonner.*"

Counseling, therapy, and special programs to encourage healthy interactions with others have been shown repeatedly to have a great impact on children who might otherwise begin to develop antisocial behaviors. Perhaps their effects can be enhanced still further if the therapies can be designed to try to correct for slight imbalances of behavior or perception introduced by particular combinations of alleles. I like to hope that our growing knowledge of the causes, genetic and otherwise, of antisocial behavior will eventually enable us to introduce *designer environments,* perhaps of the sort that I briefly suggested in the introduction. Designing these environments in such a way that they do not unfairly stigmatize the children involved in them will be very difficult. The effort will be worth it, however. Not only will the results of such environments improve our society in general, but they will be far more effective than any high-tech attempts at cures using designer genes.

In our current society, people are "written off" for a great variety of reasons. Perhaps the greatest value of our exploration of the human genome will lie in the realization that most of these reasons are wrong. No matter how battered some people are by their environments, or by at most a handful of mutant alleles that prevent them from functioning properly, all of them are in principle capable of living fuller, more productive lives. The overwhelming majority of their genes are just like the genes that the luckier members of society carry, and form an equally valuable part of our genetic heritage.

THERE ARE NO GENES FOR THE SOUL

If your background in biology is limited, you have probably had to struggle a bit with some parts of this book, despite my attempts to be as clear as I can. The technology involved in exploring human genes is very advanced. I have been forced to assume in my readers at least a smattering of chemistry, physics, and biology of the sort provided by a general undergraduate college education. Otherwise I would have had to stop for painful definitions and explanations at every step.

You must possess another requirement in order to have stuck with me this far. That is a great curiosity about the human condition. What are we made of? How do we work? Will we ever track down and comprehend the differences between ourselves and other animals?

I hope you have been able to join with me in excited anticipation of the many remarkable advances that are sure to come from exploring the genome, ranging from curing otherwise intractable diseases to understanding, accommodating, and (with Madame de Staël) forgiving those human differences that are the inevitable product of the diversity among our genomes.

You are, however, among a minority on one side of a great gulf of incomprehension. A recent Reuters report by Andrew Quinn told of a Buddhist temple in Taipei that has found a novel way to supplement its income. The Mercy Temple is one of the largest and most popular in the city. For a fee, it will ensure that the spirits of babies who have been aborted will not return to haunt their mothers and families. An undisclosed but substantial number of women have already taken advantage of the service.

Most of the Buddhists of Taipei are embarrassed by the promotional literature the temple has sent out. The literature claims that the aborted babies' souls are unable to be reincarnated, and as a result will return to the material world as angry spirits. Mainstream Buddhists claim that there is no justification for such statements in Buddhist doctrine. At the end of a 1988 speech in Los Angeles, the Dalai Lama was asked by a rather hostile member of the audience about the attitude of Buddhism to abortion. He replied that he did not see any particular conflict, that the wheel of life would continue whether or not some pregnancies were terminated. Of course, the Dalai Lama does not speak for Buddhists with the authority that the Pope once exerted over Catholics on matters of doctrine. Indeed, judging by what is happening in Taipei, the question of abortion is starting to splinter Buddhism just as it has every other major religion.

The world view of the Mercy Temple is epitomized by the little quotation with which I started this chapter. In Richard Strauss's opera, *Die Frau ohne Schatten* (The Woman Without a Shadow), a princess from the spirit world marries an emperor. She then learns that unless she can acquire a shadow the emperor will be turned to stone. She attempts to buy the shadow of a commoner, the wife of a dyer. The dyer's wife, however, discovers that if she gives up her shadow she will not be able to bear children. Tempted by the empress's inducements, she is given pause when she hears the voices of her unborn children calling to her.

Strauss and his librettist, Hugo von Hofmansthal, suggested by this charming device that the souls of children are preformed and waiting in heaven, ready to descend and inhabit mortal bodies. In doing so they tapped a wellspring of human belief. Either consciously or subconsciously, the overwhelming majority of the human race believes this to be true.

What does the chorus of unborn children have to do with the exploration of the human genome? Simply that the world view of the majority of humankind is quite different from that of the minority who are doing the actual exploring, or indeed of the slightly larger minority, including the readers of this book, who are curious enough to want to learn about it. In the course of this book we have journeyed through a high-tech world in which the human genome is slowly being laid bare. We have seen that there is nothing in principle to stop us from eventually tracking down genes that influence intelligence, skills, mental health, even behavior. But there are no genes, on any of our twenty-three chromosomes, for the soul.

The gulf between those who are concentrating on the scientific implications of this research and those who see only that it menaces the essential uniqueness and divinity of human beings is certain to grow. Bridges of education and comprehension must be built across this gulf. An essential starting point is that scientists and their allies must never assume that a human being is nothing more than just a collection of genes to be poked and manipulated, for if they do they will lose the support of the majority of their fellow human beings. The mere fact that we cannot prove the existence of a soul, or assign it a chromosomal location, does not invalidate that world view.

CHAPTER 15

Stop Press

You will recall from chapter 2 that Fred Sanger was helped by a number of collaborators as he developed the dideoxy method of sequencing DNA. One of them was Bart Barrell, who has since then followed his mentor's example in tackling a sequencing project at the outer limits of present technological capabilities. He is the head of the Cambridge group that holds the record for the largest single piece of DNA yet sequenced, the 230,000 bases that make up the genome of human cytomegalovirus. This virus causes mild flulike symptoms in most adults, but it can produce severe effects in newborn babies and people with immune system deficiencies. Scattered along its chromosome are over 200 genes and pieces of gene. It has proved to be a treasure trove of genes related to oncogenes, transport proteins, and receptor proteins, among others. The determination of the sequence, which took more than twelve person-years of effort and is now virtually complete, was accomplished entirely with standard sequencing technology. It was aided by the availability of very large amounts of pure viral DNA and by the fact that the cytomegalovirus chromosome contains no long, confusing stretches of repeated sequences.

So when Barrell recently complained in print that sequencing technology has simply not kept pace with the demands of the Human Genome Project, and that scientists are showing a notable lack of enthusiasm for sequencing long stretches of DNA that does not have an obvious function, he struck a nerve. He noted that very few really long sequences have appeared so far in the DNA data bases. This, he suggested, was a result of the fact that sequencing technologies have promised more than they can deliver, and he worried that once this was widely realized enthusiasm and funding for the project would dry up.

Without a doubt, there will be no way to accomplish the original goals of the Human Genome Project in the time allotted unless new sequencing technologies appear. Some steps are being taken to that end: in March 1991, an agreement was signed between Los Alamos National Laboratory and a company called Life Technologies, Inc., formalizing a cooperative effort to develop the single-molecule sequencing technology of Jim Jett and Richard Keller (discussed in chapter 7). Other cooperative programs are being established between industry and academia to investigate such promising technologies as scanning tunneling microscopy. Yet all these programs are some years from success, and until they come on-line, it seems likely that the Human Genome Project will continue to grow more diffuse in its goals, as most scientists chase down the implications of the genes they have already found. Very few can be expected to sequence simply for sequencing's sake.

This is not to say that the rate of discoveries concerning human genes has slowed down. Quite the contrary, as a quick scan of the events of early 1991 bears out. The data base of DNA sequences continues to double every two to three years, even though each sequence might be shorter than Barrell would like. Recall from chapter 7 Tim Hunkapiller's computer chip designed to speed searches through the data base; this invention is now in what is called beta-testing mode, meaning that several copies of it, coupled to Sun workstations, are now in use around the country. It will soon be joined by another kind of chip designed to look for similarities between pairs of sequences. Together, they will make an extremely flexible special-purpose computer.

Programs involved in the physical mapping of chromosomes around the country are progressing rapidly. By June 1991, Tony Carrano's group at Livermore had mapped about 70 percent of chromosome 19 into contigs, the largest being about two million bases long, with another nearby numbering about a million. The larger of these two contigs is known to include a gene for an autosomal muscular dystrophy, which has quite different effects from the X-linked kinds discussed in chapter 9. Possession of this set of overlapping cosmids will greatly facilitate the process of walking to this gene. The smaller contig is in a region containing some fascinating cancer-related genes, which specify proteins that while fully expressed in embryonic tissue are made in only trace amounts by normal adults. These genes are turned on again at high levels in some cancerous cells, however, making them useful for detecting cancer in its early stages and for monitoring post-treatment patients to pick up any signs of relapse.

The discovery of these *carcinoembryonic antigens,* as they are called, was only the beginning of this approach to cancer detection. Genes are being found everywhere in the genome that are turned on at high levels in various cancers. This inappropriate switching-on might simply be a reflection of the genetic mayhem common to the genomes of cancer cells (see chapter 1). As more is learned about which genes are turned on in which cancers, however, patterns are emerging, making it possible to employ these genes as specific diagnostic tools. If, as it now appears, hundreds of genes are capable of being turned on, the opportunities—and the costs—of early cancer diagnosis and improved therapy are certain to skyrocket. Nothing illustrates more vividly Dulbecco's prediction that the entire genome must be explored in order to conquer cancer. Only when it is understood *why* these genes are being turned on inappropriately will it become possible to start treating the causes of cancer and not merely the symptoms.

Both at Livermore and at Los Alamos, where the mapping of chromosome 16 continues, strenuous efforts are under way to close the gaps between the growing contigs. One particularly ingenious development involves the hybridization of probes to the nuclei of cells that are about to become sperm. In these cells the chromosomes are enormously stretched out. It is possible to hybridize pairs of probes that are separated by only 20,000 bases on the chromosome, giving two distinct fluorescent spots in the nucleus and resolving puzzling regions that other techniques cannot.

Many other groups around the world are continuing apace with mapping programs, and the competition to be the first to get an entire chromosome—or at least one arm of a chromosome—into one giant contig continues to be brisk. Hans Lehrach, for example, is now sending membranes containing thousands of clones from chromosomes X and 21 to various laboratories in order to begin his low-budget cooperative enterprise in physical mapping. We examined the rationale behind his approach in chapter 6. Just as it has been since the start of DNA sequencing, however, the great majority of effort has been concentrated on the genes, not the chromosomes.

One of these is the elusive gene for Alzheimer's, a degenerative disease of the central nervous system. Alzheimer's is found in about 5 percent of the population over age sixty-five. Sometimes, however, it can affect relatively young people, and the younger the person affected, the more likely it is that the disease will run in the person's family. Such familial Alzheimer's cases form only a small percentage of the total. Interestingly, the tendency for severe and early-onset forms of the disease to run in families is not unique to Alzheimer's. This tendency is seen in a number of other human diseases, such as cancers.

Alzheimer's is marked by large numbers of dying and degenerating neurons that form neurofibrillary tangles, and by the accumulation in the rest of the neurons of a waxy material called amyloid plaque. These same plaques and tangles are also found in people with Down syndrome, and enormous excitement greeted the finding that the gene specifying the amyloid plaque protein is located on chromosome 21, the chromosome that is found in three rather than the normal two copies in people with Down syndrome. Could this be the gene for Alzheimer's?

Probably not, for it was soon discovered that the gene or genes for familial Alzheimer's are distinct from the gene for amyloid plaque, though they are closely linked. The gene causing Alzheimer's is now thought to be a regulatory gene that controls the deposition of the amyloid protein, rather than the gene for the protein itself.

This discovery was possible only because there are excellent genetic and physical maps of chromosome 21. As a result of all this mapping activity, many probes are now available that consist of pieces of DNA from the part of chromosome 21 that harbors both the amyloid gene and the possible Alzheimer's gene. These probes have allowed researchers to ask questions about the evolution of this part of chromosome 21, questions that may at first sight appear to be nothing but "pure science" but that are now providing valuable clues to how Alzheimer's happens.

Southern blots of mouse DNA probed with human DNA from this critical part of chromosome 21 showed that the region that probably harbors the Alzheimer's gene has a homologous region on chromosome 16 of the mouse. A large group of British and American researchers headed by Stephen Dunnett decided to use this fact to try to develop a mouse model for Alzheimer's. Suppose a mouse could be made that had three copies of chromosome 16—a mouse version of human Down syndrome. Would the mouse's brain show the plaques and tangles typical of Alzheimer's or Down?

Mice could indeed be made carrying three copies of this chromosome, but unfortunately the developing young carrying the extra chromosome died about twenty days after the formation of the zygote. Dunnett's group decided to transplant fetal brain tissue from these *trisomic* mice into the brains of normal fetuses, so that these islands of transplanted tissue would survive longer. At first nothing happened, but after four months the transplanted tissue in the young mice had accumulated detectable amounts of amyloid and other proteins typical of Alzheimer's. The rest of their brain tissue was normal. These mice are difficult and expensive to make, but for the first time it has become possible to look in detail at the genetics of the disease in animals. It will soon be feasible to dissect the mouse chromosome 16 to find out which regions code for the regulatory genes that produce this simulacrum of Alzheimer's in the mouse.

The amyloid gene is regulated in very complex ways, some of which involve the production of alternate messenger RNAs from different parts of the gene. These messages are translated into different versions of the protein—some longer, some shorter, and some starting at places other than the beginning of the gene. Such *alternate transcription* is being found more and more commonly in other genes as well. For example, the dystrophin protein that is defective in the cells of muscular dystrophy patients is also expressed in the brain. You will recall from chapter 9 that about a third of muscular dystrophy patients are mentally retarded. Louis Kunkel, Alan Beggs, and their coworkers have recently shown that the brain version of the protein, which is a little shorter than the muscle protein, starts at the second exon of the gene rather than the first and is under the control of a different region of the DNA from the one that controls the muscle version of the gene. They ask a fascinating question: Since there are many cases of muscular dystrophy with no noticeable mental retardation, might there also be cases of X-linked mental retardation, caused by defects in the brain control region, that are not accompanied by musculary dystrophy? While no such cases have yet been found, the molecular tools now exist to screen families in which mental retardation seems to be inherited in a sex-linked fashion. Kunkel suggests that defects in dystrophin or its regulation might be a quite common cause of retardation.

Another very common type of X-linked mental retardation, the *fragile X syndrome,* affects about one in every 1,500 males and one in every 2,500 females. The mode of inheritance of this condition has been, until very recently, a mystery, for the X chromosome carrying the gene may be passed without effect from a normal male to his normal daughter. Only in the daughter does it somehow become "activated" to produce mental retardation in a grandson or granddaughter. The condition takes its name from the fact that, in addition to producing mental retardation, the X chromosome often breaks at a specific point, so that many cells in the affected individuals carry broken Xs. The gene that causes the retardation maps at the same point at which the chromosome breaks.

Two groups, one in France and one in Australia, have now succeeded in cloning the part of the X chromosome responsible for producing the condition, and have found the condition to be correlated with a small insertion of DNA of variable size, up to 500 bases long. The insertion can actually grow in size as the chromosome is passed from one generation to the next, and generally, though not always, the larger the insertion the more severe the condition. The sequence of this region is not yet known, but some of its properties seem to resemble those of the VNTRs that we examined in chapter 9. Of course, those VNTRs seem to have little or no effect on the people carrying them. The fragile X insertion seems to be disrupting some important gene, producing mental retardation, and on top of that somehow changing the structure of the chromosome in such a way as to make it more liable to be broken. Will it turn out to be VNTR-like? And if so, is this region a time bomb waiting to be triggered in all our X chromosomes, or do only certain people carry this modification of the X, a modification that can somehow be activated by passage through a female to produce this runaway VNTR-like activity? And does this inserted region always grow in size, or can it sometimes shrink, perhaps reversing the damage?

This and other recent work on the genetics of mental retardation reinforces the conclusion of chapter 13, that many and perhaps most cases of gene-related retardation are due not to some accidental combination of many genes, but rather to rare—and in principle curable—alterations, each having a very large effect.

Evidence is also growing to reinforce a suggestion made in chapter 6, that the overall structure of chromosomes influences the way they behave. A group headed by K. H. Choo in Australia has recently shown that parts of the regions around the *centromeres* of chromosomes 13 and 21 resemble each other strongly. The most likely explanation for this resemblance is that some kind of genetic interchange between these two otherwise very different chromosomes took place during recent evolutionary time. This might not in itself be remarkable were it not for the fact that chromosome 13, like chromosome 21, is often distributed incorrectly during the cell division called meiosis that results in the formation of eggs and sperm. As a result, infants and spontaneously aborted fetuses with an extra chromosome 13 are almost as common as those with an extra chromosome 21, although an extra chromosome 13 produces a far more severe syndrome. Is there any relationship between the fact that the centromeres of these two chromosomes resemble each other and the fact that they are often passed on to the next generation in the wrong numbers? And when did this postulated genetic interchange between them happen? It should be possible to find out by looking at the equivalent chromosomes in chimpanzees, to see whether they resemble each other in the same way. If they do not, then the interchange must have happened since the ancestors of humans and chimpanzees parted company. Yet one suspects that whatever might be causing these chromosome problems in humans, the same factors should be operating in chimpanzees, because a case of a chimpanzee with Down syndrome has been reported. The more we know about the structure of these chromosomes, the better we will be able to understand this important and devastating problem.

Many genes in the human genome continue to prove puzzlingly elusive, even though a great deal of effort is being expended in tracking them down. It is now known that the gene for Huntington's chorea, a degenerative neurological disease that first appears in middle age, is somewhere near the tip of the short arm of chromosome 4. Even though it is dominant and some extensive pedigrees are available, the gene itself had not been found as of June 1991. For years it has danced tantalizingly just beyond the reach of investigators. One reason is that the data on the location of the gene are very inconsistent, leading some workers to suggest that the gene may occupy different places on the chromosome in different people. Indeed, there are good molecular reasons why such shifts in location are more likely to occur at the tips of chromosomes than in their middles, where genes tend to behave in a more predictable fashion.

Research groups are homing in on other important genes. The gene for the neurodegenerative disease Friedreich's ataxia has been tracked down to a region that includes eleven possible candidates on a stretch of the long arm of chromosome 9. There is also much current interest in Marfan syndrome, from which Abraham Lincoln may have suffered. The syndrome is marked by connective tissue, resulting in heart defects, among other problems. The gene is now known to be somewhere near the centromere of chromosome 15, and there is a great deal of excitement about the possibility of cloning it, since the gene is suspected of being involved in many other connective tissue diseases.

Once genes have been tracked down, progress in understanding them can be astoundingly rapid. A very common dominant condition with the rather daunting name of von Recklinghausen's neurofibromatosis (NF) is found on chromosome 17. About one person in every 4,000 carries a mutation at this genetic locus. The effects of these mutations can vary widely, ranging from a few "liver spots" on the skin to multiple skin cancers, macrocephaly, and malformations of the spinal column and the lungs. The gene was cloned independently in 1990 by groups headed by Ray White and Francis Collins. They found that the NF gene resembles a gene called Ras-GAP, which is known to interact with a well-known oncogene called *ras.* Perhaps this resemblance helps to explain why skin cancers are so common in people with the disease.

Both groups began their search by cloning genes in the suspect region. They soon found, however, that these genes were not the neurofibromatosis gene, which turned out to be spread over a long region of the chromosome. Rather, they were genes that were embedded *within an intron* of the neurofibromatosis gene. At last count, three of these embedded genes had been found in one of the large introns of the NF gene, and others are likely to be discovered. These "genes within genes" carry their coding information on the DNA strand that is the antisense strand of the NF gene. Remarkably, it turned out that one of the three genes embedded in the NF gene had already been cloned and sequenced, by Daniel Mikol at the University of Chicago. This gene specifies a protein that forms part of the *myelin sheath* of nerve cells. It is at the moment quite unclear whether this embedded gene has anything to do with the NF gene or whether its location within the NF gene is simply an accident. Yet it does seem an odd coincidence that the skin tumors of NF involve the fibrous tissue that surround nerve cells, tissue that may contain myelin.

As more new genes are found and sequenced, there is a growing chance that they will be found to resemble one or more already cloned genes. Further, the story of the NF gene shows that there is even a chance that a newly discovered gene may actually turn out to *be* one of these already sequenced genes. Elimination of this overlapping effort provides a strong rationale for the early sequencing of the entire genome.

Despite our still spotty understanding of the human genome, remarkable coincidences of this kind are turning up with ever greater frequency. Stephen Friend's group at Mass General decided recently to investigate a familial condition called Li-Fraumeni syndrome, in which the victims are marked by the early onset of many different kinds of cancers. They wanted to see if people suffering from this syndrome might carry a mutant anti-oncogene that would explain the high incidence of these cancers. They considered two well-studied anti-oncogenes, and rejected the first without studying it because it is known to be involved in cancers that are never seen in the Li-Fraumeni syndrome. The second, a gene called p53 on the short arm of chromosome 17, turned out to be mutant in all seven of the patients they examined, and in some cases could be seen to be passed down from one generation to the next. It was not entirely an accident that they struck it lucky the first time. At this rate, it will not be many years before the human genome is transformed from a region of total darkness to a familiar and populous neighborhood in which educated guesses can be made about the functions of many genes with an increasing chance that they will be right.

The early gene therapy experiments of Blaese, Anderson, and Rosenberg seem at this writing to be going well. The four-year-old has received half a dozen transfusions of her genetically altered lymphocytes, and some of these appear to have colonized her immune system and to be producing ADA. A nine-year-old girl has also begun the treatment, though few data are available as yet on the course of her therapy.

Cancer genes remain just as elusive as ever, though sometimes half-digested stories can give the public the impression that a breakthrough is imminent. At the end of 1990, a group in Berkeley headed by Mary-Claire King published a study of familial breast cancer cases. They were able to show that in some of these families there was linkage of the condition to certain RFLPs in a region on the long arm of chromosome 17 (a region many millions of bases away from the p53 gene studied by Stephen Friend). The media gave much attention to this discovery, including a cover story in *Time* magazine. Yet, as the researchers were at pains to point out, only half of their cases of familial breast cancer showed linkage with this region, and, of course, familial breast cancers themselves make up only a small percentage of all breast cancers. The families that showed the linkage were those in which the cancers tended to develop early, at age fifty or less. King's group also analyzed many other families in which the tendency to breast cancer also seemed to be inherited, but in which the average age at which the cancers appeared was greater than fifty. In these families, in contrast to the early-onset families, there was no sign of linkage to the RFLPs on chromosome 17. These late-onset families might simply be families with a high incidence of breast cancer due simply to chance, with no easily traced genetic cause. Recall that in chapter 11 a similar explanation was thought perhaps to account for some of the difficulties encountered in the search for a schizophrenia gene.

Even when the "breast cancer gene" is tracked down, it will undoubtedly turn out to be only one of a number of genes that can be involved in this condition. The region of chromosome 17 that is a suspect in breast cancer is already known to be rich in oncogenes and other genes that might have some connection with the genesis of cancers. Indeed, it is beginning to look as if all the rest of our genome is rich in these genes.

As with Alzheimer's disease and many other conditions in which the genetics tends to be confusing, it seems to be easiest to track down a gene when its expression is extreme—in this case, when it leads to the early onset of the cancer. We came across the same situation in chapter 11, when we saw how genes influencing colon cancer were most easily tracked down if the cancer appeared at an early age and the expression of the precancerous state was clear.

Very little has yet been published about the category of genes dealt with in chapter 14: genes that influence personality, behavior, and abilities. This, I would venture to suggest, is merely the calm before the storm. One very interesting line of research is being pursued by Joseph Profita and his colleagues in Los Angeles and San Francisco. They are investigating the inheritance of perfect pitch, the ability to identify a note or group of notes accurately. As might be imagined, perfect pitch can be manifested in a number of ways, the most extreme of which is the ability to identify notes in isolation. Relative pitch, the ability to recognize musical intervals, is far more common. In the majority of people with perfect pitch the ability manifests itself at an early age; in a few it can be produced by intensive training. Profita has designed tests to distinguish various aspects of perfect pitch, in both musicians and the nonmusical, and has followed the character in sixty families. It appears to be inherited as an autosomal dominant, though the variable expression of the character can confuse the picture. Perfect pitch is rare: Profita has estimated on the basis of data from music teachers that its incidence is approximately one in 1,500 in the general population. And while his data show that there may be one or more major genes involved, its inheritance is likely to be complex. If such major genes are involved and if they improve the acuity of perception, one can ask why they are not more common in the population. Is there in fact some disadvantage to the character?

Perfect pitch is a human ability that is not charged with racial or class implications, so that the study of its genetics is quite "safe." It seems likely that genes for such mental characteristics and abilities will be the first to be mapped to the human genome—but they certainly will not be the last.

References

Preface

PAGE

xi The quotation by Rynders and Pueschel is from *Down syndrome: advances in biomedicine and the behavioral sciences,* ed. Siegfried M. Pueschel and John E. Rynders (Cambridge, Mass.: Ware Press, 1982).

Chapter 1: Cancer

PAGE

15 John Cairns's prescient quote, along with an excellent discussion of the problem of cancer, is from J. Cairns, The origin of human cancers, *Nature* 289 (1981): 353–57. Dulbecco's editorial suggesting a human genome project was published as R. Dulbecco, A turning point in cancer research: sequencing the human genome, *Science* 231 (1986): 1055–56.

18 The association between hepatitis B and liver cancer is a complex one. A recent review can be found in L. E. Schnipper and G. J. Bubley, Viral infections and cancer, *Comprehensive Therapy* 16 (1990): 42–48; and G. M. Dusheiko, Hepatocellular carcinoma associated with chronic viral hepatitis, *British Medical Bulletin* 46 (1990): 492–511. There have been numerous reports that healthy donor bone marrow can become leukemic, but the difficulty of tracing the donor cells with certainty in the constantly changing genetic world of the leukemia cell has recently been documented by J. Stein et al., Origins of leukemic relapse after bone marrow transplantation, *Blood* 739 (1989): 2033–40.

21 The genetics of the skin disease xeroderma pigmentosum is reviewed

PAGE

in A. R. Lehmann and P. G. Norris, DNA repair deficient photodermatoses, *Seminars in Dermatology* 9 (1990): 55–62.

22 The contributions of Michael Bishop and Harold Varmus to our understanding of oncogenes, the genes involved in the development of cancer, are reviewed in J. M. Bishop, The molecular genetics of cancer, *Science* 235 (1987): 305–11; and H. Varmus, Reverse transcription, *Scientific American* 257 (1987): 56–59.

Chapter 2: DNA

PAGE

27 Fred Sanger has traced his early career in a short autobiographical article with an attention-getting title: Sequences, sequences and sequences, *Annual Review of Biochemistry* 57 (1988): 1–28.

33 Max Delbrück referred to DNA as "that stupid substance" in a conversation with Horace Freland Judson, quoted in Judson, *The eighth day of creation: makers of the revolution in biology* (New York: Simon and Schuster, 1979), p. 59.

33 Perhaps the best discussion of the discovery of DNA and the dawn of molecular biology is given in Judson, *The eighth day of creation.* The book is superbly researched and remarkably complete.

45 Walter Gilbert has sketched out his approach to DNA sequencing in the following Nobel Prize lecture: DNA sequencing and gene structure, *Science* 214 (1981): 1305–12.

48 Sanger's definitive work on the sequence of the virus ØX174 was published in F. Sanger, A. R. Couson, T. Friedmann, et al., Nucleotide sequence of bacteriophage ØX174, *Journal of Molecular Biology* 125 (1978): 225–246. His group's paper on the complete sequence of the human mitochondrial chromosome, all 16,569 bases, appeared in S. Anderson, A. T. Bankier, B. G. Barrell, et al., Sequence and organization of the human mitochondrial genome, *Nature* 290 (1981): 457–65.

Chapter 3: Therapy

PAGE

52 Garrod's description of his early experiments on uric acid and gout can be found in Alfred Baring Garrod, *The nature and treatment of gout and rheumatic gout* (London: Walton and Maberly, 1859).

57 Many fascinating histories of gout have been written. See for example J. E. Seegmiller, Contribution of Lesch-Nyhan syndrome to the understanding of purine metabolism, *Journal of Inherited Metabolic Diseases* 12 (1989): 184–96; and William S. C. Copeman, *A short history of the gout and the rheumatic diseases* (Berkeley: University of California Press, 1964).

57 Gertrude Elion's description of her Nobel Prize work can be found

PAGE

in G. B. Elion, The purine path to chemotherapy, *Science* 244 (1989): 41–47.

59 The current status of our understanding of the human uricase gene is given by X. W. Wu, C. C. Lee, D. M. Muzny, and C. T. Caskey, Urate oxidase: primary structure and evolutionary implications, *Proceedings of the National Academy of Sciences (U.S.)* 86 (1989): 9412–16. They show that the human uricase gene has been rendered nonfunctional by mutations, while that of the baboon can function even though the baboon has little enzyme activity.

61 Consult C. T. Caskey and J. T. Stout, Molecular genetics of HPRT deficiency, *Seminars in Nephrology* 9 (1989): 162–67 for a recent review of the molecular biology of the Lesch-Nyhan syndrome.

63 The very elegant molecular biology involved in sequencing the entire HPRT gene is set out in A. Edwards et al., Automated DNA sequencing of the human HPRT locus, *Genomics* 6 (1989): 593–608. The paper also shows the reader what an entire human gene looks like when it has been sequenced, introns and all. Discussion of some of the mutants found at the HPRT locus is given in S. Singh, et al., Lesch-Nyhan syndrome and HPRT variants: study of heterogeneity at the gene level, *Advances in Experimental Medicine and Biology* 258A (1989): 151–54. Gene therapy with HPRT is summarized in J. A. Wolfe and T. Friedmann, Approaches to gene therapy in disorders of purine metabolism, *Rheumatic Disease Clinics of North America* 14 (1988): 459–77.

66 Martin Rechsteiner's letter critical of the Human Genome Project was sent to many biologists at his expense, and appears in M. Rechsteiner, The Human Genome Project: two points of view, *FASEB Journal* 4 (1990): 2941–42.

67 The work by Israeli scientists on PRPP synthetase is summarized by O. Sperling et al., Altered kinetic property of erythrocyte phosphoribosyl phosphate synthetase in excessive purine production, *Revue Européene d'Études Cliniques et Biologiques* 17 (1972): 703–6. The Guy's Hospital group's work on kidney defects and gout is in C. Calabrese, H. A. Simmonds, J. S. Cameron, and P. H. Davies, Precocious familial gout with reduced fractional urate clearance and normal purine enzymes, *Quarterly Journal of Medicine* 75 (1990): 441–50.

68 The finding of an "alcoholism gene" received wide attention in the popular press. The original paper is K. Blum, E. P. Noble, P. S. Sheridan, et al., Allelic association of human dopamine D2 receptor gene in alcoholism, *Journal of the American Medical Association* 263 (1990): 2055–60. Recent work by a group at NIH found no such association between this gene variant and alcoholism, but the patients in the NIH study were not drawn from as severely affected a group of alcoholics. See A. M. Bolos, N. Dean, S. Lucas-Derse, et al., Population and pedigree studies reveal a lack of association between the dopamine D2 receptor gene and alcoholism, *Journal of the American Medical Association* 264 (1990): 3156–60.

CHAPTER 4: POLITICS

PAGE

70 The history of the transformation of the Atomic Energy Commission into the Department of Energy can be followed in Richard T. Sylves, *The nuclear oracles: a political history of the General Advisory Committee of the Atomic Energy Commission, 1947–1977* (Ames: Iowa State University Press, 1987).

76 In addition to numerous articles in *Science, Nature,* and the popular press since the Human Genome Project's inception, a number of articles have provided insight into the intellectual genesis of the project. Notable among these is a series of articles in the journal *Genomics,* written by the people principally involved: *Genomics* 5 (1989): 385–87, 654–56, 657–60, 661–63, 952–54, 954–56. Charles DeLisi gave a prescient view of the probable shape of the project in Overview of human genome research, *Basic Life Sciences* 46 (1988): 5–10. James Watson and Charles Cantor have also written accounts in *Science.* See J. D. Watson, The Human Genome Project: past, present and future, *Science* 248 (1990): 44–49; and C. R. Cantor, Orchestrating the Human Genome Project, *Science* 248 (1990): 49–51. The OTA report is U.S. Congress, Office of Technology Assessment, *Mapping our genes: genome projects—how big, how fast?* (Baltimore: Johns Hopkins University Press, 1988). The NAS report is *Mapping and sequencing the human genome* (Washington, D.C.: National Academy Press, 1988). In this account I have also drawn from congressional transcripts, interviews with some of the principals, and information obtained from a Capitol Hill source.

CHAPTER 5: THROUGH THE GENOME WITH GUN AND CAMERA

PAGE

89 Jorge Yunis has published some of the most detailed comparisons of our chromosomes with those of our nearest relatives. Some of his pictures can be found in J. J. Yunis and O. Prakash, The origin of man—a chromosomal pictorial legacy, *Science* 215 (1982): 1525–30.

97 Tim Hunkapiller's computer chip, and the controversies surrounding it, are described in L. Roberts, New chip may speed genome analysis, *Science* 244 (1989): 576.

98 Some information about the enormous protein titin can be found in S. Labeit, D. P. Barlow, M. Gautel, et al., A regular pattern of two types of 100-residue motif in the sequence of titin, *Nature* 345 (1990): 273–76.

100 A recent article setting forth some of the difficulties in finding how proteins fold is F. M. Richards, The protein folding problem, *Scientific American* 264 (1991): 54–60. Neural networking programs that tackle the problem of protein folding are discussed in J. Kinoshita, Net

PAGE

result: folded protein. A neural network deciphers the structure of protein, *Scientific American* 262 (1990): 24–26.

103 A more elaborate tour of the living cell is given in Christian de Duve, *A guided tour of the living cell* (San Francisco: W.H. Freeman, 1984).

Chapter 6: Dismantling the Genome

PAGE

122 The history of the chromosome sorting project at the national laboratories is given in J. W. Gray, P. N. Dean, J. C. Fuscoe, et al., High-speed chromosome sorting, *Science* 238 (1987): 323–29.

133 A recent review of the remarkable advances in human gene mapping made possible by the technique of cell-cell hybridization can be found in F. H. Ruddle, A review of genomic physical mapping, *Cancer Surveys* 7 (1988): 267–94.

133 Maynard Olson's remarkable achievement in putting the human cystic fibrosis gene into yeast is detailed in E. D. Green and M. V. Olson, Chromosomal region of the cystic fibrosis gene in yeast artificial chromosomes: a model for human genome mapping, *Science* 250 (1990): 94–98.

134 A discussion of Tony Carrano's mapping method can be found in A. V. Carrano, J. Lamerdin, L. K. Ashworth, et al., A high-resolution, fluorescence-based, semiautomated method for DNA fingerprinting, *Genomics* 4 (1989): 129–36.

136 The original paper on hybridization of DNA to entire chromosomes is M. L. Pardue and J. Gall, Molecular hybridization of radioactive DNA to DNA of cytological preparations, *Proceedings of the National Academy of Sciences (U.S.)* 64 (1969): 600–4.

139 You can contrast the approach taken by the Moyzis group with that of Carrano's group at Livermore by looking at R. L. Stallings, D. C. Torney, C. E. Hildebrand et al., Physical mapping of human chromosomes by repetitive sequence fingerprinting, *Proceedings of the National Academy of Sciences (U.S.)* 87 (1990): 6218–22.

141 A note about Hans Lehrach's new approach to chromosome mapping is P. Little, Clone maps made simple, *Nature* 346 (1990): 611–12. The idea is set out in more detail in A. G. Craig, D. Nizetic, J. D. Hoheisel et al., Ordering of cosmid clones covering the herpes simplex virus type I (HSV-I) genome: a test case for fingerprinting by hybridization, *Nucleic Acids Research* 18 (1990): 2653–60.

146 Information on the incidence of Down syndrome and the effects of screening programs can be gleaned from B. A. Goodwin and C. A. Huether, Revised estimates and projections of Down syndrome births in the United States and the effects of prenatal diagnosis utilization, *Prenatal Diagnosis* 7 (1987): 261–71; J. J. Mulvihill, Medical geneticists confront ethical dilemmas: cross-cultural comparisons among 18 nations, *American Journal of Human Genetics* 46 (1990): 1200–13; C.

PAGE

Julian, P. Huard, J. Gouvernet, et al., Physicians' acceptability of termination of pregnancy after prenatal diagnosis in southern France, *Prenatal Diagnosis* 9 (1989): 77–89; N. Wilson, D. Bickley, and A. McDermott, The prevention of Down syndrome in the southwestern region of England, *West of England Medical Journal* 105 (1990): 15–17; and C. W. McGrother and B. Marshall, Recent trends in incidence, morbidity and survival in Down's syndrome, *Journal of Mental Deficiency Research* 34 (1990): 49–57.

148 Barbara Trask's observations about the range of sizes of various chromosomes 21 in the human population are published in B. Trask, G. van den Engh, B. Mayall, and J. W. Gray, Chromosome heteromorphism quantified by high-resolution bivariate flow karyotyping, *American Journal of Human Genetics* 45 (1989): 739–52.

Chapter 7: The Search for a New Sequencing Technology

PAGE

151 The story of Hood's near-replacement of Cantor as head of the Berkeley Human Genome Center is about as close to a soap opera as the project has yet provided. The pruriently minded can obtain the details from P. Selvin, Charlie Cantor gets kicked upstairs, *Science* 249 (1990): 1238–39.

152 Many of the techniques discussed here are in the early stages of development, and descriptions of them have not yet appeared in the literature.

155 Rod Balhorn's work on DNA imaging can be found in T. P. Beebe, T. E. Wilson, D. F. Ogletree, et al., Direct observation of native DNA structures with the scanning tunneling microscope, *Science* 243 (1989): 370–72. Some of the difficulties in making scanning tunneling microscope images of DNA, and the artifacts that can fool investigators, are discussed in C. R. Clemmer and T. P. Beebe, Graphite: a mimic for DNA and other biomolecules in scanning tunneling microscope studies, *Science* 251 (1991): 640–42.

156 Bustamante's work on imaging individual DNA bases is discussed in D. D. Dunlap and C. Bustamante, Images of single-stranded nucleic acids by scanning tunneling microscopy, *Nature* 342 (1989): 204–6.

159 Some of the preliminary work of Jett and Keller on sequencing individual DNA molecules is set out in J. H. Jett, R. A. Keller, J. C. Martin, et al., High-speed DNA sequencing: an approach based upon fluorescence detection of single molecules, *Journal of Biomolecular Structure and Dynamics* 7 (1989): 301–9.

161 Akiyoshi Wada's attempt to automate the sequencing of DNA has been chronicled in A. Wada, Automated high-speed DNA

PAGE

sequencing, *Nature* 325 (1987): 771–72. The saga of James Watson and the Japanese appears in L. Roberts, Watson vs. Japan, *Science* 246 (1989): 576.

Chapter 8: Sherlock Holmes Meets the Human Genome

PAGE

173 Alec Jeffreys's discovery of VNTRs, which created the whole field of forensic DNA analysis, was announced in A. J. Jeffreys, V. Wilson, and S. L. Thein, Hypervariable "minisatellite" regions in human DNA, *Nature* 314 (1985): 67–73.

175 The subhead comes from a T-shirt spotted at the Cetus Corporation.

176 Lander's article on problems with VNTR fingerprinting is E. S. Lander, DNA fingerprinting on trial, *Nature* 339 (1989): 501–5.

179 The paper showing that populations probably are in Hardy-Weinberg equilibrium for VNTR alleles is B. Devlin, N. Risch, and K. Roeder, No excess of homozygosity at loci used for DNA fingerprinting, *Science* 249 (1990): 416–20.

Chapter 9: Racing After the Killer Genes

PAGE

185 The story of B. B., the child with the X-chromosome deletion that led to the muscular dystrophy gene, is in U. Francks, A. W. Ritter, D. L. Tirschwell, et al., Minor Xp21 chromosome deletion in a male associated with expression of Duchenne muscular dystrophy, chronic granulomatous disease, retinitis pigmentosa and McLeod syndrome, *American Journal of Human Genetics* 37 (1985): 250–67.

186 Louis Kunkel and Anthony Monaco have recounted the story of their cloning of the CF gene in A. P. Monaco and L. M. Kunkel, Cloning of the Duchenne/Becker muscular dystrophy locus, *Advances in Human Genetics* 17 (1988): 61–98.

191 The variety of ways in which different animals are affected by the CF mutation is summarized in J. L. Carpenter, E. P. Hoffmann, F. C. Romanul, et al., Feline muscular dystrophy with dystrophin deficiency, *American Journal of Pathology* 135 (1989): 909–19.

193 Alan Beggs's work on the screening of muscular dystrophy mutants is reviewed in A. H. Beggs and L. M. Kunkel, Improved diagnosis of Duchenne/Becker muscular dystrophy. *Journal of Clinical Investigation* 85 (1990): 613–19.

198 The paper suggesting that Chopin might have suffered from CF is J. G. O'Shea, Was Frédéric Chopin's illness actually cystic fibrosis? *Medical Journal of Australia* 147 (1987): 586–89.

PAGE

200 The first case of autosomal linkage in humans was reported in J. H. Renwick and S. D. Lawler, Genetical linkage between the ABO and nail-patella loci, *Annals of Human Genetics* 19 (1955): 312–31.

201 The first papers to suggest the use of RFLPs to map the human genome were Y. W. Kan and A. Dozy, Antenatal diagnosis of sickle cell anemia by DNA analysis of amniotic-fluid cells, *Lancet* 2 (1978): 910–12; and D. Botstein, R. L. White, M. Skolnick, and R. Davis, Construction of a genetic linkage map in man using restriction fragment polymorphisms, *American Journal of Human Genetics* 32 (1980): 314–31.

205 The complete human map produced by Collaborative Research can be found in H. Donis-Keller, P. Green, C. Helms, et al., A genetic linkage map of the human genome, *Cell* 51 (1987): 319–37.

208 Some of the story behind the race for the CF gene is told in L. Roberts, The race for the cystic fibrosis gene, *Science* 240 (1988): 141–44; and L. Roberts, Race for the cystic fibrosis gene nears end, *Science* 240 (1988): 282–85. The three papers that detail the finding of the gene are J. M. Rommens, M. C. Ianuzzi, B. Kerem, et al., Identification of the cystic fibrosis gene: chromosome walking and jumping, *Science* 245 (1989): 1059–65; J. R. Riordan, J. M. Rommens, B. Kerem, et al., Identification of the cystic fibrosis gene: cloning and characterization of the complementary DNA, *Science* 254 (1989): 1066–73; and B. Kerem, J. M. Rommens, J. A. Buchanan, et al., Identification of the cystic fibrosis gene: genetic analysis, *Science* 245 (1989): 1073–80.

212 The papers detailing the successful "curing" of the CF defect in tissue culture cells are M. L. Drumm, H. A. Pope, W. H. Cliff, et al., Correction of the cystic fibrosis defect in vitro by retrovirus-medicated gene transfer, *Cell* 62 (1990): 1227–33; and D. P. Rich, M. P. Anderson, R. J. Gregory, et al., Expression of cystic fibrosis transmembrane conductance regulator corrects defective chloride channel regulation in cystic fibrosis airway epithelial cells, *Nature* 347 (1990): 358–63.

212 The warning posed about the dangers of CF screening can be found in C. T. Caskey, M. M. Kaback, and A. L. Beaudet, The American Society of Human Genetics statement on cystic fibrosis screening, *American Journal of Human Genetics* 46 (1990): 393.

212 The mixed success of screening programs for Tay-Sachs, sickle-cell anemia, and thalassemia is recounted in M. Modell and B. Modell, Genetic screening for ethnic minorities, *British Medical Journal* 300 (1990): 1702–4; J. Brown, Prenatal screening in Jewish law, *Journal of Medical Ethics* 16 (1990): 75–80, J. Trevelyan, Fighting Tay-Sachs disease, *Nursing Times* 85 (1989): 22–23; A. Cao, Results of programmes for antenatal detection of thalassemia in reducing the incidence of the disorder, *Blood Reviews* 1 (1987): 169–76; L. Roberts, One worked; the other didn't, *Science* 247 (1990): 18; and A. M.

PAGE

Kuliev, O. L. Mazurova, V. B. Biriukov, et al., The first case of prenatal diagnosis of sickle-cell anemia in the USSR, *Gematologiia i Transfuzologiia* 35 (1990): 8–9.

Chapter 10: The Early Days of Gene Therapy

PAGE

221 The story of Robert and his companions is set forth in K. Cornetta, R. Wieder, and W. F. Anderson, Gene transfer into primates and prospects for gene therapy in humans, *Progress in Nucleic Acid Research and Molecular Biology* 36 (1989): 311–22.

224 A summary of the many problems facing gene therapists is given in T. Friedmann, Progress towards human gene therapy, *Science* 244 (1989): 1275–81.

227 The story of Steven Rosenberg's attempts to use tumor-infiltrating lymphocytes and genetically modified cells to fight cancer is recounted in S. A. Rosenberg, Adoptive immunotherapy for cancer, *Scientific American* 262 (1990): 62–69.

229 Jeremy Rifkin's jousts with the establishment are covered in L. Roberts, Rifkin battles gene transfer experiment, *Science* 243 (1989): 734; and D. Swinbanks, Rifkin tries to block human gene transfer experiment, *Nature* 337 (1989): 398.

234 The story of smallpox variolation in the eighteenth century can be found in Derrick Baxby, *Jenner's smallpox vaccine: the riddle of vaccinia virus and its origin* (London: Heinemann Educational Books, 1981).

Chapter 11: Cancer Revisited

PAGE

239 Fanny Burney's ghastly account of her mastectomy can be found in *Eyewitness to history,* ed. John Carey (Cambridge, Mass.: Harvard University Press, 1987).

240 Eldon Gardner told his story of the discovery of familial polyposis in E. J. Gardner, Discovery of the Gardner syndrome, *Birth Defects: Original Articles Series* 8 (1972): 48–51.

242 An account of the discovery of the deletion leading to the familial polyposis gene can be found in L. Herrera, S. Kakati, L. Gibas, et al., Gardner syndrome in a man with an interstitial deletion of 5q, *American Journal of Medical Genetics* 25 (1986): 473–76. The two papers announcing the mapping of the gene are M. Leppert, M. Dobbs, P. Scambler, et al., The gene for familial polyposis coli maps to the long arm of chromosome 5, *Science* 238 (1987): 1411–13; and W. F. Bodmer, C. J. Bailey, J. Bodmer, et al., Localization of the gene for familial adenomatous polyposis on chromosome 5, *Nature* 328 (1987): 614–16.

243 The paper revealing the great range of expression of the familial

PAGE

polyposis gene is M. Leppert, R. Burt, J. P. Hughes, et al., Genetic analysis of an inherited predisposition to colon cancer in a family with a variable number of adenomatous polyps, *New England Journal of Medicine* 322 (1990): 904–8.

245 The discovery that retinoic acid could cause leukemic cells to complete their differentiation in tissue culture is in T. R. Breitman, S. J. Collins, and B. R. Keene, Terminal differentiation of human promyelocytic leukemic cells in primary culture in response to retinoic acid, *Blood* 57 (1981): 1000–4. The first use of retinoic acid in treatment of acute promyelocytic leukemia is in P. Flynn, W. Miller, D. Weisdorf, et al., Treatment of acute promyelocytic leukemia with retinoic acid: correlation with cell culture, *Blood* 60 (1982): 155a. The results of the French-Chinese collaboration can be found in S. Castaigne, C. Chomienne, M. T. Daniel, et al., All-trans retinoic acid as a differentiation therapy for acute promyelocytic leukemia: I. Clinical results, *Blood* 76 (1990): 1704–9.

246 Mapping of the retinoic acid receptor gene was accomplished by M. G. Matthei, M. Petkovich, J. F. Matthei, et al., Mapping of the human retinoic acid receptor to the q21 band of chromosome 17, *Human Genetics* 80 (1988): 186–88. A recent summary of the remarkable correlations between certain types of cancer and certain chromosome abnormalities can be found in P. C. Nowell and C. Croce, Chromosome translocations and oncogenes in human lymphoid tumors, *American Journal of Clinical Pathology* 94 (1990): 229–37.

247 The chromosome 15-17 fusion story is in H. de Thé, C. Chomienne, M. Lanotte, et al., The t(15;17) translocation of acute promyelocytic leukemia fuses the retinoic acid receptor alpha gene to a novel transcribed locus, *Nature* 347 (1990): 558–61.

249 The genes involved in the Philadelphia chromosome fusion were discovered by S. S. Clark, J. McLaughlin, M. Timmons, et al., Expression of a distinctive BCR-ABL oncogene in Ph1-positive acute lymphocytic leukemia (ALL), *Science* 239 (1988): 775–77; and by E. Fainstein, C. Marcell, A. Rosner, et al., A new fused transcript in Philadelphia chromosome positive acute lymphocytic leukemia, *Nature* 330 (1987): 386–88. The Baltimore group's work on producing cancer in retrovirus-infected mice is in G. Q. Daley, R. A. van Etten, and D. Baltimore, Induction of chronic myelogenous leukemia in mice by the P210*bcr/abl* gene of the Philadelphia chromosome, *Science* 247 (1990): 824–30.

251 A review of what is known about Burkitt's lymphoma is J. A. Goldstein and R. L. Bernstein, Burkitt's lymphoma and the role of Epstein-Barr virus, *Journal of Tropical Pediatrics* 36 (1990): 114–20.

252 The involvement of recombination-generating regions in the Burkitt's lymphoma translocation was noted by B. Shiramizu and I. Magrath, Localization of breakpoints by polymerase chain reactions in Burkitt's lymphoma with 8;14 translocations, *Blood* 75 (1990):

PAGE

1848–52. The Paris group that found the presence of these signals near the *myc* gene on chromosome 8 reported their results in O. Bernard, C. J. Larson, A. Hampe, et al., Molecular mechanisms of a t(8;14)(q24;q11) translocation juxtaposing *c-myc* and TcR-alpha genes in a T-cell leukemia: involvement of a V alpha internal heptamer, *Oncogene* 2 (1988): 195–200.

CHAPTER 12: VOICES HEARD IN A TINY PIECE OF CHROMOSOME

PAGE

253 Anne Bassett's paper reporting the chromosome 5 partial trisomy linked to schizophrenia is A. S. Bassett, B. C. McGillivray, B. D. Jones, and J. T. Pantzar, Partial trisomy chromosome 5 cosegregating with schizophrenia, *Lancet* 1 (1988): 799–800.

254 A discussion of the Soviet approach to schizophrenia can be found in Sidney Bloch, *Psychiatric terror: how Soviet psychiatry is used to suppress dissent* (New York: Basic Books, 1977).

258 The paper finding linkage of schizophrenia to chromosome 5 is R. Sherrington, J. Brynjolfsson, H. Petursson, et al., Localization of a susceptibility locus for schizophrenia on chromosome 5, *Nature* 336 (1988): 164–67.

261 Papers *not* finding this linkage are J. L. Kennedy, L. A. Guiffra, H. W. Moises, et al., Evidence against linkage of schizophrenia for markers on chromosome 5 in a northern Swedish pedigree, *Nature* 336 (1988): 167–70; D. St. Clair, D. Blackwood, W. Muir, et al., No linkage of chromosome 5q11-q13 markers to schizophrenia in Scottish families, *Nature* 339 (1989): 305–9; S. D. Detera-Wadleigh, L. R. Goldin, R. Sherrington, et al., Exclusion of linkage to 5q11-13 in families with schizophrenia and other psychiatric disorders, *Nature* 340 (1989): 391–93.

269 A survey of tomato genetics can be found in C. M. Rick and J. I. Yoder, Classical and molecular genetics of tomato: highlights and perspectives, *Annual Review of Genetics* 22 (1988): 281–300.

271 Steven Tanksley's approach to quantitative gene mapping in tomato is detailed in A. H. Paterson, E. S. Lander, J. D. Hewitt, et al., Resolution of quantitative traits into Mendelian factors by using a complete linkage map of restriction fragment length polymorphisms, *Nature* 335 (1988): 721–26.

275 The difficulties and complexities of this mapping are explored in A. H. Paterson, S. Damon, J. D. Hewitt, et al., Mendelian factors underlying quantitative traits in tomato: comparison across species, generations and environments, *Genetics* 127 (1991): 181–97.

Chapter 13: Fanfare for the Common Person

PAGE

277 A discussion of the history and mission of CEPH is in J. Dausset, H. Cann, D. Cohen, et al., Centre d'étude du polymorphisme humain (CEPH): collaborative genetic mapping of the human genome, *Genomics* 6 (1990): 575–77.

284 A selection of polemics on the subject of genetics and IQ is, in approximate order from left to right: Richard C. Lewontin, Stephen Rose, and Leon J. Kamin, *Not in our genes: biology, ideology and human nature* (New York: Pantheon, 1984), Leon J. Kamin, *The science and politics of I.Q.* (New York: Halsted, 1974), Hans J. Eysenck and Leon J. Kamin, *The intelligence controversy* (New York: Wiley, 1981), Richard J. Herrnstein, *I.Q. in the meritocracy* (Boston: Little, Brown), and Hans J. Eysenck, *The IQ argument: race, intelligence and education* (New York: Library Press, 1971).

285 A fascinating life of Francis Galton, more accessible than his autobiography or the forbidding biography by Karl Pearson, is Derek W. Forrest, *Francis Galton: the life and work of a Victorian genius* (New York: Taplinger, 1974).

285 Galton's wonderful study is Francis Galton, *Hereditary genius, an inquiry into its laws and consequences* (Cleveland: Meridian, 1962).

286 The Bach family tree is traced in Karl Geiringer and Irene Geiringer, *The Bach family: seven generations of creative genius* (London: Allen and Unwin, 1959).

292 The early history of the Mensa Society can be found in Victor Serebriakoff, *IQ: a Mensa analysis and history* (London: Hutchinson, 1966). Terman's IQ study is detailed in Lewis W. Terman, *Genetic studies of genius* (Stanford, Calif.: Stanford University Press, 1947) and discussed in May V. Seagoe, *Terman and the gifted* (Los Altos, Calif.: W. Kaufmann, 1975).

293 Nigel Hunt's little book is *The World of Nigel Hunt* (New York: Garrett, 1967).

294 The standard deviation of the bell-shaped distribution of smart and dumb alleles is given by $\sqrt{2npq}$, where n is the number of genes and p and q are the average proportions of smart and dumb alleles at those genes. Since we have specified that these proportions are always 0.5, the standard deviation is the square root of 10,000 times 0.5 times 0.5, which is 50. About 95 percent of such a bell-shaped population lies within two standard deviations of the mean, so that most of the population will have between 4,900 and 5,100 smart alleles.

298 The unpleasant discovery about Cyril Burt's data is detailed in Kamin, *The science and politics of I.Q.* The Minnesota study is T. J. Bouchard, Jr., D. T. Lykken, M. McGue, et al., Sources of human psychological differences: the Minnesota study of twins, *Science* 250 (1990): 223–28.

PAGE

299 Burckhardt's book is Jacob Burckhardt, *The civilization of the Renaissance in Italy* (New York: Oxford University Press, 1937).

299 Something about the remarkable life of Leon Battista Alberti can be gleaned from Franco Borso, *Leon Battista Alberti* (Oxford: Phaidon, 1977) and Joan Kelly, *Leon Battista Alberti: universal man of the early Renaissance* (Chicago: University of Chicago Press, 1969).

Chapter 15: Stop Press

PAGE

316 Bart Barrell's complaint about the state of DNA sequencing is B. Barrell, DNA sequencing: present limitations and prospects for the future, *FASEB Journal* 5 (1991): 40–45.

318 A recent survey on the confusing genetics of Alzheimer's is K. H. Buetow, R. Shiang, P. Yang, et al., A detailed multipoint map of human chromosome 4 provides evidence for linkage heterogeneity and position-specific recombination rates, *American Journal of Human Genetics* 48 (1991): 911–25.

319 Hans Lehrach's first moves on his mapping program are detailed in D. Nizetic, G. Zehetner, A. P. Monaco, et al., Construction, arraying, and high-density screening of large insert libraries of human chromosomes X and 21: their potential use as reference libraries, *Proceedings of the National Academy of Sciences (U.S.)* 88 (1991): 3233–37.

319 Dunnett's work on the mouse model for Alzheimer's is S. J. Richards, J. J. Waters, K. Beyreuther, et al., Transplants of mouse trisomy 16 hippocampus provide a model of Alzheimer's disease neuropathology, *EMBO Journal* 10 (1991): 297–303.

320 Kunkel's investigation of the muscular dystrophy protein in the brain is F. M. Boyce, A. H. Beggs, C. Feener, and L. M. Kunkel, Dystrophin is transcribed in the brain from a distant upstream promoter, *Proceedings of the National Academy of Sciences (U.S.)* 88 (1991): 1276–80.

320 The papers dealing with the cloning of the fragile region of the X chromosome are I. Oberle, F. Rousseau, D. Heitz, et al., Instability of a 500-base pair DNA segment and abnormal methylation in fragile X syndrome, *Science* 252 (1991): 1097–1102; and S. Yu, M. Pritchard, E. Kremer, et al., Fragile X syndrome characterized by an unstable region of DNA, *Science* 252 (1991): 1179–81.

321 The characterization of the centromeric regions of chromosomes 13 and 21 is reported in B. Vissel and K. H. Choo, Four distinct alpha satellite subfamilies shared by human chromosomes 13, 14, and 21, *Nucleic Acids Research* 19 (1991): 271–77.

322 An update on the search for Huntington's chorea and Alzheimer's genes can be found in J. F. Gusella, The search for the genetic defects in Huntington's disease and familial Alzheimer's disease, *Research*

PAGE

Publications—Association for Research in Nervous and Mental Disease 69 (1991): 75–83.

323 The discovery of the function of one of the "genes within genes" in the NF region is reported in D. Viskochil, R. Cawthon, P. O'Connell, et al., The gene encoding the oligodendrocyte-myelin glycoprotein is embedded within the neurofibromatosis type 1 gene, *Molecular and Cellular Biology* 11 (1991): 906–12.

323 The molecular lesion involved in the Li-Fraumeni syndrome is reported in D. Malkin, F. P. Li, L. C. Strong, et al., Germ line p53 mutations in a familial syndrome of breast cancer, sarcomas, and other neoplasms, *Science* 250 (1990): 1233–38.

324 The recent advance in breast cancer genetics can be found in J. M. Hall, M. K. Lee, B. Newman, et al., Linkage of early-onset breast cancer to chromosome 17q21, *Science* 250 (1990): 1684–89.

325 Some early data on the inheritance of perfect pitch are given by J. Profita and T. G. Bidder, Perfect pitch, *American Journal of Medical Genetics* 29 (1988): 763–71.

GLOSSARY

ADA The enzyme adenosine deaminase. When this enzyme is missing in the cells of the body, a product builds up that kills the T cells of the immune system.

Adenine A purine base; an important component of DNA, RNA, and many other compounds in the cell.

Adenosine deaminase deficiency One form of severe combined immune deficiency (SCID), caused in this case by the absence of ADA.

Allele One form of a particular gene. For example, people who have different ABO blood types carry different allelic forms of a gene that has a locus, or location, near the tip of the long arm of chromosome 9. Within a species, most alleles may differ by one or a few bases, producing no differences or small differences in the resulting protein. Some, however, show larger differences. Often one allele will replace another in the course of evolution.

Alpha helix This is a helical structure formed by some of the chains of amino acids in proteins, and should not be confused with the double helix of DNA. Some proteins are made up primarily of alpha helices, while others have different kinds of structures; most are a mixture of alpha helices and other types of structures.

Amino acids These small molecules can act as both acids and bases. The acidic part of one amino acid can link to the basic part of another, forming the long chains of proteins. There are twenty different common types of amino acids, which vary greatly in their chemical properties, and which can link up to produce millions of different proteins with extremely variable properties.

Antenatal diagnosis See *prenatal diagnosis.*

Anti-oncogene A gene that counteracts the cancer-inducing effects of a mutated oncogene. One of the steps leading to the appearance of a cancer is the inactivation of anti-oncogenes.

Antisense strand The noncoding strand of the DNA double helix; the strand that does not carry a message.

Bases Each of the building blocks of DNA and RNA, called nucleotides, consist of phosphoric acid, a sugar, and a molecule called a nitrogenous base, or base for short. In every DNA molecule there are four types of base, which pair with the corresponding bases on the other DNA strand according to specific rules. Adenine always pairs with thymine, guanine with cytosine. The sequence of bases specifies the genetic code. Since the sequence of bases along the *coding* strand of the DNA is not constrained by these pairing rules, the process of evolution can produce a great variety of messages written in the four-letter language of the bases.

Burkitt's lymphoma A cancer associated with both Epstein-Barr virus and a translocation between chromosomes 8 and 14.

cDNA library A *plasmid* library made from DNA copies of the messenger RNAs found in the cell. The copies are made using reverse transcriptase, and the library is rich in genes that happen to be turned on in the kind of cells from which the library was made.

Chromosome sorter A device for separating large numbers of a particular chromosome from all the chromosomes in the cell.

Chromosome Most human cells contain forty-six chromosomes, twenty-three derived from one parent and twenty-three from the other. Each chromosome consists of a long and complexly folded strand of DNA, which in the case of the largest human chromosome is almost 300 million bases long. The genes are scattered rather unequally along the chromosome, tending to be bunched at the ends, while many points on the chromosome contain long stretches of highly repetitious DNA. When the cell is active, chromosomes stretch out, allowing the genes to be transcribed easily into *RNA.* When the cell is dividing, the chromosomes coil up and contract.

Clone A hugely multiplied set of copies of a piece of DNA.

Codon One of sixty-four three-letter code words making up the genetic code. Sixty-one of these code for the twenty amino acids, with between one and six code words for each amino acid. The remaining three are stop signals, indicating that the end of a protein has been reached.

Contig A region of the physical map of a chromosome in which the pieces of DNA are known to overlap, providing an unbroken sequence of information for that region of the chromosome.

Cosmid A vector, able to multiply inside a bacterium, which can carry a fairly large piece of foreign DNA. Cosmids can be used to clone pieces of DNA up to 50,000 bases long.

Cystic fibrosis A disease in which the transport of chloride ions across cell membranes is affected, damaging the pancreas and the lungs. It is the most common severe genetic disease among Caucasians.

Cytoplasm The region of the cell lying outside the nucleus, where proteins are synthesized.

Cytosine A *pyrimidine* base; one of the bases coding information in DNA and RNA.

Dideoxynucleotides Analogues of the normal building blocks of DNA that the DNA polymerase can insert by mistake into a growing chain of DNA as it is being synthesized, bringing the synthesis of the chain to a stop at the point of the insertion. These analogues are an essential part of Frederick Sanger's method of DNA sequencing, allowing a set of nested fragments to be produced.

DNA fingerprinting The use of unique combinations of DNA sequences, usually *VNTRs,* to identify an individual. If enough of these sequences can be examined, the individual can be identified with virtual, though not absolute, certainty.

DNA polymerase An enzyme that makes a new strand of DNA by using an existing strand as a template. The enzyme can only work in the presence of a primer—a short stretch of double-stranded DNA somewhere on the single strand, which it can lengthen. It cannot turn a completely single-stranded piece of DNA into a double-stranded piece because it is unable to begin; it needs to be primed.

DNA Deoxyribonucleic acid, the molecule that carries the genetic information of humans and most other organisms on the planet. DNA takes the form of a double helix, resembling a twisted ladder. The information is encoded on one of its two strands—the coding strand—in the form of a sequence of bases.

Dominant A genetic character that is expressed even if it is only one of the two alleles at that locus (in contrast to a *recessive* trait). For example, in humans the allele for brown eyes is (usually) dominant over the allele for blue.

Down syndrome The syndrome, or collection of symptoms, expressed by a person carrying three copies of chromosome 21 rather than the usual two. The syndrome is marked by mental retardation and various organ malformations.

Dystrophin The protein that is altered in the disease muscular dystrophy.

E. coli See *Escherichia coli.*

Electrophoresis A technique by which molecules, such as proteins or DNA fragments, can be separated by the application of an electric current. The molecules are forced by the current through a thin layer of gel. The gel can be designed so that its pores are ideal for the separation of molecules within a given range of sizes and shapes.

Enzyme A protein molecule that catalyzes, or aids, a biochemical reaction. Most reactions in the living cell would proceed with infinitesimal slowness were it not for the aid provided by a huge array of different enzymes.

Epstein-Barr virus The virus associated with Burkitt's lymphoma; also the cause of mononucleosis.

Escherichia coli A common bacterium that lives in the guts of many animals, and that is ideal for making many copies of cloned genes because it can readily harbor plasmids and cosmids carrying genes from other organisms.

Exons The fragments into which the coding part of the gene is divided, separated by long, mysterious noncoding regions called introns. As a result, most genes are much longer than they apparently need to be.

ϕ174 A single-stranded DNA virus that multiplies in bacteria. The chromosome of this organism was the first to be sequenced completely.

GenBank The largest computer repository of DNA sequence information.

Gene pool The group of genes available to any given species. Reproductive barriers prevent the interspecies exchange of genes. All human groups are capable of interbreeding, however, so the human gene pool is very large. Five billion humans carry ten billion copies of each of the hundred thousand human genes making up our gene pool. This means that at each genetic locus tens of thousands of new mutations are arising in people throughout the world each generation, adding to the variation already present in our species.

Gene therapy The addition of a functioning gene to cells or tissues, resulting in the alleviation of the symptoms of a genetic disease. It is hoped that gene therapy will eventually lead to gene surgery: the removal of a damaged gene or portion of a gene, and its precise replacement with a functioning one.

Gene Strictly speaking, a stretch of DNA that codes for a protein. As we have learned more about the genome, however, this definition of a gene has become a little fuzzy around the edges. Most genes code for proteins, but nearly all of them have introns that break them up and regulatory regions nearby that govern when they turn on and off. Some genes, like the immunoglobulins, are pieced together from smaller parts scattered along the chromosome. Some of the regions that regulate the expression of genes do not code for proteins, but instead interact with proteins that are made by genes elsewhere in the genome.

Genetic screening The process of surveying a population to find out which individuals are heterozygous for a recessive mutant allele. One particularly successful screening program searches for the mutant alleles leading to Tay-Sachs disease.

Genome All the genes of an organism, along with all the other DNA of the chromosomes. In humans, only about 8 percent of the genome is made up of genes; the rest consists of introns, flanking regions between the genes, copies of retroviruses, and other more mysterious pieces of DNA, as well as long stretches of highly repeated DNA of uncertain function.

Gout A disease in which uric acid accumulates in the blood and crystallizes in the joints. Gout has many genetic and nongenetic causes.

Guanine A purine base; one of the four bases that code information in DNA and RNA.

Hardy-Weinberg Law A simple law that allows a researcher to predict the proportion of homozygotes and heterozygotes in a population, provided that the allele frequency is known or can be inferred and that the members of the population marry at random with respect to the allele. (Other restrictions on the law apply, but these are the most important.)

Heterozygote An organism with two different alleles at a particular genetic locus. For example, a person with blood type AB is a heterozygote at the ABO locus, having received an A allele from one parent and a B allele from the other. We are all heterozygous at some of our genetic loci, homozygous at others.

HLA Human lymphocyte antigens: a very polymorphic set of genes located on chromosome 6.

Homozygote An organism with two identical alleles at a particular genetic locus. For example, a person with blood type O is homozygous at the ABO locus, having received identical alleles from each parent. This does not prevent him or her from being heterozygous at other genetic loci, however.

HPRT An enzyme that is missing in individuals suffering from Lesch-Nyhan syndrome, and that is found only in trace amounts in some sufferers from familial gout.

Hybridization of DNA A procedure in which two single-stranded pieces of DNA meet and match up. When the double helix of DNA is broken apart into single strands by heat or chemicals, the bases that are normally in the interior of the molecule are exposed. This molecule now has astounding specificity; mixed together with an enormous number of DNA molecules, only a few of which are complementary to it, it will eventually bump into and match with one of its complementary molecules and recreate the double helix. The experimenter has the option of altering the conditions, making them more or less stringent. Under more stringent conditions, the molecule will only match with a precisely complementary sequence. Under less stringent conditions, it can match with a sequence that is not its precise complement. Under stringent conditions, human genes will match only with their complementary human strands. When the stringency is lowered a little, they will match the equivalent genes of a monkey. When it is lowered still further, they will be able to match with the equivalent genes of a bird, or a reptile. This allows the experimenter to search for genes in different organisms that have an evolutionary relationship to each other, even a remote one.

Insulin A small hormone that regulates the transport of sugar into cells. It is the first protein for which the complete amino acid sequence was determined.

Integration The insertion of foreign DNA into a chromosome. In human cells, integration can be carried out by a retrovirus or retroposon that carries the foreign DNA.

Interleukin 2 One of the many growth factors made by tissues in the body. This one stimulates the growth of T cells, the cells of the immune system that can recognize and destroy foreign cells.

Introns Parts of a gene that are transcribed into RNA in the nucleus, but are removed before the RNA travels to the cytoplasm and is translated into protein. Most human genes have many introns scattered along their lengths, which have the effect of enormously lengthening the genes. Introns appear to be involved in gene regulation, but why there should be so many of them remains a puzzle.

Jumping along a chromosome The process of cloning part of a chromosome and then using it as a jumping-off point to clone another part a known distance away, without having to clone all the intervening regions. This allows an experimenter to travel rapidly from a known region of a chromosome toward a gene of interest.

Lesch-Nyhan syndrome A recessive genetic disorder marked by mental retardation and an uncontrollable tendency toward self-mutilation. It results from the absence of the enzyme HPRT, causing the levels of purines to reach poisonous heights.

Library of DNA A random collection of bits of the DNA of an organism, cloned into plasmid or cosmid vectors. In a well-constructed library, it should be possible to find a clone carrying any one of an organism's genes.

Linkage The tendency for two genes close to each other on a chromosome to be passed together to the next generation. The process of recombination exchanges pieces of maternal and paternal chromosomes each generation, but the more tightly linked two genes are, the less likely it is that they will be separated by recombination in a given generation.

Locus As its name implies, the locus is a location: In this case, it is the location of a gene on the chromosome. Geneticists speak of *alleles* of a *gene* at a *locus* on a *chromosome* in the same way that one would speak of tenants in an apartment at an address in a city.

Marek's disease A highly infectious cancer of chickens, triggered by a DNA virus.

Messenger RNA An RNA molecule that carries a copy of the information in a gene out into the cytoplasm, where it can be translated into protein.

Muscular dystrophy A wasting muscular disease, caused by a defect in the protein dystrophin. Duchenne and Becker muscular dystrophy are, respectively, more and less severe mutations at the muscular dystrophy locus. MD mutations are recessive, but because the gene is on the X chromosome most of the disease's victims are male.

Mutation So many kinds of mutations are now known that any definition must be very broad. I define it here, as I have elsewhere, as any inherited alteration in the *genome* of an organism, aside from the regular mixing of genes produced by the process of genetic recombination.

Nested fragments Fragments of DNA, RNA, or protein produced by experimenters interested in sequencing these molecules. A set of nested DNA fragments will consist of all the possible fragments of that piece of DNA that end in a particular base.

Nucleus A region inside the cell, bounded by a membrane, that contains the chromosomes.

Oncogene A gene that, when mutant, can help to trigger a cancer. Normally, oncogenes have a variety of functions in the cell. Many of them code for growth factors, so that when they are turned on permanently as a result of a mutation they can lead to the uncontrolled growth that is the hallmark of cancer.

PCR Polymerase chain reaction, an elegant new technique that allows an experimenter, starting from a DNA molecule, to clone large quantities of any known piece of DNA.

Plasmid A small piece of circular DNA that can multiply inside a bacterium and be used to clone any fairly short piece of DNA inserted into it.

Polygenes Also known as genes for quantitative characters, these are genes that individually have very small effects on a character, but that may be quite numerous and are usually scattered around the chromosomes. Most polygenes probably have their major effect on some other character, and their effect on the character being measured is thus an incidental one.

Polymerase An enzyme that polymerizes or joins together long chains of subunits. The polymerases talked about in this book are those responsible for building chains of DNA or RNA.

Polymorphism Greek for "many forms"; thus, a gene that is polymorphic comes in many different allelic forms. No individual can have more than two of these, though there may be more than two alleles in the population. For example, there are three common alleles—A, B, and O—at the ABO blood group locus, but nobody can carry all three. Many genetic loci are polymorphic, and many of these polymorphic alleles in turn contribute to the genetic variation in the human population.

Prenatal diagnosis The process of determining whether a fetus carries a harmful genotype. Most prenatal diagnoses are carried out using small numbers of the fetal cells that float in the amniotic fluid in which the fetus is bathed. Sampling of this fluid cannot be carried out safely before the fourteenth week of pregnancy at the earliest, restricting the number and complexity of tests that can be done in any one pregnancy.

Primer A short, single-stranded piece of DNA that bonds to its complementary region on a longer single-stranded piece of DNA, forming a short region of double helix. This primes the reaction carried out by DNA polymerase, allowing it to complete the job of synthesizing double-stranded DNA from the single strand.

Probe A piece of DNA that is labeled radioactively or with a fluorescent dye and used to seek out and mark its complementary DNA strand on a gel or on a chromosome.

Proto-oncogene A gene that has the potential to become a cancer-triggering oncogene if it is mutated.

Purine In DNA, the bases adenine and guanine. These molecules are larger than the pyrimidines (in DNA, cytosine and thymine), so adenine always pairs with thymine and guanine with cytosine on the complementary strand. As a result, the backbones of the two DNA strands remain a constant distance apart.

Pyrimidine See *purine.*

Recessive An allele that is masked if a dominant allele is present. About a third of human disease genes are recessive, and will show their effect only if the individual is homozygous for them.

Recombination, genetic The process by which alleles are exchanged each generation between the maternally derived and paternally derived chromosomes. Were it not for this process, all the genes on any given chromosome would always be inherited as a unit.

Replisome A structure made up of a collection of enzymes, prominent among which are the DNA polymerases, which are involved in the replication of DNA.

Restriction enzyme An enzyme that will recognize a specific sequence of bases on a DNA molecule and cut the molecule only there. Some restriction enzymes cut the DNA cleanly, leaving a blunt end; others make staggered cuts on the two strands, producing complementary "sticky ends." If a given enzyme recognizes a short DNA sequence, it will tend to cut the DNA often; if it recognizes a longer sequence, it will cut the DNA into larger pieces.

Retinoic acid A molecule related to vitamin A that is important in the differentiation of cells as they take up various functions in the developing organism. A gene coding for a protein that recognizes retinoic acid has been found to be involved in a cancer that can be treated with this compound.

Retroposon or transposable element Elements having some of the properties of retroviruses, and able to move DNA around from one part of the genome to another. They cannot, however, leave the cell and travel elsewhere, as retroviruses can. Much more is known about the activities of retroposons in organisms like yeast and fruit flies than in humans.

Retrovirus An RNA virus that carries a gene for the important enzyme reverse transcriptase. Retroviruses include many cancer viruses and the AIDS virus. These viruses, very important for gene therapy, can carry foreign DNA into the nuclei of cells and insert it into their chromosomes.

RFLP Restriction fragment length polymorphism. RFLPs are the many places where a restriction enzyme will recognize a site in the DNA of one person but not recognize the same region in the DNA of another person because the sequence is slightly different. Once they have been discovered, RFLPs can then be followed in families. They are enormously valuable tools in gene mapping.

Ribosomes Complex collections of proteins and pieces of RNA, found in the cytoplasm, that form the sites at which the messenger RNA is translated into protein.

RNA This molecule acts as a go-between to carry the message of DNA to the cytoplasm of the cell where proteins are made. It is like a single-stranded DNA, and can hybridize to the DNA strand that is complementary to it.

Rous sarcoma virus A virus causing cancer in chickens, discovered by Peyton Rous early in this century. The first cancer virus to be found.

Scanning tunneling microscope A device for examining the surfaces of materials at a level sufficiently detailed to pick up images of individual atoms.

SCID Severe combined immune deficiency, a genetic disease in which the T cells of the immune system have been destroyed. About a third of the cases are due to the absence of the enzyme ADA.

Sense strand The strand of DNA that carries the genetic information.

Sickle-cell anemia A recessive genetic condition, common among peoples from tropical Africa, that causes hemoglobin molecules to join together in long chains, distorting the shape of the red blood cells. Heterozygotes for this allele are protected against the severest effects of a malaria common to the region.

Southern blot Named for its inventor, E. M. Southern; a method of transferring electrophoretically separated fragments of DNA from the soft, fragile electrophoresis gel to a more permanent material such as a nylon membrane. The DNA is then bound firmly to the membrane, where it can be probed repeatedly with different radioactive probes. Once the location on the membrane of the DNA that is complementary to a probe has been determined, the probe can be washed off, leaving the original DNA still firmly bound to the membrane and ready to be hybridized to a different probe.

Tay-Sachs disease A recessive genetic disease, common among Ashkenazi Jews and some other groups, that is marked by neurological degeneration. The condition is invariably fatal in the first year or two of life.

Thalassemia A collection of recessive blood diseases found in the Mediterranean and the Far East, and marked by a lack of one or another of the protein subunits of hemoglobin. Heterozygotes for the thalassemias, like heterozygotes for sickle-cell anemia, are protected against the severer effects of malaria, which appears to be why these diseases are so common.

Thymine One of the pyrimidine bases of DNA. In RNA, thymine is replaced by uracil.

Titin The largest known protein, some 30,000 amino acids long. It is found in muscle cells.

Transcription The process by which the genetic information on the sense strand of DNA is transcribed onto a complementary messenger RNA. Transcription is carried out by the enzyme RNA polymerase.

Transfer RNA A small molecule that binds to an amino acid and carries it to the ribosome. Each of the thirty-two types of transfer RNA also possesses an anticodon that is complementary to the codon of the messenger RNA that specifies that amino acid.

Transformation The book mentions two types of transformation. *Bacterial* transformation is the process by which bacteria take up DNA from outside the cell. The DNA may be in the form of a plasmid that can live inside the bacterial cell, replicating independently of the bacterial chromosome, or it may be in the form of DNA that is incorporated directly into the bacterial chromosome, changing its genetic makeup permanently. *Cellular* transformation is the quite different process by which human or other animal cells in tissue culture take on cancerlike qualities, growing without limit and piling up in mounds on the surface of a petri dish.

Translation The process by which the genetic information on messenger RNA is translated by the ribosome into protein.

Translocation The attachment of a piece of one chromosome onto another, different chromosome. In the familial schizophrenia case examined in chapter 12, a piece of chromosome 5 had been translocated into chromosome 1. Translocation occurs seldom and is an abnormal event. It should not be confused with the quite normal process of genetic recombination, which is a more regular interchange of material between the two copies of a particular kind of chromosome—such as chromosome 5—in the cell.

Uric acid A breakdown product of purines in the cell. High levels of uric acid can lead to gout.

Vector A plasmid, cosmid, or retrovirus that can be modified by the experimenter to carry foreign DNA into a cell.

VNTR Variable number of tandem repeats, another hideous acronym indicating short stretches of repeated DNA sequences that are scattered everywhere on the chromosomes. Different individuals carry different numbers of repeats in these stretches. In effect, each VNTR has many alleles, differing from each other by their size. It is now possible to measure the sizes of several of these VNTRs in an individual, providing a genetic fingerprint.

Walking along a chromosome This is a less athletic exercise than chromosome jumping. The experimenter wishing to walk along a chromosome starts with a cloned piece of DNA and uses it to probe for a longer piece of the organism's DNA that includes that clone. The end of the longer piece can then be cloned in turn. The next step is to look for another long piece of DNA that includes *that* clone and then repeat the process, thus gradually moving along the chromosome. If it is not clear at the outset which way to proceed, it may be necessary to walk in both directions along the chromosome until one reaches some recognizable landmark.

X chromosome One of the sex-determining chromosomes. A female carries two X chromosomes, while a male carries one X and one Y.

X linkage A gene that is located on the X chromosome, like the gene for HPRT, is said to be X-linked. Recessive X-linked characters, such as hemophilia or muscular dystrophy, are much commoner in males than in females, because males carry only one X chromosome. The Y chromosome has few genes in common with the X, so genes on the Y cannot mask the effects of genes on the X.

X-ray crystallography When a beam of X rays is passed through a crystal, the way the beam is scattered provides clues to the structure of the molecules making up the crystal. This technique has been important in determining the structures of many proteins and of DNA.

Xeroderma pigmentosum A condition marked by skin lesions and ultimately by skin cancers, in which the cells are unable to repair damage to their DNA caused by ultraviolet light.

Y chromosome The other sex chromosome. In humans, the Y chromosome determines male characteristics; a person with a single X and no Y develops as a female, though she will be sterile.

YAC A yeast artificial chromosome. It is possible to insert a stretch of up to a million bases of DNA into one of these chromosomes and then insert this artificially constructed chromosome into a yeast cell. There, it will behave like any other chromosome. This allows very large pieces of DNA to be cloned, since yeast cells carrying a YAC can be grown easily to enormous numbers.

Zygote The cell that is formed when a sperm and egg fuse. In humans, both sperm and egg carry twenty-three chromosomes, so the resultant zygote carries forty-six. All the cells of the body are descended from the zygote.

INDEX